5G Networks

5G Networks

Planning, Design and Optimization

Christofer Larsson

ACADEMIC PRESS
An imprint of Elsevier

Academic Press is an imprint of Elsevier
125 London Wall, London EC2Y 5AS, United Kingdom
525 B Street, Suite 1650, San Diego, CA 92101, United States
50 Hampshire Street, 5th Floor, Cambridge, MA 02139, United States
The Boulevard, Langford Lane, Kidlington, Oxford OX5 1GB, United Kingdom

Notices

Knowledge and best practice in this field are constantly changing. As new research and experience broaden our understanding, changes in
research methods, professional practices, or medical treatment may become necessary.

Practitioners and researchers must always rely on their own experience and knowledge in evaluating and using any information, methods,
compounds, or experiments described herein. In using such information or methods they should be mindful of their own safety and the
safety of others, including parties for whom they have a professional responsibility.

To the fullest extent of the law, neither the Publisher nor the authors, contributors, or editors, assume any liability for any injury and/or
damage to persons or property as a matter of products liability, negligence or otherwise, or from any use or operation of any methods,
products, instructions, or ideas contained in the material herein.

Library of Congress Cataloging-in-Publication Data
A catalog record for this book is available from the Library of Congress

British Library Cataloguing-in-Publication Data
A catalogue record for this book is available from the British Library

ISBN: 978-0-12-812707-0

For information on all Academic Press publications
visit our website at https://www.elsevier.com/books-and-journals

Working together
to grow libraries in
developing countries

www.elsevier.com • www.bookaid.org

Publisher: Mara Conner
Acquisition Editor: Tim Pitts
Editorial Project Manager: Leticia Lima
Production Project Manager: Surya Narayanan Jayachandran
Designer: Mark Rogers

Typeset by VTeX

To the memory of my father

Contents

Preface

This book is in some sense a sequel to *Design of Modern Communication Networks* (2014) having the aim of describing some of the most important concepts and capabilities of 5G from a mathematical point of view. I would like to see this as an idea book; its content is influenced by my own understanding of 5G and challenges that operators and solution vendors are facing.

5G networks are inherently complex and their planning requires a wider scientific scope than traditional network design. Thus, some attention is paid to big data and machine learning techniques as well. Indeed, optimization and machine learning can be seen as dual disciplines, and big data occur as a natural part of network management.

The content focuses on design and optimization tasks in design and operations that can be formulated as tractable optimization problems. Many of these problems may seem overwhelmingly complex, but appropriate problem formulations and novel optimization methods have been used to show amazing results and provide new insight into network planning matters.

The overall emphasis is on resource efficiency, whether the primary objective is capacity, coverage, latency or energy consumption. It can be shown that proper planning and design lead to substantial gains in several performance parameters. The achievable improvement with proper optimization is often substantial, possibly 10–30% compared with "traditional" methods, and the gain can often be noticed in several aspects simultaneously.

It has been the aim to present technology-independent content as far as possible. That is the reason why there are few details on actual network technologies, protocols, and functionality. There is plenty of literature available on these topics, and the algorithms described in this book can rather easily be translated into technology-specific situations.

The main idea permeating the methods described in this book is randomization. Instead of trying to solve problems exactly, we aim at finding a good solution with high probability. Most design tasks are combinatorial in their nature. In addition, the number of potential optimization parameters increases as new functionality is introduced and complex dependencies must be resolved. Most problems discussed throughout the text can therefore be characterized as "hard".

In general, combinatorial optimization is difficult, and diverse problems need different approaches, though the ideas underpinning many solution methods are still fairly simple. It should be possible to implement the described algorithms in any programming language of choice with reasonable effort.

A large part of the content of the book presents methods, findings and conclusions drawn from industrial or academic projects I have taken part in. As mathematical prerequisites, the reader is probably familiar with some combinatorics, optimization, fundamental probability theory, queueing theory, statistics and analysis, that is, a little bit of everything, and an open mind.

I am grateful to the many researchers and scientists who have made their interesting papers freely available on the Internet.

Christofer Larsson
Bratislava
June 2018

CHAPTER 1

Concepts and Architectures in 5G

5G networks are often described by a set of strict performance criteria. To cater for vastly diverse traffic and quality of service requirements of various use cases, the general demands on the networks are

- Massive system capacity,
- Very high data rates,
- Minimal latency,
- Extremely high reliability and availability,
- Energy-efficient and secure operation.

To achieve this, not only need networks be upgraded with high-performance elements, but planning, design and management have to be carried out according to optimal engineering methods to maintain performance and cost efficiency.

A fairly new paradigm in communication networking is *network elasticity*, which can be interpreted as a dynamic and optimal utilization of resources and – as far as possible – a demand-driven allocation of capacity. The justifications for flexible resource allocation are at least twofold: from a capacity point of view, resource pooling is always more efficient than distributed resources in terms of utilization; secondly, dynamic functional allocations through network function virtualization (NFV) improves the resilience approaching that of a distributed logic. To achieve these goals cost-effectively, planning for high resource utilization and resilience are imperative.

A consequence of virtualized architectures is the need for a novel approach to traffic engineering, which must be able to cater for a wide range of diverse traffic sources from applications such as massive sensor networks, telemedicine, and virtual reality.

It is well known that aggregation of many types of traffic leads to self-similar behavior, often leading to starvation of some traffic types by others. Furthermore, user mobility has been compensated for by resource overprovisioning in traditional networks, which becomes wasteful when traffic demands and mobility increase further.

1.1 Software-Defined Networking (SDN)

The concept of software-defined networking (SDN), which originated in the mid-1990s, is a framework to manage network behavior dynamically via programmable applications using

5G Networks. https://doi.org/10.1016/B978-0-12-812707-0.00006-1

open interfaces to the network. This allows for a flexible utilization of network resources, in contrast to the static allocation of resources in conventional networks.

Traditionally, the control of packet flows is realized by nodes forwarding packets to the best destination according to the packet header information and a set of static rules. In effect, routers try to accommodate the quality of service on a per-flow basis, mostly agnostic of the states in other parts of the network. The control of the single packet flow and the interaction between flows are intrinsically linked together. As a result, some parts of the network may be overloaded whereas others may be underutilized.

SDN allows decoupling of the control from the flow, resulting in two separate domains referred to as the control plane and the data plane, respectively. A consequence of this decoupling is the possibility to create resource and flow control logic that is based on more information than is contained in individual flows. The logic can then be based on the state of different parts of a network, the characteristics and interaction of various flows, or external factors, such as dynamic reservation of resources.

As a matter of fact, the logical separation into control plane and data plane was implemented for synchronous digital hierarchy (SDH), that is, connection-oriented networks, already in the mid-1990s. The implementation was facilitated by the fact that the routes in such networks already were semistatic, and networks consisted of relatively few nodes [1].

There are several incentives for an SDN-based implementation. Firstly, traffic characteristics change much more rapidly today than in the past through the fast deployment of new services. Secondly, traffic flows depend on an increasingly cloud-based service distribution. Thirdly, with rapidly increasing traffic volumes and big data, resources need to be managed more cost-efficiently.

Centralized and Distributed Control

In principle, network control functions can be either centralized or distributed. The advantage of a centralized control is the ability to apply a control policy taking into account the state of the entire network. The main disadvantages are delays in receiving information updates and applying control actions and the risk of overload or failure of a single centralized control function.

At present, network control is typically decentralized and residing in the routers due to its fast reaction time and resilience against failure. The control actions, however, are only based on network state information in a small neighborhood of the node itself. Distributed control functions are also based on rather simple logic, which may not lead to flow control which is optimal for the entire network.

In SDN, control functions can be separated into fast flow control and routing functions, which are decentralized and implemented locally in routers and switches, and longer-term control strategies residing in a network entity we refer to as an SDN orchestrator.

Network Function Virtualization (NFV)

NFV is essentially decoupling of software from hardware. It provides a layer between the hardware with its operating system and software applications, which are known as hypervisors. Thus, NFV represents the physical implementation of functions as virtual machines or virtual servers on general-purpose hardware platforms.

To distinguish between SDN and NFV, we can note that SDN refers to the decoupling of network logic from data transport, whereas NFV is the rapid provisioning and control through virtualization [2].

The resilience concern of a centralized logic is addressed by NFV; it can be seen as resource sharing on the platform level (Platform-as-a-Service). It constitutes an important framework to ensure resilience of the function of the central logic. Furthermore, the SDN/NFV concept supports open interfaces, such as OpenFlow, reducing the implementational complexity of a centralized intelligence in complex networks.

OpenFlow

OpenFlow specifies the communication between the control plane and the data plane. It implements one of the first standards in SDN enabling the SDN controller to interact directly with the forwarding nodes – switches and routers – in a network.

The control logic triggers queue settings and forwarding tables through OpenFlow. It is also possible to collect traffic traces and destination addresses from the nodes. The queues in the nodes are assumed to be configurable in terms of bandwidth, which could be specified as a minimum/maximum rate pair [3]. Version 1.3 of OpenFlow supports slicing, that is, multiple configurable queues per port. Similarly, the buffer sizes assigned to the queues should be configurable. In principle, it is sufficient to be able to specify a maximum buffer size per queue. This is particularly important to limit the impact of self-similar traffic on the rest of the network [4].

The update of packet forwarding tables in switches and routers is a standard operation, where globally optimal routes are determined by the central logic. We also assume that traffic traces are available, either via OpenFlow or some other means. Full traces rather than statistics are required for reasonable accuracy on short time scales.

OpenFlow implements the Orchestrator-Forwarder interface for manipulating flow tables and requests traffic statistics and network state information from the nodes to model the network topology and for load monitoring. For large networks, we expect to have multiple orchestrators, and OpenFlow is used to share information between these entities.

We will need to classify traffic efficiently, and ideally use lower-order protocol header fields to extract this information. Examples are traffic class header field in MPLS, the type of service field in IPv4, traffic class in IPv6, and possibly the source IP address and source and/or destination ports.

1.2 IT Convergence

5G is associated with many emerging IT trends, such as machine learning, artificial intelligence and Big Data. We touch upon these vast topics throughout the book with the justification that these technologies pertain both to network management and planning based on use cases on a high level. We refer to some advanced algorithms as machine learning techniques (avoiding the somewhat ambiguous term artificial intelligence). Most data can be classified as Big, due to its increasing granularity, inhomogeneity and detail.

Big Data

With the term Big Data, we mean data arriving in a streaming fashion at very high speeds, possibly containing different formats, for which traditional off-line analysis is inadequate in terms of processing speed and memory requirement. Problems related to Big Data occur naturally in high-speed networks. In Gbps optical links, data arrives at nanosecond time scales. We still need to extract useful information in almost real-time. Handling of high frequency data requires new approaches and specially designed algorithms.

Edge Computing

Edge computing refers to assigning some of the computational burden to the "edge" of the network, separate from clouds and central logic. The edge is typically closer to the data source and edge computing functions often include pre-processing, analytics and transformations, reducing the required transmission bandwidth. This is expected to be an integral part of many Internet of Things and cloud applications.

Security and Integrity

There are many concerns on how to ensure security, data protection and user integrity. These are difficult topics for many reasons, such as social, legal and commercial. Although not directly connected to design, these issues present a challenge of growing importance.

Resilience and robustness are characteristics somewhat connected to security, whereas integrity is more of an operational concern, or more precisely, policy setting and policing. Some protective techniques are based on game theory, where the idea is to formulate optimal strategies using as much known information about the adversary as possible.

Energy Efficiency

Energy efficiency of IT systems is increasingly becoming a concern. Datacenters consume huge amounts of energy, and the cryptocurrency Bitcoin has shown an exponential energy demand.

We note that resource optimization, including routing, assignment and scheduling often lead to energy savings as well as performance and other cost benefits. However, apart from wireless sensor networks, there is still relatively little investigation into the impact of optimization on energy efficiency. We simply estimate the energy savings by a quantity proportional to the efficiency ratio for other resources.

1.3 Building Blocks

The architectural evolution towards general purpose hardware, efficient transmission and open software has changed how networks are built and operated. These changes also put forward challenges in terms of interoperability and warranty.

Optical Fiber

In fiber-optic communications, wavelength-division multiplexing (WDM) is a technology for multiplexing a number of optical carrier signals onto an optical fiber by using different wavelengths (or "colors") of laser light. The technique enables bidirectional communications over a single fiber, and facilitates expansion of capacity by proper planning or by upgrading the interfaces at the fiber ends.

Channel plans vary, but a typical WDM system uses 40 channels at 100 GHz spacing or 80 channels with 50 GHz spacing. New technologies are capable of 12.5 GHz spacing (sometimes called ultra dense WDM). New amplification options enable the extension of the usable wavelengths more or less doubling the capacity. Optical fiber systems that can handle 160 concurrent signals over 100 Gbps interfaces result in a single fiber pair carrying over 16 Tbps.

SD-WAN

Software-defined networking in a wide area network (SD-WAN) simplifies the operation of a WAN by decoupling networking hardware from its control logic. The concept is similar to software-defined networking and virtualization technology in transport networks and datacenters.

A main application of SD-WAN is to implement high-performance networks using ubiquitous Internet access, partially or completely replacing more expensive connection technologies such as MPLS.

A centralized controller is used to set policies and prioritize traffic. The SD-WAN uses these policies and availability of network bandwidth to route traffic. The goal is to ensure that application performance meets given service level agreements.

Features of SD-WAN include resilience, security and quality of service (QoS) functions, such as real time detection of outages and automatic switch over to working links.

It supports quality of service by application level awareness, giving priority to critical applications through dynamic path selection, or splitting an application between two paths to deliver it faster. SD-WAN communication is usually secured using IPsec.

Open Source Software

Open source software is often perceived as being equivalent to free software, that is, being free to distribute and modify, and not carrying any license charges. While the two categories essentially are synonymous, there is a difference in the values and beliefs behind their creation.

Open source is a development and distribution paradigm. The principle is that the source code, the actual computer program, is made publicly available to inspect and modify. For software to be considered "free", it should allow the "four freedoms":

- To make use of the program for any purpose,
- To access the source code and study and change it as you wish,
- To freely redistribute copies of the software,
- To distribute freely modified versions of the software.

These types of software have been very successful in their ability to quickly deliver innovation. Following the principles of open exchange, rapid prototyping, and community-oriented development, open source solutions have been developed to fulfill most business needs.

These range from end user applications, operating systems, and libraries, with many commonly used in the industry, such as Linux, MySQL, WordPress, PHP, and Apache.

Built-in security is a concern for businesses choosing software. Many may consider free or open-source solutions less secure than proprietary solutions. However, free and open source software communities are contributing to cybersecurity systems, such as Apache Spot for example.

1.4 Algorithms and Complexity Classes

This book is about algorithms for network design problems, almost all of which are *computationally hard*, so it is important to have some way to evaluate their performance and thus be able to compare different algorithms for the same problem. We will therefore have a brief look at *algorithm analysis* and *computational complexity classes*.

In algorithm analysis, we imagine that we have a fictitious computer (often called the *random access machine* (RAM) model) with a central processing unit (CPU) and a bank of memory cells. Each cell stores a *word*, which can be a number, a string of characters, or a memory address. We assume further that the CPU performs every *primitive operation* in a constant number of steps, which we assume do not depend on the size of the input. We use high-level primitive operations such as performing an arithmetic operation, assigning a value to a variable, indexing into an array, comparison of two numbers, and calling a method. By counting the number of basic operations in the algorithm, we obtain an estimate of the running time of the algorithm. (The actual execution time is obtained by multiplying the total number of steps by the CPU execution time for basic operations.)

It is common to analyze the worst-case running time of an algorithm, both because it is usually much easier than finding the average running time, but also because the worst-case running time is a more useful measure, since it provides an upper bound for all possible inputs. This is particularly important for the classes of problems that we consider in this book.

Let the *input size*, n, of a problem be the integer number of bits used to encode an input instance. We also assume that characters and each numeric type use a constant number of bits.

We can usually avoid rigorous technical arguments and will be satisfied with an algorithm construction if we have "done our best", or that we have taken "reasonable precautions" to make it as efficient as possible. What we mean by that is – in very imprecise terms – that we

Table 1.1: Common running times of algorithms.

Constant	$O(1)$
Logarithmic	$O(\log(n))$
Linear	$O(n)$
Near-linear	$O(n\log(n))$
Polynomial	$O(n^k), k \geq 1$
Exponential	$O(a^n), a > 1$

try to find a representation as short and efficient as possible for the problem instances. We also assume that any evaluation of a solution to the problem is performed efficiently.

Define the worst-case running time of an algorithm A to be the time A runs with an input instance of size n, where the worst case is taken over all possible inputs having an encoding with n bits.

We will express the worst-case algorithm running times using the "big-Oh" notation. Suppose we have a function $f(n)$, representing the number of steps of an algorithm. Then we say that $f(n)$ is $O(g(n))$ of another function $g(n)$, if there exist a real constant $c > 0$ and an integer constant $n_0 \geq 1$ such that $f(n) \leq c \cdot g(n)$ for all $n \geq n_0$. This should be interpreted as "$f(n)$ is bounded above by a constant c times $g(n)$ for large enough n", and the constants need not to be specified when stating the upper bound.

Complex algorithms are typically broken down into subroutines. Therefore the following fact is useful in the analysis of composite algorithms. The class of polynomials is closed under addition, multiplication, and composition. That is, if $p(n)$ and $q(n)$ are polynomials, then so are $p(n)+q(n)$, $p(n) \cdot q(n)$, and $p(q(n))$. Thus, we can add, multiply, or compose polynomial time algorithms to construct new polynomial time algorithms.

Any sum of functions is dominated by the sharpest increasing one for large n. Therefore, we have, for example, $\log(n) + \log(\log(n)) = O(\log(n))$, $2^{10} = O(1)$, and $n + n^2 = O(n^2)$.

An algorithm is called *efficient* if it runs (at most) in time $O(n^k)$ of the input size n and some $k > 0$, that is, if it is bounded above by a polynomial. Table 1.1 lists some common names and running times used to describe algorithms.

Optimization Problems

Since most of the problems we encounter in relation to networks are hard to solve, we cannot expect to find efficient algorithms for any of them. In practical terms, that also means that any *exact* algorithm or method will have a running time which is exponential in the input size n, and we will never find any polynomial upper bound to the worst-case running time. We will formalize that now.

Most problems in network design are *optimization problems*, where we search for an *optimal value*, such as cost. For a discussion on computational complexity, however, it is useful to think of the optimization problems as *decision problems*, that is, problems for which the output is either *true* or *false* only.

We can convert an optimization problem into a decision problem by introducing a parameter k and ask if the optimal value is as most k (or at least k). If we can show that a decision problem is hard, then its related optimization formulation must be hard too.

Example 1.4.1. *Suppose we have a network and want to find a shortest path from a node i to a node j. Formulating this as a decision problem, we introduce a constant k and ask if there is a path with length at most k.*

In order to define some important complexity classes, we refer to the class of decision problems as a *language L*. An algorithm A is said to *accept* a language L, if, for each $x \in L$, it outputs the value true and false otherwise (for improperly formulated x). We assume that if x is in an improper syntax, then the algorithm given x will output false.

The *polynomial time* complexity class $\mathcal{P}$ is the set of all decision problems L that can be accepted in worst-case polynomial time $p(n)$, where $p(n)$ is a polynomial and n is the size of x. The *nondeterministic polynomial time* complexity class $\mathcal{NP}$ is a larger class which includes the class $\mathcal{P}$ but also allows for languages (decision problems) that may not be in $\mathcal{P}$.

In the complexity class $\mathcal{NP}$, an algorithm can perform nondeterministic modifications of $x \in L$, such as it ultimately outputs true, where the verification is done in (polynomial) time $p(n)$, where $p(n)$ is a polynomial and n is the size of x. In other words, the time of asserting that x is true is polynomial, and the generation of such an x may require a polynomial number of nondeterministic modifications. Still there is no guarantee that such a solution will be found. We can only *guess* a solution and verify it in polynomial time. If we try to explore all possible modifications of x in the algorithm, this procedure would become an exponential time computation, since the time required increases very rapidly with the size of the input.

No one knows for certain whether $\mathcal{P} = \mathcal{NP}$ or not. Most computer scientists believe that $\mathcal{P} \neq \mathcal{NP}$, which means that there are no efficient algorithms for solving any $\mathcal{NP}$ problem.

Example 1.4.2. *In network design, there is no way to efficiently calculate how many links and which ones should be included in an optimal design. Selecting the links is therefore a nondeterministic procedure. Verification, that is, summing up edge costs and comparing them, is, however, fast.*

We can note that there may be little difference in the formulation of two problems, which nevertheless will happen to fall into different complexity classes.

Example 1.4.3. *Finding the shortest path in a network is easy, while finding the longest path in a network is hard.*

Showing Problem Hardness

Given a problem, how do we know whether an efficient algorithm exists for it or not? It is an important question, because if a problem belongs to the class of known hard problems, we would not have to spend time on trying to find an exact solution. This question is answered by the theory of $\mathcal{NP}$-*completeness*, which rests on a foundation of automata and language theory.

There are some problems that are at least as hard as every problem in $\mathcal{NP}$. The notion of *hardness* is based on the concept of polynomial time *reducibility*. A problem L is polynomial time-reducible to another problem M if there is a function f, computable in polynomial time, such that $x \in L \iff f(x) \in M$. The problem M is said to be $\mathcal{NP}$-*hard* if every other problem L in $\mathcal{NP}$ is reducible to M in polynomial time. A problem is $\mathcal{NP}$-*complete* if it is $\mathcal{NP}$-hard and in the class $\mathcal{NP}$ itself. It is then one of the hardest problems in $\mathcal{NP}$. If one finds a deterministic polynomial time algorithm for even one $\mathcal{NP}$-complete problem, all $\mathcal{NP}$-complete problems can be solved in polynomial time. This would mean that $\mathcal{P} = \mathcal{NP}$. Note that there are problems believed to be in $\mathcal{NP}$ but not be $\mathcal{NP}$-complete. Also most $\mathcal{NP}$-hard problems are complete, but not all.

In order to show that a problem is $\mathcal{NP}$-complete, we need to have at least one $\mathcal{NP}$-complete problem. Such a problem is *satisfiability* (SAT) of a logical expression. It has been proven that satisfiability is $\mathcal{NP}$-complete. The proof is complicated, but it shows that satisfiability is at least as hard as any other problem in $\mathcal{NP}$. A variant of the satisfiability problem is the restricted *3-SAT*, restricted to clauses with three literals.

In SAT problems, we have a set of Boolean variables $\mathcal{V} = \{v_1, \ldots, v_n\}$ and a set of clauses (or subexpressions) $\mathcal{C}$ over $\mathcal{V}$. The expression is a combination of the logical operations AND (denoted by $\cdot$) and OR (denoted by $+$). We denote by a bar the complement of a variable such that if v_i is true then $\bar{v}_i$ is false and vice versa. The problem is to find a combination of values of the variables in $\mathcal{V}$ such that each clause evaluates to true, and therefore the full expression evaluates to true.

The 3-SAT problem is a variant of SAT which takes three variables per clause. For example, the following formula could be an instance of 3-SAT. The clauses are the parentheses containing the three variables. We have

$$(\bar{v}_1 + v_2 + \bar{v}_7)(v_3 + \bar{v}_5 + v_6)(\bar{v}_2 + v_4 + \bar{v}_6)(v_1 + v_5 + \bar{v}_2). \tag{1.1}$$

We note that 3-SAT is in $\mathcal{NP}$, for we can construct a nondeterministic polynomial-time algorithm that takes an expression with three variables per clause, guesses an assignment of Boolean values for these variables, and then evaluates it to see if it equals true by inserting the values into the clauses. 3-SAT is $\mathcal{NP}$-complete. An interesting fact is that the 2-SAT, expressions with only two variables in each clause, is in $\mathcal{P}$. We can use 3-SAT and reduction to show that some other problem in $\mathcal{NP}$ is $\mathcal{NP}$-complete, which is illustrated in Example 1.4.4.

Example 1.4.4. *Show that integer programming is $\mathcal{NP}$-complete. Consider the following integer programming example:*

$$\max x_1 + 2x_2 \tag{1.2}$$
$$subject\ to \tag{1.3}$$
$$x_1 \geq 2 \tag{1.4}$$
$$x_2 \geq 0 \tag{1.5}$$
$$x_1 + x_2 \leq 4 \tag{1.6}$$
$$x_1, x_2 \quad integers. \tag{1.7}$$

First we formulate this problem as a decision problem. Introduce a constant to compare the objective function with, say, $k = 5$. The decision problem reads: is there a pair of values (x_1, x_2) such that the decision problem

$$x_1 + 2x_2 \geq k \tag{1.8}$$
$$subject\ to \tag{1.9}$$
$$x_1 \geq 2 \tag{1.10}$$
$$x_2 \geq 0 \tag{1.11}$$
$$x_1 + x_2 \leq 4 \tag{1.12}$$
$$x_1, x_2 \quad integers \tag{1.13}$$

outputs true?

The IP problem is in $\mathcal{NP}$, because we can guess a pair of values (x_1, x_2), verify that the side conditions are satisfied, and, if so, calculate the value of the objective function.

To show that IP is $\mathcal{NP}$-complete we use reduction from 3-SAT. Recall the form of 3-SAT,

$$(v_1 + v_2 + \bar{v}_3)(\bar{v}_1 + v_4 + v_5)... \tag{1.14}$$

If we can solve IP in polynomial time and 3-SAT can be formulated as an IP in polynomial time, then we can also solve 3-SAT in polynomial time, and as a consequence, $\mathcal{P} = \mathcal{NP}$.

Table 1.2: **The complexity of some network design problems on graphs of order $n = 15$.**

Problem	Upper bound	Relative execution time
Connected networks	$4.05 \cdot 10^{31}$	$1.29 \cdot 10^{31}$ years[a]
Shortest rings	$4.36 \cdot 10^{10}$	4360 seconds
Number of spanning trees	$1.95 \cdot 10^{15}$	62 years

[a] 283 million times the age of Earth.

Make the integer variables correspond to Boolean variables and have constraints serve the same role as the clauses. The IP will have twice as many variables as the SAT instance, one for each variable and one for its complement.

A Boolean variable v_i can be expressed, letting $x_i = v_i$ and $y_i = \bar{v}_i$, as

$$1 \geq x_i \quad \geq \quad 0, \tag{1.15}$$
$$1 \geq y_i \quad \geq \quad 0, \tag{1.16}$$
$$x_i + y_i \quad = \quad 1. \tag{1.17}$$

A clause (v_1, v_2, v_3) is then expressed as $x_1 + x_2 + x_3 \geq 1$. The objective function is unimportant; we can simply let $k = 0$. Thus, IP is $\mathcal{NP}$-complete.

When one fails to prove that a problem is hard, there is a good chance that there is an efficient algorithm. We will not prove that problems in this book are $\mathcal{NP}$-hard or $\mathcal{NP}$-complete, but just convince ourselves that they are and select solution methods accordingly.

Algorithms for Hard Problems

Most problems where we want to find a set of links such that a network possesses some specified property, such as path diversity or resilience, are $\mathcal{NP}$-hard. Network links can be expressed as a pair of nodes (u, v), and one may be deceived to think that the effort to solve design problems are proportional to n^2 as well. This, however, is incorrect since the number of possible configurations of a network of order n is $2^{n(n-1)/2}$, that is, exponential in n. This essentially means that the effort required to solve such a problem in general grows exponentially with its order (and size). We illustrate some network problems and their complexity in Example 1.4.5.

Example 1.4.5. *The total possible network configurations having n nodes grows very rapidly with n. Table 1.2 shows the upper bounds for common problems for $n = 15$. Supposing that evaluation of each configuration takes 1 μs on a computer, the corresponding times to evaluate all configurations are listed in the third column.*

Usually, an exact solution cannot be found with reasonable effort, even when we know what we mean by a solution being optimal. Instead, we have to resort to clever heuristics or approximations. For networks of realistic sizes, the choice of solution method is very important. The choice of algorithm is usually a trade off between computational effort and accuracy. The additional time required to find a more "accurate" solution may not be worthwhile spending. The effects of statistical variations and measurement errors of traffic may far exceed the gain of trying to find a more accurate solution.

We also need to pay attention to verification of our results. A good approach, if possible, is to solve a problem using different algorithms and comparing the results. If the results are similar, the likelihood of the solutions being close to the optimum increases, provided that the design criteria are correctly formulated.

Loosely speaking, an algorithm is a set of rules that defines a sequence of operations. Its is desirable that an algorithm is efficient and well defined, so that any input data leads to a defined state and that the algorithm does not "freeze" in an infinite loop. We therefore require that the number of steps the algorithm needs to complete a task is finite.

Algorithms may be classified in various ways, which also reflect the way they may be implemented on a computer (and the choice of programming language). An algorithm may evaluate an expression *directly*, *iteratively*, by calling itself repeatedly with a value previously determined, or *recursive*, where successive calls to the algorithm determine successively smaller problems. A recursive algorithm must have a base case which can be determined without a recursive function call and a recursive part which specifies how the algorithm calls itself with successively smaller problem sizes.

A simple example is the recursive evaluation of the factorial $f(n) = n!$. If $n \leq 1$, then $f(n) = 1$, otherwise $f(n) = nf(n-1)$. Since we usually are interested in evaluating $n!$ for $n > 1$, we simply take $f(1) = 1! = 1$. One example is the following: $(4!) = 4 \cdot (3!) = (4 \cdot 3) \cdot (2!) = (4 \cdot 3 \cdot 2) \cdot (1!) = (4 \cdot 3 \cdot 2 \cdot 1)$. The factorial can also be viewed as an iteration. Starting from $f(1) = 1$, we get $f(2) = 2f(1)$, $f(3) = 3f(2)$ and $f(4) = 4f(3) = 4 \cdot 3 \cdot 2 \cdot 1$.

There are a number of important approaches for constructing algorithms, of which we mention a few. This is not intended to be an exhaustive or mutually exclusive set of classes, but rather a list of principles that are useful in solving hard problems.

Brute force

The method of sequentially listing and evaluating all possible configurations does certainly determine an optimal solution – eventually. It is only a reasonable method for very small problems, but is also useful in testing algorithm correctness.

Analytical methods

An analytical method can be evaluated exactly or numerically (that is, approximately with arbitrarily small error). The main analytical methods used are mathematical programming (analytical optimization), combinatorics, and probability calculus. Such methods are not available for a general $\mathcal{NP}$-hard problem.

Sometimes, however, hard problems can be made analytically tractable by letting some group of parameter tend to infinity or zero (or any other suitable limit). For example, the flow in some networks can be described analytically when the size of the network grows to infinity. These bounds are very useful in analyzing hard problems.

Approximations

An approximative method gives a result close to the exact value and with a bound on its error – or equivalently, the distance between the approximation and the exact value. There are many ways to construct approximations. A problem is often simplified so that it becomes tractable by other methods. Limits and asymptotics are often used.

Heuristics

Heuristic methods differ from approximations in that there is no guarantee on the error limit. Another view is that a heuristic method is relying on a principle which may serve as a first "best guess". The optimality of the result remains unprovable, but heuristics often give reasonable results and are easy to program. One of the best known heuristics is the *greedy principle* or (greedy heuristic). It is used in many various contexts, in exact methods as well as in approximations.

The greedy method is applicable to optimization problems, that is, problems that involve searching through a set of configurations to find one that minimizes or maximizes an objective function defined on these configurations. The general formula of the greedy method could not be simpler. In order to solve a given optimization problem, we proceed by a sequence of choices. The sequence starts from some well-understood starting configuration and then iteratively makes the decision that seems best from all of those that are currently possible.

Problem restriction

An efficient way of making problems tractable is by restriction of the search space. Possibly some parameters are kept fixed while others are allowed to vary. Reduction of the state space is usually the method for deriving bounds. In combinatorial optimization we may use the technique of *branch-and-bound*, where an integer optimization problem is replaced by its continuous counterpart, which is relatively easy to solve, and parameter values are successively refined based on these.

Divide and conquer

The divide and conquer principle is a method to solve a complex problem by dividing it into subproblems of smaller size, recursively solving each subproblem, and then merging the solution parts to produce a solution to the original problem. Decompositional methods can be classified as relying on this principle. Another, related, technique is dynamic programming. It is similar to divide and conquer in that it is very general, and it often produces efficient algorithms for otherwise hard problems.

Randomization

An algorithm can have deterministically controlled or random transition between states and have fixed or random input data. When the state transitions are deterministic, the algorithm is called deterministic, and an algorithm with stochastic state transitions is called a randomized algorithm. Randomization is a very general and important technique. Randomized algorithms include simulation, Monte Carlo methods, and metaheuristics, or any method that is dependent on random numbers. The method class relies on random numbers of high quality, and it is worthwhile to study and implement algorithms for the generation of reproducible and uniformly distributed random numbers.

Metaheuristics are numerical optimization methods that mimic a physical system or biological evolution. Two commonly used methods are *simulated annealing* and *evolutionary algorithms*. The two methods represent two groups of techniques: local search and population-based search. The strength of these methods is that they are very general and often easy to formulate. Monte Carlo methods and simulation are also indispensable tools for estimation and verification.

Simulation should be used with care, however. It is not a panacea for solving hard problems. It can be misleading if some parameters are incorrectly set or conditions improperly formulated.

CHAPTER 2

Network Modeling and Analysis

A network is commonly modeled as a graph, a structure consisting of vertices (or nodes) and edges (or links) between the vertices. A vertex is a point where an edge originates or terminates, and an edge can therefore be uniquely described by a pair of vertices (u, v). A graph is denoted $G = (V, E)$, where V is the set of vertices and E the set of edges. We use the terms network and graph somewhat synonymously. Strictly speaking, a network can be regarded as a physical system of nodes and links connecting them, while a graph is the model of such a construction. Thus, in practical applications, we will mostly use the terms nodes and links, since these more closely reflect the physical components of a network.

The graph concept is very versatile. We may model a transport network by letting the nodes represent routers that link the fibers or wireless connections between them. In social networks, we can let the nodes represent users and links symbolize users who are friends. Similarly, it can be used to describe collaboration or business networks, certain biochemical processes, atomic lattices, etc.

2.1 Basic Properties

Some basic properties can be defined on graphs (see Table 2.1). The number $n = |V|$ of nodes is called the *order* of the graph, and the number $m = |E|$ of links is its *size*. We often associate some properties with the components of a network, particularly links. The links can be thought of as carrying a flow of some commodity between nodes, or more generally as symbolizing a relationship between nodes. If the commodity transported by the network is only allowed to flow in one direction, the graph is called *directed*, whereas if flows are allowed in any of the two directions, it is *undirected*.

Most networks we study are undirected, following the principle of full duplex links in transport networks. Furthermore, problems defined on directed graphs are often associated with intricacies not present in undirected networks, so the solution methods may be quite different. In particular, for certain problems the existence of a solution may be guaranteed in an undirected graph, but not necessarily so if it is directed.

We may also associate numerical values to the edges, representing for example their capacity, cost, distance, or latency. Collectively, numerical values assigned onto the edges are known as *weights*, and the sum of all weights is the weight of the graph. Vertices can also be assigned

5G Networks. https://doi.org/10.1016/B978-0-12-812707-0.00007-3

Table 2.1: Basic properties of graphs.

Property	Symbol	Description
Order	$n, |V|$	The number of nodes (vertices) in a graph.
Size	$m, |E|$	The number of edges (links) in a graph.
Weight	w_e	A property assigned to (usually) an edge e, such as cost or distance.
Degree	$\deg(v)$	The number of edges incident to a vertex v. In directed graphs, we distinguish between in-degree (edges pointing towards v) and out-degree (edges pointing out from v).
Diameter	D	The maximum path length between any two nodes u, v.
Average path length	ℓ	The average path length between any two nodes u, v.
Clustering coefficient	C_c	The number of edges between neighbors of a given node v.

weights. For any vertex, the number of edges originating from it is the *degree* of the vertex. We will find that the degree is an important characteristic of graphs and a useful measure in network analysis and design. Fig. 2.1 depicts an undirected graph of order 7 and size 10 with edge and vertex weights.

2.2 Graph Representations

There are various ways to represent a graph. A particularly convenient representation is the $n \times n$ *adjacency matrix* A, which also can be used in algebraic equations. For an undirected graph it is defined as a matrix with a positive one in the positions (u, v) and (v, u) whenever there is an edge connecting vertices u and v, and zeros elsewhere. The adjacency matrix is in this case symmetric. The adjacency matrix is the most commonly used graph representation in applications.

Another representation, suitable for directed graphs, is the vertex-edge $n \times m$ incidence matrix J where an entry (v, e) is a positive one in the position of the origin of the directed edge, a negative one in the position of its termination, and zeros elsewhere. The vertex-edge incidence matrix can also be defined for undirected graphs, where all nonzero entries are positive ones. This matrix is useful in representing many flow problems since each row represents a vertex's incoming and outgoing edges. When a vertex-edge matrix of a directed graph is multiplied by a flow vector $\mathbf{f}$, the result is the inflow minus outflow to and from each vertex.

It is immediately clear that we can define n^2 pairs of vertices altogether. If we require that the vertices be different in defining an edge (there are no self-loops) and if we let the edge associated to the same pair of vertices be independent of the order of the vertices (that is, an undirected graph), $(u, v) = (v, u)$, there can be a total of $n(n-1)/2$ different edges. If we let each graph of order n be identified by its edge configuration, where any edge may or may not be present, we have $2^{n(n-1)/2}$ different graphs.

So far, we have identified edges by a pair (u, v) of vertices. In computer implementations of most algorithms, however, it is convenient to have a single parameter indexing an edge. Suppose that n vertices are indexed by the natural numbers. A simple mapping between a pair of vertices and a edge index is obtained by introducing two variables l and s, the largest and the smallest of the two vertex indices, respectively,

$$l = \max\{u, v\},$$
$$s = \min\{u, v\}.$$

Then we can assign a single number k to any edge $e_{uv} = e_k$ by

$$k = l \cdot (l - 1)/2 + s. \tag{2.1}$$

This indexing simplifies storing and retrieving data in vectors. The converse transformation, given an edge index k, is performed by searching for the largest number l such that

$$k \leq l \cdot (l - 1)/2,$$

which is the largest of the vertex indices, and the smallest is $s = k - l(l - 1)/2$, recalling that $s < l$ since we do not have any self-loops.

By indexing all the $s - t$ paths and the edges in the network, it is also possible to create a matrix by mapping paths between s and t against edges, with a positive one in a position where an edge is part of the path, and zeros elsewhere. The path index is arbitrary, and the edge index can be based on Eq. (2.1), with nonexisting edges removed and the index adjusted to represent existing edges only. This representation is called an *edge-path* (or edge-chain) incidence matrix. This matrix, denoted by K, can be useful in flow problems, but it tends to be large.

2.3 Connectivity

One of the simplest but also one of the most fundamental problems in graph theory is to determine whether two vertices in a graph are connected or not. The problem may seem trivial enough for small graphs, but in large graphs, a systematic approach becomes necessary. We formally define connectivity as follows.

Definition 2.3.1 (Connectivity). *In an undirected graph $G = (V, E)$, two vertices $u, v \in V$ are said to be* connected *if G contains a path from i to j. Otherwise, they are said to be dis-connected. A graph is called* connected *if every pair of distinct vertices in V can be connected through some path. Otherwise it is called* disconnected. •

The connectivity problem can be solved by simply traversing all possible paths in the graph until either the destination vertex has been found, or all paths have been traversed without finding the destination vertex. There are two strategies for searching through a graph: depth-first search and breadth-first search. The problem of determining vertex connectivity can be solved efficiently using any of these strategies. A simple algorithm to ascertain the connectivity of a graph G is as follows.

Algorithm 2.3.1 (Graph connectivity).

Given a graph $G = (V, E)$.

STEP 0: Begin at an arbitrary node of the graph G.
STEP 1: Proceed from that node using either the depth-first or the breadth-first search, counting all nodes reached.
STEP 2: Once the graph has been entirely traversed, if the number of nodes counted is equal to the number of nodes of G, the graph is connected; otherwise it is disconnected.

Output TRUE or FALSE.   •

Depth-First Search

Suppose we are given a starting vertex s and we would like to know which other vertices s is connected to. The *depth-first search* (DFS) method can be used to determine the set of connected vertices. We label vertices as "unvisited" and "visited" to be able to keep track of where we are in the search. Initially we let all vertices except s be marked as unvisited. Now, s is referred to as the current vertex. Whenever we find an unvisited neighbor vertex, we continue as far as we can by marking the unvisited vertex as visited, and we set the new vertex as the current vertex. When arriving at a point where no more unvisited neighbors can be found, we *backtrack* along the path until an unvisited vertex can be found again. If we arrive back to s without finding any such neighbors, the algorithm terminates; all vertices that can be reached have been visited. Of course, we can stop the search earlier if we are looking for a particular vertex t that has been visited before backtracking back to s. We can note that each edge is being traversed twice, once in the forward direction, once when backtracking.

Example 2.3.1. *If we perform a depth-first search on the network in Fig. 2.1 starting from $s = 1$, we would for the first probe get the sequence of vertices $s, 2, 3, 4, 6, 7$. Next, we backtrack to vertex 2 from where we reach vertex 5, omitting vertex 6, which we have already been visited. Therefore, we backtrack again, but this time all the way to $s = 1$ and stop, since we have in fact visited all vertices.*

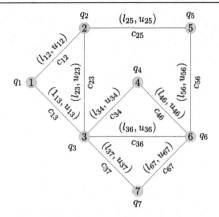

Figure 2.1: Graph of order $n = 7$ and size $m = 10$ representing a network with edge and vertex weights. The edges are labeled with capacity limits in brackets, (l_{uv}, u_{uv}), and costs, c_{uv}. The vertices are assigned demands q_u.

Breadth-First Search

The alternative to DFS is BFS in which all the neighbors of the current one are visited before moving on to the next level. Consequently, there is no need for backtracking. Again we start from a vertex s. When the current vertex has unvisited neighbors, these vertices are visited in turn. If no further unvisited vertices can be found, we move on to the next vertex in the order they have been visited. Continuing in this manner, we will eventually reach a vertex which has no unvisited neighbors and there is no vertex to move on to, and then the algorithm terminates.

Example 2.3.2. *Using BFS on the example in Fig. 2.1, starting from $s = 1$, the first scan gives 2, 3. Moving on to vertex 2, we reach vertex 5. Proceeding to vertex 3, we reach vertices 4, 6 and 7. Now there is no visited vertex left with unvisited neighbors, so the search is completed.*

2.4 Shortest Paths

Finding the *shortest path* in a graph is a fundamental problem of immense importance in graph theory as well as in traffic engineering of communication networks. More complex problems can often be reformulated as a sequence of shortest-path problems. The shortest-path distances or costs can be defined arbitrarily to model various aspects that can then be solved by a shortest-path algorithm, or by using shortest-path problems as subroutines. In communication networks, the shortest path is a basic *routing principle* used, for example, in the *Open Shortest Path First* (OSPF) and the *Intermediate System to Intermediate System* protocols. An efficient routing principle leads to decreasing transport distance which lowers propagation delay in the network and also the transport cost per flow unit, which can be seen as a measure of the cost effectiveness of the network. Shortest-path algorithms can be formulated both for undirected and directed graphs, where by a directed graph we mean that each edge in an undirected graph is given an orientation.

Dijkstra's Algorithm

One of the most widely used algorithms for shortest paths in applications is *Dijkstra's algorithm*. It works on undirected graphs and the weights (distances) need to be nonnegative for the algorithm to work. Note that even if we only want the shortest path between the vertices s and t, there is no better way (in the worst case) than to determine the shortest paths from s to all other vertices.

Starting from a source s, the algorithm scans its neighbors to find the distances to them. Next, the neighbor which is closest is chosen and is scanned for its neighbors' distances in turn. It continues selecting scanned neighbors at increasing distances and progresses throughout the network until the destination t has been reached. This process is referred to as the *greedy principle*, as it always chooses the closest vertex for subsequent scanning.

The algorithm works by using a distance "best estimate" $l(v)$ from the source s to every other vertex v in the network, initially set to ∞. The estimate $l(v)$ is successively updated as the algorithm progresses. Let $d(s, v)$ be a function representing the actual distance from s to some vertex v. The algorithm keeps track of vertices to which the actual distance has been determined in a set S of *visited* vertices. In determining the actual distance to a yet unvisited vertex u neighboring S, the relation

$$d(s, v) = \min_{u \in S}(d(s, u) + w_{uv}) \tag{2.2}$$

is used, where w_{uv} is the direct distance from u to v (or more generally, the weight on edge (u, v)). This means that, knowing the actual distances from s to all vertices $v \in S$, the actual distance to a new vertex u is the smallest sum of the distance from s to any vertex v in S and the direct distance from v to u. When scanning for neighbors from S, the algorithm chooses the closest unvisited neighbor to S in each step. Therefore the relation (2.2) is guaranteed to work; there cannot be a shorter way to reach u.

Algorithm 2.4.1 (Dijkstra).

Given an (undirected) graph $G = (V, E)$, nonnegative edge costs $c(\cdot)$, and a starting vertex $s \in V$.

STEP 0:

   Set $S := \{s\}, l(s) := 0, l(v) = \infty, v \in V, v \neq s$

STEP 1 to $|V| - 1$:

   while $S \neq V$ **do**

   find a closest unvisited neighbor u

   for all $v \in V - S$ **do**

   $l(v) := \min\{l(v), l(u) + w_{uv}\}$

   set $S := S \cup \{u\}$

   end (while)

Output the shortest distances from s to all $u \in V$ in the vector **l**.   •

Algorithm 2.4.1 shows the steps in Dijkstra's algorithm. In its basic form, the algorithm does not return the shortest path, only its distance. It is, however, easy to reconstruct the path by adding the preceding vertex to the estimated distance label when it is updated via (2.2). Every time a vertex is added to S, the edge corresponding to the vertex and its preceding vertex is recorded as part of the path. When the destination t has been found and processed, we can backtrack to s following the recorded vertices to find the actual path.

We can estimate the running time of the algorithm as follows. Each iteration requires processing of vertices that have not yet been visited, which is at most $|V|$. Since there are $|V|$ iterations (including the initialization), the algorithm has a running time of $O(|V|^2)$. This running time refers to a "standard" implementation. By using clever data structures, faster implementations can be achieved.

Example 2.4.1. *The steps in Dijkstra's algorithm are shown in Fig. 2.2. Initially, the set S consists of the source s only and all labels are set to ∞. The two neighbors of s, vertices 2 and 3, are labeled with their distances from s. The closest vertex (3) is then added to S. Reaching vertex 2 from vertex 3 gives a total distance of 4, so the previous label of vertex 2 remains unchanged. Then, vertex 2 is added to S. The new closest neighbor of S, vertex 5, can be reached from s via vertex 3 at a distance of 2. The only unvisited neighbor of S is vertex 4 at distance 5 from s. In the following iteration, vertex 5 is added to S. Its neighbors, 4 and t, can now be labeled with the estimated distances 4 and 7, respectively. Next, vertex 4 is added to S, and its neighbor t can be reached at a total distance of 6 from s. The sink t can now be added to S, and the algorithm terminates.*

Note that in the third iteration, we have a tie between the distance to vertices 2 and 5. It does not matter which of the two vertices we visit; the final result is the same. Note that Dijkstra's algorithm does not produce the actual shortest path. In order to find this, we can add the previous vertex to the label and backtrack from t once the algorithm terminates.

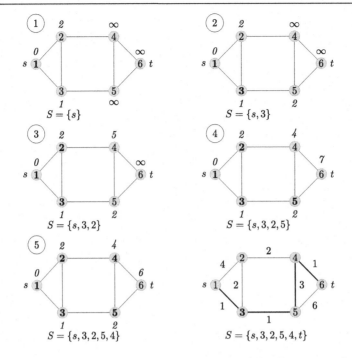

Figure 2.2: Dijkstra's algorithm used to find the shortest path.

The Bellman–Ford Algorithm

Another algorithm for finding shortest paths is the Bellman–Ford algorithm. It is similar to Dijkstra's algorithm in that it computes successive refinements of the vertex distances, but contrary to Dijkstra's algorithm, it processes all vertices in each iteration. It is slower than Dijsktra's algorithm, but has the advantage of being capable of handling negative edge weights. The algorithm is based on the following principle, called *Bellman's* equations:

$$
\begin{aligned}
l(s) &= 0, \\
l(v) &= \min_{u \in S}\{l(u) + w_{uv}\}, \quad v \neq s,
\end{aligned}
\tag{2.3}
$$

where w_{uv} is the distance (or, more generally, the weight) of the edge (u, v). For a source s, the distance estimates $l(s)$ is initially set to zero and all other estimates are set to ∞. The estimates are successively improved by iterating (2.3) until no further improvement is achieved. It turns out that in the worst case $|V| - 1$ iterations are sufficient to reach the final result. Eq. (2.3) is also called a dynamic programming equation. It is called the *necessary optimality condition of dynamic programming*.

This so-called *principle of optimality* can be summarized as follows. Suppose we know that the shortest path from s to t is known to pass through vertex u. Then the path from u to t must

Table 2.2: Three iterations of the Bellman–Ford algorithm for the shortest-path problem. After the third iteration, no further improvements are achieved. The length of the shortest path is the value of u_6^5.

i	u_i^0	u_j^0	d_{ij}	u_i^1
1	0	–	–	0
2	∞	0	2	2
3	∞	0	1	1
4	∞	∞	–	∞
5	∞	∞	–	∞
6	∞	∞	–	∞

i	u_i^1	u_j^1	d_{ij}	u_i^2
1	0	–	–	0
2	2	0	2	2
3	1	0	1	1
4	∞	2	3	5
5	∞	1	1	2
6	∞	∞	–	∞

i	u_i^2	u_j^2	d_{ij}	u_i^3
1	0	–	–	0
2	2	0	2	2
3	1	0	1	1
4	5	2	2	4
5	2	1	1	2
6	∞	5	2	7

i	u_i^3	u_j^3	d_{ij}	u_i^4
1	0	–	–	0
2	2	0	2	2
3	1	0	1	1
4	4	2	2	4
5	2	1	1	2
6	7	4	2	6

be the shortest path from u to t. The principle works "backwards" in that, starting from vertex u closer to t, the shortest path from u to t is "unaware" of how to get from s to u. Should it happen that the shortest path from s to t passes through u, then the u–t shortest path must be a segment of the s–t shortest path.

The algorithm uses *successive approximation*, starting from an initial approximation for v. After a sufficient number of iterations using (2.3), a point is reached where no further improvements can be made. Such a point is called a *fixed point*. As initial approximation we take $l(s) = 0$ and $l(v) = \infty$, $v \in V$, $v \neq s$.

In the same way as in Dijkstra's algorithm, we can record the immediate predecessor when updating the distances and, when reaching the fixed point, backtrack to get the actual path. Even if the algorithm can handle negative distances, it will fail in the presence of a negative distance cycle (a cycle whose distances sum to a negative value). The Bellman–Ford algorithm runs in $O(|V| \cdot |E|)$ time, or since $O(|E|) = O(|V|^2)$, in $O(|V|^3)$ time. Compared to Dijkstra's algorithm it is therefore slower, but, being capable of handling negative weights, more robust.

Example 2.4.2. *Using the Bellman–Ford algorithm on the same problem as in Example 2.4.1, summarized in Table 2.2, we begin by letting $u_s^0 = 0$ and all $u_s^0 = \infty$ for $i \neq s$. In the first iteration, we have only vertex s with $u_s < \infty$, so $u_2^1 = \min\{\infty, 0 + 2\} = 2$ and $u_3^1 = \min\{\infty, 0 + 1\} = 1$. For all other vertices, the distances are trivially $u_i = \min\{\infty, \min_{k \neq j}\{\infty + d_{ki}\}\} = \infty$. The second iteration gives $u_2^2 = \min\{2, 2\}$ with the second argument equal to $2 = \min\{2, 1 + 3, \infty\}$ for $k = s$ ($k = 3$ gives $1 + 3 = 4$ and all other* *vertices give* ∞). *Continuing with vertices 4 and 5 gives the distances 5 and 2, respectively. Continued iteration successively improves the distance estimations, until no improvement can be made (after iteration 5).*

The fixed point is reached after four iterations, equaling the number of nodes traversed to reach t from s: $\{s, 3, 5, 4, t\}$. *In the worst case, the algorithm converges after up to* $|V| - 1$ *iterations. The first vertex s is the starting point, so a maximum of* $|V| - 1$ *vertices can be traversed by the shortest path.*

The Bellman–Ford algorithm is used for example in the *Routing Information Protocol* (RIP) in autonomous systems (subsystems like IP networks owned by an operator). Each node in the network calculates the distances to all neighboring nodes within the autonomous system and sends its tables to the other nodes. When a node receives a table from a neighbor, it recalculates its distance table based on the information received.

2.5 Minimum Spanning Trees

In network design, we want to balance two – often conflicting – objectives. Since the network cost is dependent on the number of links, we would like as few links as possible. The cost of transportation of a flow in terms of resources used in the network, however, is related to the number of links on the path of the flow, and we desire this number to be as low as possible. The second objective is minimized when there is a large number of links in the network. The most cost-efficient network topology is likely to be somewhere in between these two extremes.

Sparseness of Graphs

In order for a network to have a low cost, its corresponding graph should be *sparse*. Intuitively there should not be any "unnecessary" links present, where the importance of a link is a balance between the conflicting goals of the cost of the *network* and the cost of *routing a flow* (the longer the path a flow has to take to reach its destination, the more expensive it is). The *sparseness* of a graph is measured by its size and its weight.

As a matter of fact, it is very straightforward to find the sparsest possible subgraph of a general weighted graph $G = (V, E)$. A tree T which contains every vertex of a connected graph $G = (V, E)$ is called a *spanning tree* for G. Suppose G has a weight w_{ij} assigned to its edges (for example, length). The tree T which contains all vertices in G and minimizes the sum of the weights of the edges of T,

$$w(T_{\min}) \leq w(T) = \sum_{(u,v)\in T} w_{uv}, \quad \text{for all } T,$$

is called a *minimum spanning tree* (MST) for G, denoted $T_{\min}$. Note that the MST need not be unique. There are efficient algorithms to find the minimum spanning tree of a graph.

A well-known algorithm for finding spanning trees is *Kruskal's algorithm*, which builds the minimum spanning tree by greedily forming clusters. Initially, each vertex constitutes its own cluster and the set of edges of the tree $E_T = \emptyset$. The algorithm then investigates each edge in turn, where the edges are ordered by increasing weight. If an edge (u, v) connects two different clusters, then (u, v) is added to the set of edges of the minimum spanning tree E_T, and the two clusters connected by (u, v) are merged into a single cluster. Any edge (u, v) connecting two vertices that are in the same cluster is discarded. When all vertices are connected the algorithm terminates and the result is a minimum spanning tree.

Algorithm 2.5.1 (Kruskal)**.**

Given a connected weighted graph $G = (V, E)$ with n vertices and m edges.

STEP 0:   (initialize)

    for each vertex $u \in V$ **do**

      define the clusters, $C(u) \leftarrow \{u\}$

    end

    Sort E with respect to weights in increasing order. Call the list L

    Let $E_T \leftarrow \emptyset$

STEP 1:   (iterate)

    while E_T has fewer than $n - 1$ edges **do**

      starting from the top of L and progressing downwards,

      get an edge (u, v) from L

      if $C(u) \neq C(v)$ **then**

        add edge (u, v) to E_T

        merge $C(u)$ and $C(v)$ into one cluster

      end

    end

Output $T_{\min} = (V, E_T)$, a minimum spanning tree of G.   •

Kruskal's algorithm can be implemented to run in $O(m \log(m))$ time.

In the *Prim–Jarník algorithm*, we grow a minimum spanning tree starting from an initial vertex. The main idea is similar to that of Dijkstra's algorithm. We begin with a vertex s, which defines the initial cluster of vertices C. Then, in each iteration, we greedily choose a minimum weight edge (u, v) connecting a vertex u in the cluster C to a vertex v outside of C. The vertex v is then brought into the cluster C and the process is repeated until a spanning tree is formed. Since we are always choosing the edge with the smallest weight, we have a minimum spanning tree.

Algorithm 2.5.2 (Prim–Jarník)**.**

Given a connected weighted graph $G = (V, E)$ with n vertices and m edges.

STEP 0:   (initialize)

Select an arbitrary initial vertex $s \in V$ to start from

Let $E_T \leftarrow \emptyset$

Let $C \leftarrow s$

STEP 1:   (iterate)

while E_T has fewer than $n - 1$ edges **do**

select the edge of minimum weight between a tree and

a nontree vertex

add the selected edge to E_T

add the corresponding vertex to C;

end

Output $T_{\min} = (V, E_T)$, a minimum spanning tree of G.   ●

The Prim–Jarník algorithm can be implemented to run in $O(n^2)$ time.

A natural starting point in topological design is the minimum spanning tree. The problem, however, with using the MST as a network topology is that a flow may have to traverse as many as $n - 1$ links to reach its destination. The relative weighted path length in a subgraph $G' = (V, E')$ of a complete graph $G = (V, E)$, where $E' \subseteq E$, is defined by its *stretch factor*.

Definition 2.5.1. *Let* $G = (V, E)$ *be a complete graph with weight function w and let* $G' = (V, E')$, $E' \subseteq E$, *be a subgraph of G. Let* $d_{G'}(u, v)$ *denote the* shortest weighted path *between u and v in G', that is, the minimum of* $w(p)$ *over all paths p from u to v in G'. Then the* stretch factor *(or* dilation*) of G' is*

$$\text{stretch}(G') = \max_{u,v \in V} d_{G'}(u, v)/d_G(u, v).$$

Definition 2.5.1 allows us to define a class of graphs known as *spanners*, a class of graphs with low size, weight, and stretch, which can be used in topological design.

Example Topologies

Some simple generic topologies are often used in practice, such as the minimum spanning tree, topologies made up by the *complete graph* (often referred to as a "full mesh" topology), the *maximum-leaf spanning tree* (also known as the "star" topology), and the 2-circulant (the "ring" topology). The geometric distance is used as weight in Fig. 2.3, which is proportional to the edge distances depicted.

We can use these examples for comparison of quality measures for topologies. The properties of these topologies are summarized in Table 2.3. The example topologies have the same order, but different size and weight, which usually determine the cost of the network. They also have different degrees, diameter, and stretch, which are related to network performance.

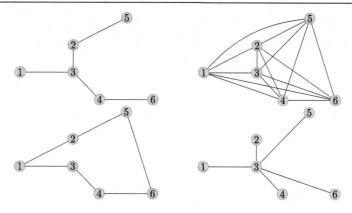

Figure 2.3: Some common network topologies: the minimum spanning tree (top left); the complete graph (top right); the shortest ring (bottom left); the minimum-weight maximum-leaf spanning tree (bottom right).

Table 2.3: Some characteristics of the sample topologies in Fig. 2.3.

Property	MST	Mesh	Ring	Star
Size (links)	5	15	6	5
Relative weight (cost)	1.00	4.82	1.51	1.20
Minimum degree	1	5	2	1
Maximum degree	2	5	2	5
Diameter	4	1	3	2
Stretch	2.13	1	4.3	2.35

The Traveling Salesman Problem

A common network topology, especially for high-speed networks, is the ring topology. In Table 2.3 we can see that rings possess some desirable properties, such as low size and low stretch, and provide path diversity with their minimum degree of 2. The problem of constructing such a ring while minimizing the cost is known as the *traveling salesman problem* (TSP). As the name suggests, the objective is to optimize the route (distance) of a "traveling salesman" that needs to visit n locations and return to the starting point so that the total distance of the trip is minimized.

The TSP is a classical $\mathcal{NP}$-complete problem because of its general usefulness and its intuitive interpretation. There are many methods for the TSP available. Some simple heuristics are easy to implement, but the result may be rather far from the optimum. Better results can be obtained by employing more sophisticated algorithms or combining different methods to improve an initial approximation. Two heuristics that can be used as first approximations are the nearest neighbor, incremental insertion, and k-optimal methods (such as the Kernighan–Lin algorithm) (see [5]). A solution or approximate solution to the TSP is called a *tour*.

Since a ring is a special type of network, its hardness implies the hardness of topological design in general. The problem can be described as follows. Let some points in the plane be the locations of cities. If there are n cities, a solution to the TSP can in theory be found by evaluating the weight of each $(n-1)!$ possible orderings of the cities. However, the number of orderings grow very quickly with the size n of the problem. When letting the cities be represented by points in the two-dimensional plane, we also assume that the *triangle inequality* is satisfied, that is,

$$d(u, v) \leq d(u, k) + d(k, v), \quad \text{for all } u, v, k \in V.$$

In other words, the distance of a direct link is always equal to or less than a path going through any other point in the plane.

The Nearest Neighbor Algorithm

One of the most straightforward methods to solve the traveling salesman problem is the *nearest neighbor* algorithm. Starting from an arbitrary vertex s, we connect to its nearest neighboring vertex and continue until vertex s is reached again. The algorithm, being a heuristic, does not guarantee an optimal solution. A straightforward improvement is to try the n different starting vertices and take the shortest of the n paths.

Incremental Insertion

Incremental insertion represents another type of heuristic where vertices are inserted into a tour. Starting from a single vertex s, it selects the farthest vertex not yet connected and forms a shortest path to determine the edges to connect this vertex with other vertices on this path. A vertex far from s is an attempt to add edges by decreasing cost. If s, u, and v are vertices and V_T is the set of vertices in the tour T, insert vertex j such that

$$\max_{v \in V} \left\{ \min_{u \in V_T} \{ d(s, u) + d_{uv} \} \right\}.$$

The minimization ensures that we insert the vertex in the position that adds the smallest amount of distance to the tour, while the maximum ensures that we pick the worst such vertex first. This works rather well, because when we add distant vertices first, paths to closer vertices are optimized better.

k-Optimal Methods

Whereas the methods described so far provide a more or less accurate approximation, the k-optimal (also known as the Kernighan–Lin algorithm) provides a way of improvement by

Figure 2.4: **The 2-optimal method can be used to improve the solution to the traveling salesman problem, where the initial solution is given by the nearest neighbor heuristic.**

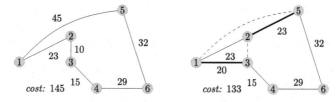

Figure 2.5: **The best solution given by the nearest neighbor algorithm can be improved by the 2-optimal method.**

local search. The method randomly selects $k \geq 2$ edges from an initial tour, and then the corresponding nodes are reconnected by k new edges. If the cost decreases, an improvement has been found. Otherwise, the solution is discarded. The procedure is repeated until no further improvement can be made or a predefined number of iterations has been reached. Usually, 2-optimal or 3-optimal methods are used. For $k > 3$, it is likely that the computation time becomes too long to be of practical use. The 2-optimal method in the context of topological design is also known as the *branch exchange* algorithm, illustrated in Fig. 2.4. Fig. 2.5 shows an initial solution to a TSP obtained by the nearest neighbor algorithm, followed by an improvement using the 2-optimal method.

When searching for edges to transform, we usually use a random selection method. It is important, however, that the edges are disjoint, i.e., do not have any vertices in common. If we search for vertices, then we also must assure that the edges actually exist in the input graph and that the transformed edges do not exist in the input graph. If they do, the algorithm produces a degenerate solution. There are other practical considerations, such as need for testing the graph for connectivity.

2.6 Network Resilience

In network design, we focus much of our discussion on aspects of reliability and resilience. Firstly, we need to define what exactly is meant by "reliability", as well as some way to measure it. Measures that are studied in network reliability are $\{s, t\}$- or *two-terminal reliability*, *k-terminal reliability*, and *all-terminal reliability*. Using the undirected graph $G = (V, E)$ as a topological model, these measures correspond to the restriction of the connectivity analysis to subsets of the vertex set V of cardinality 2, $2 < k < |V|$, and $|V|$, respectively.

In other words, we select the appropriate measure depending on whether we want to study the reliability between two specific vertices, the mutual reliability in a k-subset of the vertices, or the mutual reliability between all vertices. We will use all-terminal reliability measures in this text, since these are the most natural measures in a core network design context.

Network Cuts

The connectivity of two vertices in a graph, defined in Definition 2.3.1, is discussed in Section 2.3. Two vertices u and v in a graph are connected if there is a path from u to v. Depth-first search or breadth-first search can be used to determine whether two vertices are connected or not. Connectivity (between two vertices) is central to reliability, and this concept can be generalized to more than two vertices. However, rather than using the term "connectivity" to denote a Boolean variable (being connected or not), it is here used to mean "degree of connectivity". In a graph, two basic reliability measures are the rather intuitive concepts of *connectivity* and *cuts*.

Definition 2.6.1. A *vertex cut* for two vertices u and v is a set of vertices such that when removed, u is disconnected from v. The *two-terminal vertex connectivity* $\kappa(u, v)$ is the size of the smallest vertex cut disconnecting u and v. The *(all-terminal) vertex connectivity* $\kappa(G)$ of a graph G is the smallest vertex-cut that disconnects G. $\kappa(G)$ equals the minimum of $\kappa(u, v)$ over all pairs of distinct vertices u and v.   •

The "removal" of a vertex should here be understood so that the vertex and its incident edges are "blocked". Note that a complete graph with n vertices has no vertex cuts at all, but by convention it is set to $n - 1$. Similarly, an *edge cut* is defined.

Definition 2.6.2. An *edge cut* for two vertices u and v is a set of edges such that when removed, u is disconnected from v. The *two-terminal edge connectivity* $\lambda(u, v)$ is the size of the smallest edge cut disconnecting u and v. The *(all-terminal) edge connectivity* $\lambda(G)$ of a graph G is the smallest edge cut of G that disconnects G. Here $\lambda(G)$ equals the minimum of $\lambda(u, v)$ over all pairs of distinct vertices u and v.   •

The following relations hold for the above measures in a graph G with n vertices and m edges:

$$\kappa(G) \leq \lambda(G) \leq \delta(G) \leq \frac{1}{n} \sum_{i=1}^{n} d_i = 2m/n, \tag{2.4}$$

where $\delta(G)$ is the *minimum degree* of the graph. Note that the minimum degree of the graph is easily verified, but it is only an upper bound to the more precise and useful connectivity measures $\kappa(G)$ and $\lambda(G)$. A stronger reliability measure can be defined by requiring that the paths of connection are *disjoint*.

Definition 2.6.3. *Given a connected undirected graph $G = (V, E)$ and two distinct vertices $u, v \in V$, a set of paths from u to v is said to be* vertex-disjoint *if no vertex other than u and v themselves is on more than one of the paths. Similarly, the set is called* edge-disjoint *if none of the paths share an edge with any other path.* •

Denote the greatest number of vertex-disjoint paths between two vertices u and v by $\kappa'(u, v)$ and the greatest number of edge-disjoint paths between u and v by $\lambda'(u, v)$. Then the following relations hold.

Theorem 2.6.1 (Menger's theorem for vertex connectivity). *Let G be a finite undirected graph and u and v two nonadjacent vertices. Then the size of the minimum vertex-cut for u and v (the minimum number of vertices whose removal disconnects u and v) is equal to the maximum number of pairwise vertex-independent paths from u to v, that is, $\kappa(u, v) = \kappa'(u, v)$.*

Theorem 2.6.2 (Menger's theorem for edge connectivity). *Let G be a finite undirected graph and u and v two distinct vertices. Then the size of the minimum edge-cut for u and v (the minimum number of edges whose removal disconnects u and v) is equal to the maximum number of pairwise edge-independent paths from u to v, that is, $\lambda(u, v) = \lambda'(u, v)$.*

Menger's theorems asserts that the two-terminal vertex connectivity $\kappa(u, v)$ equals $\kappa'(u, v)$ and the two-terminal edge connectivity $\lambda(u, v)$ equals $\lambda'(u, v)$ for every pair of vertices u and v. This fact is actually a special case of the *max-flow min-cut theorem*, which can be used to find $\kappa(u, v)$ and $\lambda(u, v)$. The vertex connectivity and edge connectivity of G can then be computed as the minimum values of $\kappa(u, v)$ and $\lambda(u, v)$, respectively. These theorems thus allow a redefinition of connectivity as the sizes of the smallest vertex-cuts $\kappa(G)$ and edge-cuts $\lambda(G)$, respectively.

A graph is called *k-vertex-connected* if its vertex connectivity is k or greater. Similarly, it is called *k-edge-connected* if its edge connectivity is k or greater.

Example 2.6.1. *Consider the cube graph in Fig. 2.6. It has a vertex-cut of 3 and is therefore, according to Theorem 2.6.1, 3-vertex-connected. It is also, according to (2.4), 3-edge-connected (both $\kappa(G)$ and $\delta(G)$ equal 3).* □

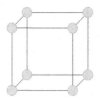

Figure 2.6: A 3-vertex-connected graph (the cube graph).

The Deletion-Contraction Principle

Let $G = (V, E)$ be a connected graph and $e = \{u, v\} \in E$. The *deletion* $G - e$ is the graph G with edge e removed, that is $(V, E - \{e\})$. The *contraction* G/e is obtained by merging the two end vertices u and v of e together. For a contraction to make sense, we require that the edge e is not a self loop.

Proposition 2.6.3. *Let $e \in E$ and let $\mathcal{T}(G)$ be the set of spanning trees of G. Then, we have the bijections*

$$\{T \in \mathcal{T}(G) | e \notin T\} \leftrightarrow \mathcal{T}(G - e)$$

and

$$\{T \in \mathcal{T}(G) | e \in T\} \leftrightarrow \mathcal{T}(G/e).$$

That is, the spanning trees of G that do not contain e are in bijection with the spanning trees of the deletion $G - e$, and the spanning trees of G that do contain e are in bijection with the spanning trees of the contraction G/e.

The first bijection is quite obvious: the set of spanning trees of G not containing e equals the set of spanning trees of G with e removed. In the second bijection, the edge e is made "permanent" by merging the two nodes. Thus, we have

$$\tau(G) = \tau(G - e) + \tau(G/e)$$

for any graph G and edge e. This gives a recursive algorithm to compute $\tau(G)$ for any graph. Unfortunately it is computationally inefficient to do so – the complexity of the algorithm is $O(2^n)$. It is, however, an important theoretical tool.

Example 2.6.2. *To illustrate how the deletion-contraction principle works, consider the four-node network in Fig. 2.7. The first component is the graph with the diagonal from the upper left-hand corner to the bottom right-hand corner removed. The second component is the contraction: the two corresponding vertices are merged and the edges (apart from the deleted one) are preserved.*

Figure 2.7: The first deletion-contraction step.

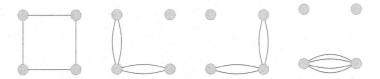

Figure 2.8: The second deletion-contraction step.

In the second contraction-deletion step, shown in Fig. 2.8, these two components are decomposed further. The second diagonal is now removed from both previous components, giving the four components shown. It is now easy to see that each component has exactly four spanning trees, so the total number of spanning trees for the graph is 16.

Edge contraction can also be used to find minimum cuts in the following randomized algorithm, presented in [8]. An edge $e \in E$ in the graph $G = (V, E)$ is chosen randomly and is contracted, that is, the vertices at each end are merged. Any edge between the two merged vertices are removed, so that there are no edges that are loops. All other edges are retained. Each contraction reduces the number of vertices of G by one.

The edge contraction does not change the size of the cut between two separate vertices u and v, since no edge between them has been contracted. Had it been, u and v would have been merged. As a result, a cut between u and v in a graph after some contraction step is also a cut in the original graph. The contractions are repeated until only two vertices remain. The remaining number of edges between these two vertices is the size of the cut.

The algorithm finds the size of any *cut. By randomization, we can search for a minimum cut and determine the probability that one has been found. If the algorithm is run $N^2/2$ times, the probability of not finding a minimum cut is*

$$\left(1 - \frac{2}{N^2}\right)^{N^2/2},$$

which can be made arbitrarily small by using sufficiently large N. □

CHAPTER 3

Network Science

The graph as a network model is a powerful tool to understand phenomena of many large networks. We invariably think of social media as prime examples hereof, but any structure where relations between entities can be identified can be thought of as a network. Examples of such structures include author collaborations in publications, biochemical reactions and spreading of diseases. Using random graphs as our main tool, we study the characteristics and evolution of large networks. The presentation loosely follows findings published by Barabási et al. [15] and Albert and Barabási [16] and references therein.

3.1 The Small-World Phenomenon

The term small-world phenomenon refers to the average path length between any two nodes in a large, densely connected network; any node is assumed to be directly linked to nodes far away with nonnegligible probability. The distance L between any two nodes, expressed in the average number hops (traversed nodes), is related to the total number of nodes n as

$$L \propto \log n.$$

The small-world phenomenon means that the minimum path between far-away nodes is relatively short. This is in essence the foundation of the popular expression "six degrees of separation", although misleading by the use of the term "degree", which has an entirely different meaning in graph theory.

Even though random graph theory has been around for many decades, it has attracted renewed interest as a means to model and analyze large communication networks, in particular the Internet and social networks.

3.2 The Erdős–Rényi Model

Analysis of the internet has shown the presence of the small-world property, that is, the average path length is proportional to the logarithm of the network order n. The small-world property, measured by average path length, is therefore one of three main characteristics we expect to be present in a network model.

5G Networks. https://doi.org/10.1016/B978-0-12-812707-0.00008-5

The Erdős–Rényi model [17][18], denoted $G(n, p)$, takes n nodes, and for each node pair $(u, v), u, v = 1 \ldots, n$, a link is formed with probability p, independently from every other node pair. If the number of links in a so-generated graph is m, determined by

$$m = E(n) = p(n(n - 1)/2),$$

each link in such a graph has equal probability

$$p_e = p^m (1 - p)^{n(n-1)/2 - m}$$

of being present.

Social networks also show formation of cliques – groups of friends or acquaintances where each member knows (is linked to) every other member. The degree to which cliques tend to form is measured by the clustering coefficient. Let us fix a random node v with degree k_v, that is, having k_v neighbors. If these k_v neighbors form a clique, there are $k_v(k_v - 1)/2$ links between them. The clustering coefficient for a node v is simply the fraction of the actual number of links E_v between the neighbors of v to the maximum number possible, or

$$C_v = \frac{2E_v}{k_v(k_v - 1)}.$$

The clustering coefficient for an entire network is the average cluster coefficient over all nodes v,

$$C = \sum_{i=1}^{n} C_i / n.$$

In an Erdős–Rényi random graph links are created with probability p, so that the clustering coefficient is $C = p$. For large networks, p must be small for a generated network of order n and size m. In real networks of the same order and size, however, the clustering coefficient is much larger. A successful model of large networks therefore needs to capture a larger degree of clustering in the network.

We will also pay attention to how the node degree k is distributed throughout the graph. Let $p(k)$ be the probability function that a node has degree k.

Following Erdős–Rényi, links are created randomly, and node degrees therefore tend to be close to the average node degree $\langle k \rangle = 2m/n$. As a matter of fact, the node degree in such a random network is distributed according to a Poisson distribution with parameter $\langle k \rangle$. (We use the notation $\langle \cdot \rangle$ for average values throughout this chapter.)

In real networks, however, the degree distribution tends to follow distributions rather different from Poisson. For many networks, such as the World Wide Web and the Internet, the node degrees follow a power law distribution,

$$p(k) \sim k^{-\gamma}, \tag{3.1}$$

where $\gamma > 0$ is a scaling parameter. An important property of the power law (3.1) is that scaling the argument k by a constant c gives a proportionate scaling of $p(k)$: $p(c \cdot k) \sim c^{-\gamma} k^{-\gamma} = c^{-\gamma} p(k)$, which is the reason why the power law distribution also is called *scale-free*. Networks where the degree distribution follow a power law are consequently called scale-free networks.

Graph Evolution

The objective of random graph theory is to study properties of the properties associated with graphs with n nodes as $n \to \infty$. The construction process of a graph, the creation of links on a set of n nodes, is called an evolution.

Assuming a set of n nodes and randomly added links, the connection probability p corresponding to the evolving graph increases as it approaches the fully connected graph in which $p \to 1$. It is of particular interest to determine at which probability p some particular property of the graph is likely to appear.

It turns out that many properties of graphs appear quite suddenly as p increases. For example, a random graph of order n and a given connection probability p is either almost always connected or almost always disconnected. For many properties of graphs, there is therefore a critical probability $p_c(n)$ dependent on the number of nodes n.

We are often interested in finding the critical probability $p_c(n)$ of the formation of special structures in a graph, such as trees, cycles, and cliques. A cycle in a graph is a subgraph such that there is a path from a node that leads back to the node itself. A tree is a graph without any cycles. In general terms, we may ask if there is a critical probability at which a subgraph with q nodes and l links appear.

More formally, consider a random graph $G(n, p)$ and a small subgraph F with q nodes and l links representing a pattern. Firstly, we would like to determine how many subgraphs F, not necessarily isolated, there are in G. A subgraph F has q nodes which can be chosen from the total of n nodes in $C_n^q = \begin{pmatrix} q \\ n \end{pmatrix}$ ways and l edges created with probability p^l. The q nodes can also be permuted to obtain a maximum of $q!$ new subgraphs. Corrected for graph isomorphisms, the actual value is $q!/a$, with a being the number of isomorphic graphs. Then, the expected number X of subgraphs F is

$$E(X) = C_n^q \frac{q!}{a} p^l \cong \frac{n^q p^l}{a}. \tag{3.2}$$

For random graphs, it can subsequently be shown that for a subgraph $F \subset G$ of order q, the critical probabilities are

(a) tree: $p_c(n) = bn^{-q/(q-1)}$,
(b) cycle: $p_c(n) = bn^{-1}$,
(c) clique: $p_c(n) = bn^{-2/(q-1)}$,

for some constant b. A fundamental structural change occurs when the average degree $\langle k \rangle$ passes the critical value $\langle k \rangle_c = 1$: for $\langle k \rangle < 1$ the largest cluster is a tree, whereas for $\langle k \rangle_c = 1$ it contains approximately $n^{2/3}$ of the nodes. This is known as a giant cluster. Other clusters tend to be rather small, typically trees. As the average degree increases further, the smaller clusters grow together and connect to the giant cluster.

Degree Distribution

In a random graph with connection probability p, the degree k_v of a node v follows a binomial distribution with parameters $N-1$ and p, that is,

$$p(k_i = k) = C_{N-1}^k p^k (1-p)^{N-1-k}, \tag{3.3}$$

where C_{N-1}^k is the binomial coefficient. That is, we take the probability of having k edges which is p^k, the probability of the absence of additional edges which equals $(1-p)^{N-1-k}$, and there are C_{N-1}^k ways of selecting these k edges.

We are interested in the degree distribution of large graphs, both as a characteristic affected by graph evolution and because the number of nodes with high (or low) degree is of importance in itself. Let X_k denote the number of nodes with degree k. The degree distribution is the probability $\mathbf{P}(X_k = r)$ that X_k equals the number r. From Eq. (3.3), we take the expectation,

$$\mathbf{E}(X_k) = N\,\mathbf{P}(k_1 = k) = \lambda_k,$$

where

$$\lambda_k = NC_{N-1}^k p^k (1-p)^{N-1-k}.$$

The distribution of X_k, determined by $\mathbf{P}(X_k = r)$, approaches a Poisson distribution with parameter λ_k as $n \to \infty$,

$$\mathbf{P}(X_i = r) = e^{-\lambda_k} \frac{\lambda_k^r}{r!}.$$

The longest path in a graph is called its diameter, that is the largest distance between any two nodes. In a disconnected graph, the diameter is infinite by definition. However, when connected subgraphs are forming, the diameter of a disconnected graph can also be defined as the largest diameter of its subgraphs.

In a random graph, diameters tend to be small unless the connectivity probability is very small. It turns out that the diameter D is proportional to $\ln(N)/\ln(\langle k \rangle)$, where $\langle k \rangle$ is the average node degree,

$$D = \frac{\ln(N)}{\ln(\langle k \rangle)} = \frac{\ln(N)}{\ln(pN)}.$$

We note that for small p where $\langle k \rangle = pN < 1$, a random graph is composed of isolated trees with diameter equal to the diameter of a tree. Whenever $\langle k \rangle > 1$, a giant cluster appears, and the diameter of the graph is proportional to $\ln(N)/\ln(\langle k \rangle)$.

Rather than using the diameter of a graph, we can describe its spread by calculating the average path length ℓ between all distinct node pairs. Following the result for the diameter, we might suspect that ℓ scales similarly to the number of nodes, so that

$$\ell \sim \frac{\ln(N)}{\ln(\langle k \rangle)}. \tag{3.4}$$

From Eq. (3.4) we see that we can plot $l \ln(\langle k \rangle)$ as a function of $\ln(N)$ to obtain a straight line with slope 1.

Clustering Coefficient

Clustering is clearly discernable in real networks, and the degree and manner of how an evolving graph coalesces, or forms clusters, is therefore of particular interest. We may interpret *clustering* as the probability that two neighbors are connected, which in random networks is equal to the probability that any two nodes are connected,

$$C_c = p = \frac{\langle k \rangle}{N}. \tag{3.5}$$

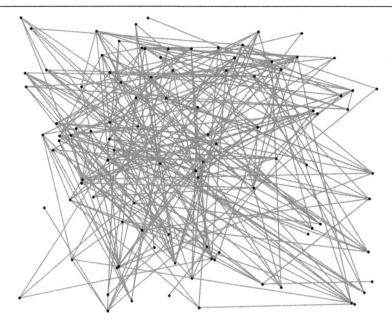

Figure 3.1: An Erdős–Rényi random graph of order $n = 100$ and $p = 0.05$.

Example 3.2.1. *Fig. 3.1 shows a simulated Erdős–Rényi random graph of order $n = 100$ and connectivity probability $p = 0.05$, so that $\langle k \rangle = 5$, in agreement with the measured value 5.09. The theoretical and measured degree distributions are shown in Fig. 3.2. The measured average path length is $\ell \approx 3.32$ compared to $\ln(N)/\ln(\langle k \rangle) \approx 2.86$, and the clustering coefficient is $0.043 \approx p$.*

3.3 Scale-Free Networks

The Erdős–Rényi random graph fails to model some characteristics observed in real networks. The degree distribution in business and social networks turns out to follow power law (Pareto) distributions rather than a binomial distribution. That is, we expect to have $\mathbf{P}(k) \sim k^{-\gamma}$ for some scaling parameter γ. We therefore need a modified model for such networks.

The Barabási–Albert Model

Investigations into real networks, including the World Wide Web and social networks, have shown power law distributions of the node degrees. This effect can be shown to be generated in a growing network, where nodes and links are constantly added and where new links are added in proportion to the degrees of already present nodes.

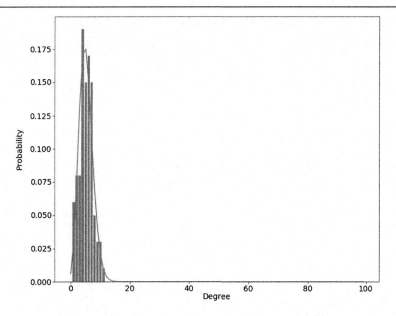

Figure 3.2: Degree distributions of the random graph.

The Barabási–Albert model mimics an evolving network, which is constantly growing by added nodes and links, similarly to members and connections in a social network. Connections are not completely random, but rather proportional to the number of existing links a node has. This assumption is rather natural since a node's affinity to other nodes tends to propagate out from the node through its connections. This is an example of preferential attachment: a document will likely include links to well-known publications with high degrees, which in turn are easy to find.

The model is initiated from a nucleus – a small set m_0 of connected nodes from which the network can grow. Then, for each added node, having m initial links, the connection probability to other nodes is proportional to the degrees of these nodes.

Algorithm 3.3.1 (The Barabási–Albert model).

Initiate with a connected nucleus of m_0 nodes, fix a number $m \leq m_0$ of initial edges for added nodes, and let t be a stop time.

STEP 0:   (initialize)
      Connect the m_0 nodes of the nucleus
STEP 1:t (iterate)

Add a node and connect its m edges by preferential attachment

The probability Π that a new node will be connected to node i depends on the degree k_i of node i, such that

$$\Pi(k_i) = \frac{k_i}{\sum_i k_i}. \tag{3.6}$$

Output generated graph G. ●

As the model simulates a growing network, we can index the network states by a time variable t. After t time steps the model has generated a network with $N_t = t + m_0$ nodes and mt links. Numerical simulations show that the model generates a scale-invariant network where the probability that a node has k links follows a power law with an exponent $\gamma_{BA} = 3$. The scaling exponent is independent of m, the only parameter in the model. It can be shown that the dynamic growth process and the preferential attachment together lead to the scale-free property of the generated network. We take a closer look at average path length, node degree distribution, and the clustering coefficient.

The average path length ℓ is shorter in a Barabási–Albert network than in the Erdős–Rényi network of the same order and size. This suggests that the scale-free property also brings nodes closer than randomly distributed edges. It can be shown that the average path length of the Barabási–Albert model increases approximately logarithmically with N, which can be described by the generalized logarithmic form

$$\ell = a \ln(N - b) + c, \tag{3.7}$$

for some constants a, b and c. Bollobás and Riordan [19] found that the diameter of the Barabási–Albert is asymptotically $\ln(N)/\ln(\ln(N))$.

In random graph models where edges are distributed randomly, the node degrees are uncorrelated. In the Barabási–Albert model, however, a correlation between degrees tends to develop [20]. To this end, we let the degree k_v of a given node v in a large network be a continuous variable of the growth rate t of the network. Here, t denotes the evolution of the network, that is, the changes when adding one node at a time. With preferential attachment, the rate of change of k_v with t is proportional to the connection probability $\Pi(k_v)$. We have

$$\frac{\partial k_v}{\partial t} = m\Pi(k_v) = m\frac{k_v}{\sum_{i=1}^{N-1} k_i}.$$

The sum of all degrees in the network (except the newly added one) is $\sum_i k_i = 2mt - m$, which gives

$$\frac{\partial k_v}{\partial t} = \frac{k_v}{2t},$$

which has the solution

$$k_v(t) = m \left(\frac{t}{t_v}\right)^\beta \quad \text{with } \beta = \frac{1}{2}, \tag{3.8}$$

using the initial condition that each node v at the time of its creation t_v has degree $k_v(t_v) = m$. Eq. (3.8) shows the scale-free property of the model by the power law node degree distribution.

Asymptotically as the network grows ($t \to \infty$), the degree distribution $P(k)$ can be shown to satisfy

$$P(k) \sim 2m^{1/\beta} k^{-\gamma} \quad \text{with } \gamma = \frac{1}{\beta} + 1 = 3 \tag{3.9}$$

with the exponent γ independent of m. This result, that $P(k)$ is independent of t and N, shows that $P(k)$ reaches a limiting scale-free state, which is in agreement with empirical and numerical findings.

Letting t be the evolution in time of the model, where a time step is equivalent to the addition of a node and m links, we derive an expression for the growth of a special graph element – a connected node pair (u, v), where $\deg(u) = k$ and $\deg(v) = l$.

Let N_{kl} denote the number of node pairs where one node has degree k, the other has degree l and the nodes are connected by an edge.

Without loss of generality, we can assume that the node with degree k was added later than the node with degree l, so that $k < l$ according to Eq. (3.8), that is, older nodes have a higher probability of having a high degree than younger nodes. Then,

$$\frac{\partial N_{kl}}{\partial t} = \frac{(k-1)N_{k-1,l} - kN_{kl}}{\sum_k kN(k)} + \frac{(l-1)N_{k,l-1} - lN_{kl}}{\sum_k kN(k)} + (l-1)N_{l-1}\delta_{k1}.$$

The terms on the right-hand side contribute to the change in N_{kl}. The first term signifies an increase by adding an edge to a node previously having degree $k - 1$, and deducting the case of adding an edge to a node previously with degree k, resulting in degree $k + 1$. The second term represents the corresponding changes to nodes with initial degree l. The last term corrects for the case that $k = 1$ where the added edge is the same as the edge already connecting the nodes. Here, δ_{k1} equals one if $k = 1$ and zero otherwise. The correction term is necessary since in the first term $k - 1 = 0$ for $k = 1$. Note also that $k < l$, so only the case $k = 1$ needs to be considered.

Investigations show that the clustering coefficient C_c is about five times higher in the scale-free graph compared to the random graph (the Erdős–Rényi model). The clustering coefficient of the Barabási–Albert model can be approximated by a power law distribution

$$C_c \sim N^{-0.75}.$$

However, this coefficient decays slower with N than the clustering coefficient for random graphs $C_c = \langle k \rangle / N$.

3.4 Evolving Networks

Analyses of real networks show that even when the degree distribution is well described by a power law distribution

$$P(k) \sim k^{-\gamma},$$

their measured exponents γ vary between 1 and 3. In contrast, the Barabási–Albert model yields a power law with fixed exponent. Expressing the connectivity probability as

$$\Pi(k) \sim k^{\alpha},$$

the constant α may vary compared to empirical findings for real networks. For the internet, for example, $\Pi(k)$ appears to depend linearly on k, that is, $\alpha \cong 1$, as predicted in the Barabási–Albert model. In other networks, the dependence of the connectivity on k is sublinear, with $\alpha = 0.8 \pm 0.1$.

In real networks, it can be noticed that there is a nonzero probability that a new node attaches to an isolated node v ($\deg(v) = 0$), so that $\Pi(0) \neq 0$. To generalize the ansatz for $\Pi(k)$, we let

$$\Pi(k) = a + k^{\alpha}, \tag{3.10}$$

where $a > 0$ caters for the nonzero initial connection probability, or *attractiveness* of the node. When $a = 0$, a node having $k = 0$ will never be able to attract any edges. In real networks, however, new nodes have a nonzero probability of being connected to, such as a new publication that is cited for the first time.

Modeling $\Pi(k)$ by Eq. (3.10) does not destroy the scale-free property of the degree distribution. As a matter of fact, it can be shown that

$$P(k) \sim k^{-\gamma} \quad \text{with } \gamma = 2 + a/m.$$

Another phenomenon in real networks is that the average degree $\langle k \rangle$ does not remain constant, but increases in time. Recall that the average degree in the Barabási–Albert model is $\langle k \rangle = 2|E|/|V| = 2mt/t = 2m$, that is, constant. The conclusion is that in many real networks, the number of edges grows faster than the number of nodes, leading to *accelerated growth*.

We have seen that in the Barabási–Albert model, new nodes connect to b nodes already in the system with preferential attachment

$$P_i = b \frac{k_i}{\sum_j k_j}.$$

Moreover, at every time step a linearly increasing number of edges are distributed between the nodes, with the probability of an edge connecting nodes i and j being

$$P_{ij} \frac{k_i k_j}{\sum_{s,l,s \neq l}^{k_s k_l}} N(t) a.$$

Here $N(t)$ is the number of nodes in the system and the summation goes over all nonequal values of s and l. As a result of these two processes the average degree of the network increases linearly in time, following $\langle k \rangle = at + 2b$. The continuum theory predicts that the time-dependent degree distribution displays a crossover at a critical degree,

$$k_c = \sqrt{b^2 t} (2 + 2at/b)^{3/2},$$

such that for $k << k_c$, $P(k)$ follows a power law with exponent $\gamma = 1.5$ and for $k >> k_c$, the exponent is $\gamma = 3$. This result explains the fast decaying tail of the degree distribution, and it indicates that as time increases the scaling behavior with $\gamma = 1.5$ becomes increasingly visible.

Another important metric is the rich club index, which measures the extent to which nodes with high node degrees connect to each other. Networks with high rich club indices are said to show the rich club effect. It is defined as

$$\phi(k) = \frac{2E_{k'>k}}{N_{k'>k}(N_{k'>k} - 1)},$$

where $E_{k'>k}$ is the number of edges between nodes with a degree $k' > k$, and $N_{k'>k}$ is the number of such nodes.

Example 3.4.1. *Fig. 3.3 shows a simulated Barabási–Albert graph of order $n = 100$ and the initial number of edges per node $m = 3$, so that $\langle k \rangle = 6$ (ignoring the initiation cluster) and the degree distribution is as shown in Fig. 3.4. The average path length is $\ell \approx 2.6$ (compared to the theoretical 3.0) and the clustering coefficient is 0.18 (compared to the theoretical 0.21). The evolution illustrated by the weights of three nodes is shown in Fig. 3.5, which can be interpreted as the "first mover advantage". The rich club index is depicted in Fig. 3.6.*

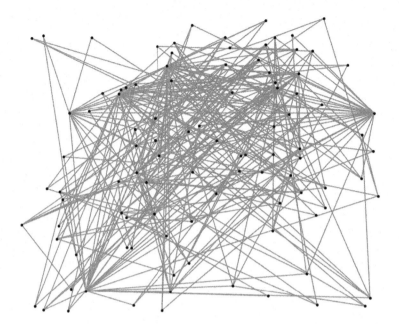

Figure 3.3: A Barabási–Albert graph of order $n = 100$ and $m = 3$.

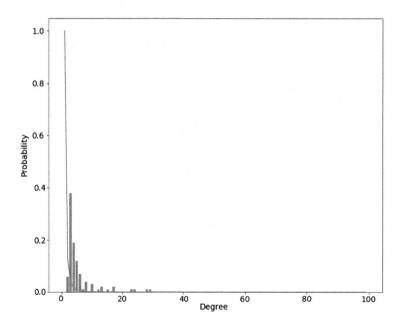

Figure 3.4: Degree distribution in the Barabási–Albert graph.

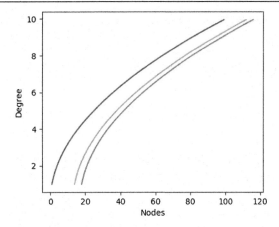

Figure 3.5: Evolution in the Barabási–Albert graph.

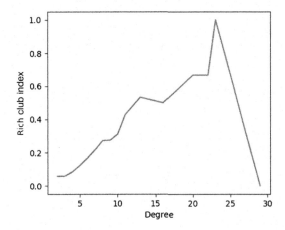

Figure 3.6: Rich club index in the Barabási–Albert graph.

The Barabási–Albert model incorporates only a single process for network growth – addition of new nodes connecting to nodes already in the system. In real systems, however, many other events affect the network on different time scales. Such events are based on the fundamental operations of addition or deletion of nodes and addition and deletion of edges.

An edge modification process is the following edge rewiring. After a node is added, it does not remain passive, but existing nodes and connections can be expected to change over time. We may therefore select an edge randomly (or rather, one of its endpoints) and reconnect (rewire) to a node proportionally to its degree. We can also allow creation of a number c of new edges at every time step, connecting two nodes with probability proportional to the product of their degrees. When $c > 0$ the network is called developing, and when $c < 0$ it is called decaying. In many real networks nodes and edges have a finite lifetime, or finite edge or node capacity as well.

In the Barabási–Albert model all nodes that have increasing degrees follow a power law time dependence with the same dynamic exponent $\beta = 1/2$. This implies that older nodes have statistically higher degrees. In real networks the degree does not only depend on a node's age, but nodes grow in different paces.

Bianconi and Barabási propose a modification of the Barabási–Albert model, in which each node is assigned a constant fitness parameter η_i [16]. This represents a competitive aspect, as each node has an intrinsic ability to compete for edges at the expense of other nodes. Whenever a new node v with fitness η_v is added to the system, where η_v is chosen from a distribution $\rho(\eta)$, each such node connects with m edges to the nodes already present in the network with probability

$$\Pi_i = \frac{\eta_i k_i}{\sum_j \eta_j k_j}.$$

This model allows a new node with a few edges to acquire edges at a high rate if it has a high fitness parameter. The rate of change of node v is

$$\frac{\partial k_v}{\partial t} = m \frac{\eta_v k_v}{\sum_j \eta_j k_j}.$$

Recalling the degree evolution (3.8) with a fitness-dependent $\beta(\eta)$, we have

$$k_{\eta_v}(t, t_v) = m \left(\frac{t}{t_v} \right)^{\beta(\eta_v)},$$

where the exponent satisfies

$$\beta(\eta) = \frac{\eta}{C} \quad \text{with } C = \int \rho(\eta) \frac{\eta}{1 - \beta(\eta)} d\eta.$$

Example 3.4.2. *Fig. 3.7 shows a simulated Bianconi–Barabási graph of order $n = 100$, initial number of edges per node $m = 3$, and uniform fitness, so that $\langle k \rangle = 6$ (ignoring the kernel) and with a degree distribution as shown in Fig. 3.8. The average path length is $\ell \approx 2.5$ (compared to the theoretical 3.0) and the clustering coefficient is 0.27 (compared to the theoretical 0.21). The evolution illustrated by the weights of three nodes is shown in Fig. 3.9, which can be interpreted as "fit-gets-richer". The rich club index is shown in Fig. 3.10.*

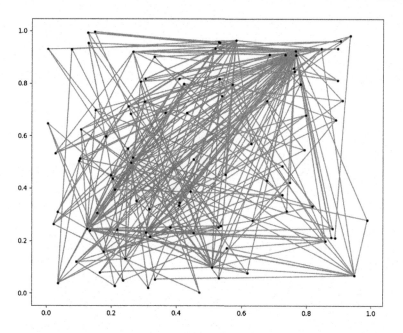

Figure 3.7: A Bianconi–Barabási graph of order $n = 100$ and $m = 3$.

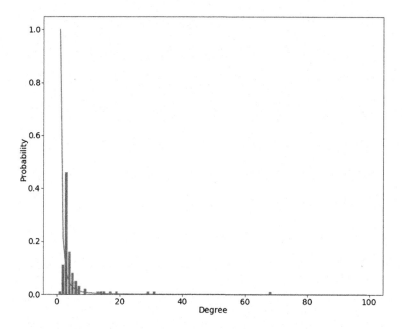

Figure 3.8: Degree distribution in the Bianconi–Barabási graph.

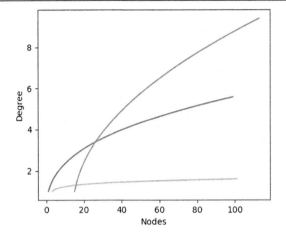

Figure 3.9: Evolution of the Bianconi–Barabási graph.

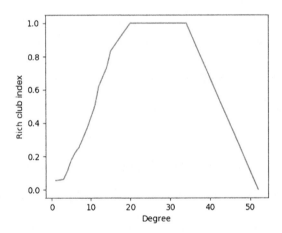

Figure 3.10: Rich club index in the Bianconi–Barabási graph.

3.5 Degree Correlation

By studying real networks, we can get an idea of degree correlation, which also captures the way nodes connect to each other. We can roughly classify networks according to their degree correlation. Referring to nodes with high edge degree as hubs, we call a network

- *assortative* when hubs tend to link to each other,
- *neutral* when nodes connect to each other according to expected random probabilities,
- *disassortative* when hubs tend to avoid linking to each other.

Let e_{jk} be the probability to find a node with degree j and degree k at the two ends of a randomly selected edge, so that

$$\sum_{jk} e_{jk} = 1, \tag{3.11}$$

$$\sum_{j} e_{jk} = q_k, \tag{3.12}$$

where q_k is the probability of finding a node having degree k at the end of a link. Then we have

$$q_k = \frac{k p_k}{\langle k \rangle}. \tag{3.13}$$

The tendency of a node to connect to more well connected nodes is expressed by $q_k = C k p_k$, which after normalization gives Eq. (3.13). In the absence of degree correlations, we have

$$e_{jk} = q_j q_k. \tag{3.14}$$

Average Next Neighbor Degree

To measure degree correlations, we define the *average next neighbor degree* $k_{nn}(k)$ as the average degree of the closest neighbors of nodes with a specified degree k. In terms of the adjacency matrix A, we derive the expressions

$$k_{nn}(k_i) = \frac{1}{k_i} \sum_{j=1}^{N} A_{ij} k_j, \tag{3.15}$$

$$k_{nn}(k) = \sum_{k'} k' P(k'|k). \tag{3.16}$$

Then, for a neutral network, we have

$$k_{nn}(k) = \sum_{k'} k' q_{k'} = \frac{\langle k^2 \rangle}{\langle k \rangle}. \tag{3.17}$$

For an assortative network, the correlation implies that the higher the degree k of a given node is, the higher the average of its nearest neighbors is, so that $k_{nn}(k)$ increases with k. In a disassortative network, hubs tend to link to low-degree nodes, so that $k_{nn}(k)$ decreases with k. The advantage of $k_{nn}(k)$ as a degree measure is that it is only a function of k and is easy to interpret.

The Correlation Coefficient

We have used the correlation coefficient to express the magnitude of the degree correlations, which can be written

$$\sum_{jk} jk(e_{jk} - q_j q_k).$$

The correlation coefficient is expected to be positive for assortative networks, zero for neutral networks, and negative for disassortative networks. To make the measure universal, we normalize it by the maximum for a perfectly assortative network having $e_{jk} = q_k \delta_{jk}$, that is,

$$\sigma_r^2 = \max \sum_{jk} jk(e_{jk} - q_j q_k) = \sum_{jk} jk(q_k \delta_{jk} - q_j q_k),$$

so that

$$r = \frac{\sum_{jk} jk(e_{jk} - q_j q_k)}{\sigma_r^2}, \quad -1 \leq r \leq 1,$$

where $r \leq 0$ indicates a disassortative, $r = 0$ a neutral, and $r \geq 0$ an assortative network. Among real networks, social networks tend to be assortative, whereas technical networks (such as the Internet and the World Wide Web) are disassortative.

To see the relation between $k_{nn}(k)$ and r, we set

$$k_{nn}(k) = a \cdot k^\beta.$$

Solving for a gives

$$a = \frac{\langle k^2 \rangle}{k^{\beta+1}},$$

so that

$$\begin{aligned} \beta < 0 &\quad \rightarrow \quad r < 0, \\ \beta = 0 &\quad \rightarrow \quad r = 0, \\ \beta > 0 &\quad \rightarrow \quad r > 0. \end{aligned}$$

Structural Cut-Off

If we allow only one link between two nodes in a finite network, there may eventually not be enough hubs to satisfy assortativity. Consider the two sets of nodes having degree k and k', respectively, and let N_k and $N_{k'}$ be the corresponding number of nodes. Then the maximum number of edges between these sets is

$$m_{kk'} = \min\{kN_k, k'N_{k'}, N_kN_{k'}\}.$$

We can see this by observing that there cannot be more links between the two groups than the overall number of edges joining the nodes with degree k and k', respectively. Also, only allowing simple edges, we cannot have more links between the two sets than if we connect every node with degree k to every node with degree k' once. Letting the number of edges connecting the sets with degrees k and k' be

$$E_{kk'} = e_{kk'}\langle k \rangle N,$$

where $e_{kk'}$ is the proportion of edges between the two sets, we define the ratio

$$r_{kk'} = \frac{E_{kk'}}{m_{kk'}} \leq 1.$$

Let k_s and be the degree such that $r_{k_s k_s} = 1$. Then we refer to this limit as the structural cut-off of the network, which depends on the network size N.

For neutral networks we have

$$m_{kk'} = \min\{kN_k, k'N_{k'}, N_kN_{k'}\},$$
$$m_{k_s k_s} = k_s N_{k_s} = k_s N p_{k_s}$$
$$r_{k_s,k_s} = \frac{\langle k \rangle N k_s^2 p_{k_s}^2}{\langle k \rangle^2 k_s p_{k_s} N} = \frac{k_s p_{k_s}}{\langle k \rangle} = q_{k_s} < 1 \quad \forall k_s,$$

leading to the structural cut-off

$$r_{k_s k_s} = \frac{\langle k \rangle N k_s^2 p_{k_s}^2}{\langle k \rangle^2 N^2 p_{k_s}^2} = \frac{k_s^2}{\langle k \rangle N} = 1,$$
$$k_s(N) = \sqrt{\langle k \rangle N}.$$

Therefore, in a neutral network of order N restricted to single links between nodes, it is impossible to have nodes with degree larger than $k_s(N)$. This implies that if there are nodes such that $k > k_s(N)$, the network cannot be neutral, but tends to be disassortative.

For a scale-free network of order N, we have the expected maximum degree $k_{\max}$, which is such that

$$\int_{k_{\max}}^{\infty} P(k)\mathrm{d}k \approx \frac{1}{N},$$

since the probability cannot exceed one node having such a degree. At the same time, using the power law probability of node degrees,

$$\int_{k_{max}}^{\infty} P(k)dk = (\gamma - 1)k_{min}^{\gamma-1} \int_{k_{max}}^{\infty} k^{-\gamma} dk = \frac{k_{min}^{\gamma-1}}{k_{max}^{\gamma-1}},$$

$$k_{max} = N^{\frac{1}{\gamma-1}}.$$

For $\gamma < 3$, a common value in real networks, k_{max} diverges faster than $k_s \sim \sqrt{N}$, which is the case in a neutral network.

It should be noted that situations exist where $k_{max} > k_s$ can arise in a neutral network due to structural limitations, thereby showing disassortativity. Therefore, the structural cut-off should be investigated when calculating the degree correlation. Examples of assortative networks are the internet and scientific collaboration networks. Power grids tend to be neutral, whereas the World Wide Web and e-mail networks show disassortative tendencies.

3.6 Importance

It is often of interest to find the "importance" of nodes in a network, called *centrality*. In a directed graph modeling a social network, a node's indegree indicates how popular you are (such as the number of followers), whereas the outdegree show how many people you know.

We define the *degree centrality* as the normalized node degree,

$$C^D(i) = \frac{k_i}{N-1}.$$

This, however, is a local measure. By inspection of a graph, we may also be interested in a node's position relative to groups. Such nodes are important for the propagation of information in a network.

We therefore define the *betweenness centrality* measure as

$$\tilde{C}^B(i) = \sum_{j<k} \frac{d_{jk}(i)}{d_{jk}},$$

where d_{jk} is the number of shortest paths between j and k and $d_{jk}(i)$ is the number of shortest paths between j and k going through i. Normalizing gives

$$C^B(i) = \frac{\tilde{C}^B(i)}{(N-1)(N-2)/2}.$$

Another measure, the *closeness centrality*, aims at capturing how close a node is to all other nodes. It is defined as

$$\tilde{C}^C(i) = \left(\sum_{j=1}^{N} d(i,j)\right)^{-1},$$

that is, the reciprocal of the sum of all shortest path distances from node i to all other nodes. This can also be normalized to obtain

$$C^C(i) = \frac{\tilde{C}^C(i)}{N-1}.$$

We note that in the Erdős–Rényi model, different nodes tend to score high in different centrality measures, whereas in the Bianconi–Barabási model a few nodes score high in all three measures.

3.7 Robustness

The term *robustness* refers to the ability of a network to operate under node failures. We distinguish between random failures, where nodes fail with a certain probability independently of their degrees, and attacks, where nodes fail with a probability proportional to their degrees. It has been shown that many networks exhibit a scale-free topology. Such networks turn out to be robust with respect to random failures. The reason for this is that when the probability of failure is equal for each node, the large nodes – or hubs – are relatively unlikely to fail.

We view a network as a collection of clusters of connected nodes. A fully connected network therefore consists of a single (giant) cluster. A network failure can therefore be viewed as the situation where the network breaks down into smaller clusters, that is, locally connected network fragments. For an initially connected network, suppose we remove a fraction f of the nodes. It appears that there is a critical rate f_c at which the connectivity in the sense of the existence of a giant cluster is destroyed. We can characterize the network topology in terms of the size of f as

- $f = 0$ All nodes are part of the giant cluster,
- $0 < f < f_c$ The network largely remains connected in a giant cluster,
- $f > f_c$ The network is fragmented into many small clusters.

Failure of a node cause the disabling of its connecting links as well, thereby causing more damage than a single link failure. Clearly, robustness is strongly dependent on network topology: scale-free networks are more robust than random networks, but are more susceptible to attacks targeting highly connected nodes.

Starting from a connected network from which we successively remove a node, a number of links are disabled with each node which annihilate some paths between other nodes. The breakdown of an initially connected network subject to node removal can be studied by determinating the relative size of the largest connected cluster S and the average path length ℓ within this cluster, as a function of the fraction f of nodes removed. Not unexpectedly, the largest cluster size decreases, whereas the average path length increases as links are disabled through the node removals.

For a giant cluster to exist each constituent node must be connected to at least two other nodes on average. This implies that we can expect the average degree k_i of a randomly chosen node i that is part of the giant cluster to be at least two. We let $\mathbf{P}(k_i | i \leftrightarrow j)$ denote the joint probability that a node in a network with degree k_i is connected to a node j that is part of the giant cluster. Using this conditional probability the expected degree of node i can be expressed as

$$\langle k_i | i \leftrightarrow j \rangle = \sum_{k_i} k_i \, \mathbf{P}(k_i | i \leftrightarrow j) = 2.$$

We can write the probability as

$$\mathbf{P}(k_i | i \leftrightarrow j) = \frac{\mathbf{P}(k_i, i \leftrightarrow j)}{\mathbf{P}(i \leftrightarrow j)} = \frac{\mathbf{P}(i \leftrightarrow j | k_i) p(k_i)}{\mathbf{P}(i \leftrightarrow j)}$$

using Bayes' theorem. For a network without degree correlations, we have

$$\mathbf{P}(i \leftrightarrow j) = \frac{2L}{N(N-1)} = \frac{\langle k_i \rangle}{N-1} \text{ and } \mathbf{P}(i \leftrightarrow j | k_i) = \frac{k_i}{N-1}$$

that is, we have $N - 1$ nodes to connect to, each with probability $1/(N-1)$ and this is multiplied by the degree k_i. Simplifying, we obtain

$$\sum_{k_i} k_i \, \mathbf{P}(k_i | i \leftrightarrow j) = \sum_{k_i} k_i \frac{\mathbf{P}(i \leftrightarrow j | k_i) p(k_i)}{\mathbf{P}(i \leftrightarrow j)} = \sum_{k_i} k_i \frac{k_i \, p(k_i)}{\langle k \rangle}$$

This reasoning shows the *Molloy–Reed criterion* for the condition of having a giant cluster, that is

$$\kappa = \frac{\langle k^2 \rangle}{\langle k \rangle} > 2.$$

Networks with $\kappa < 2$ do not have a giant cluster, but are fragmented into many disconnected components. The Molloy–Reed criterion connects a network's connectivity with the first two moments of the node degrees $\langle k \rangle$ and $\langle k^2 \rangle$. The result is valid for any degree distribution p_k. The critical fraction f_c is given by

$$f_c = 1 - \frac{1}{\frac{\langle k^2 \rangle}{\langle k \rangle} - 1}.$$

This implies that when $f < f_c$ the network remains connected, that is, there is a giant cluster with high probability, whereas when $f > f_c$ the giant cluster is likely to break down and render the network disconnected. Scale-free networks are robust under random failures, since the likelihood of a hub failing is relatively small. Since

$$f_c = 1 - \frac{1}{\kappa - 1}$$

we have

$$\kappa = \frac{\langle k^2 \rangle}{\langle k \rangle} = \left| \frac{2 - \gamma}{3 - \gamma} \right| \times \begin{cases} K_{\min} & \gamma > 3 \\ K_{\max}^{3-\gamma} K_{\min}^{\gamma-2} & 3 > \gamma > 2 \\ K_{\max} & 2 > \gamma > 1, \end{cases}$$

where

$$K_{\max} = K_{\min} N^{\frac{1}{\gamma-1}}.$$

When $\gamma > 3$, κ is finite, and so the network breaks apart at a finite f_c, that depends on $K_{\min}$. When $\gamma < 3$, κ diverges in the $N \to \infty$ limit due to the positive exponent of $K_{\max}$, so that $f_c \to 1$. This means that in the limit $N \to \infty$, all nodes need to be removed for a network to break down. When the network is finite, the breakdown criterion can be written

$$\kappa \cong 1 - CN - \frac{3 - \gamma}{\gamma - 1} \tag{3.18}$$

The breakdown of the giant cluster is indicated by a faster than linear decrease in S as f. At the same time, ℓ increases up to the critical fraction f_c where it peaks, and thereafter decreases as the giant cluster rapidly breaks down. This behavior is illustrated in Figs. 3.11–3.12. For small f, random graph theory indicates that ℓ scales as $\ln(SN)/ln(\langle k \rangle)$, where $\langle k \rangle$ is the average degree of the largest cluster. Since the number of links decreases faster than the number of nodes in operation, $\langle k \rangle$ decreases faster with increasing f than SN, and consequently ℓ increases. When $f \approx f_c$, the average path length ℓ no longer depends on $\langle k \rangle$ and decreases with S.

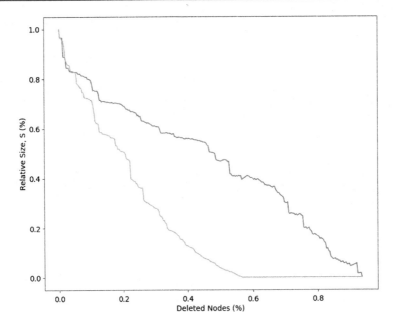

Figure 3.11: Largest cluster size S as function of the fraction f of removed nodes for a random network and a scale-free network.

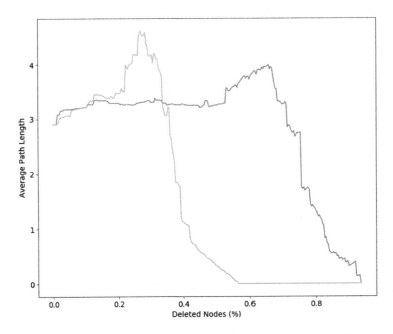

Figure 3.12: Average path length ℓ as function of the fraction f of removed nodes for a random network and a scale-free network.

In contrast, scale-free networks show a remarkable robustness subjected to random node failures. Compared to a random network, the decreasing cluster size reaches zero at a higher f, and ℓ increases slower to a less prominent peak. As the size of a scale-free network increases, the threshold $f_c \to 1$.

The degree distribution $P(k)$ changes with node removals as follows. Suppose a node has an initial degree k_0 chosen from a distribution $P(k_0)$. After random removal of a fraction f of the nodes, the probability of the node having the new degree $k \leq k_0$ is $C_{k_0}^k (1-f)^k f^{k_0-k}$ and the new degree distribution becomes

$$P(k) = \sum_{k_0=k}^{\infty} P(k_0) C_{k_0}^k (1-f)^k f^{k_0-k}.$$

Thus, the average degree and its second moment in the new network follows $\langle k \rangle = \langle k_0 \rangle (1-f)$ and $\langle k^2 \rangle = \langle k_0^2 \rangle (1-f)^2 + \langle k_0 \rangle f (1-f)$, leading to the critical fraction

$$f_c = 1 - \frac{1}{\frac{\langle k_0^2 \rangle}{\langle k_0 \rangle} - 1}, \tag{3.19}$$

where $\langle k_0^2 \rangle$ and $\langle k_0 \rangle$ are determined from the original degree distribution.

In a Erdős–Rényi graph, we have $k_0 = pN$ and $k_0^2 = (pN)^2 + pN$. Eq. (3.19) then gives the critical fraction as $f_c = 1 - 1/(pN)$. If initially $\langle k_0^2 \rangle / \langle k_0 \rangle = 2$, we have $pN = \langle k \rangle = 1$ and $f_c = 0$, so that any number of nodes removed leads to the fragmentation of the network. The higher the original degree $\langle k_0 \rangle$, the more robust the network is, that is, a larger number of nodes can be removed without disconnecting the network. In scale-free networks, the degree distribution follows a power law

$$P(k) = ck^{-\gamma}, \quad k = m, m+1, \ldots, K,$$

where m and $K \approx mN^{1/\gamma-1}$ are the smallest and largest node degrees, respectively. Using a continuum approximation (valid for $K >> m >> 1$), we obtain

$$\frac{\langle k_0^2 \rangle}{\langle k_0 \rangle} \to \frac{|2-\gamma|}{|3-\gamma|} \times \begin{cases} m & \text{if } \gamma > 3 \\ m^{\gamma-2} K^{3-\gamma} & \text{if } 2 < \gamma < 3 \\ K & \text{if } 1 < \gamma < 2. \end{cases}$$

It follows that for $\gamma > 3$ the ratio is finite and the critical fraction is

$$f_c = 1 - \frac{1}{\frac{\gamma-2}{\gamma-3} m - 1}.$$

When $\gamma < 3$ the ratio diverges with K. In the range $2 < \gamma < 3$,

$$f_c = 1 - \frac{1}{\frac{\gamma-2}{3-\gamma}m^{\gamma-2}K^{3-\gamma} - 1},$$

and $f_c \rightarrow 1$ when $N \rightarrow \infty$. This remains true for $\gamma < 2$. Thus, infinite networks with $\gamma < 3$ do not fragmentize under random failures since a spanning cluster exists for arbitrarily large f.

3.8 Attack Tolerance

We model an attack on a network by removing hubs, that is, nodes with high degrees, with high probability. Thus, let f be the fraction of removed hubs. This operation will change the degree distribution $P(k)$ in general and the maximum degree of the network K_{max} in particular. The attack problem can actually be reduced to the robustness problem by appropriately modifying K_{max} and $P(k)$. In a targeted attack, the network breaks down faster than under random node removals with a much lower critical fraction f_c.

In this case, the nodes with the highest number of links are removed in each step. It follows that scale-free networks break down faster than random networks due to the presence of a few highly connected nodes. If we remove a fraction f of the hubs, the maximum degree changes from, say, K_{max} to $\tilde{K}_{max}$. Then we have

$$\int_{\tilde{K}_{max}}^{K_{max}} P(k)\,\mathrm{d}k = f,$$

where

$$\int_{\tilde{K}_{max}}^{K_{max}} P(k)dk = (\gamma - 1)K_{min}^{\gamma-1} \int_{\tilde{K}_{max}}^{K_{max}} k^{-\gamma}dk = \frac{\gamma - 1}{1 - \gamma}K_{min}^{\gamma-1}(K_{max}^{1-\gamma} - \tilde{K}_{max}^{1-\gamma}).$$

Since $\tilde{K}_{max} \leq K_{max}$ and the initial maximum degree is the denominator, we can ignore this term and get

$$\left(\frac{K_{min}}{\tilde{K}_{max}}\right)^{\gamma-1} \approx f$$

or

$$\tilde{K}_{max} = K_{min}f^{\frac{1}{1-\gamma}},$$

which is the new estimate of the maximum degree after removing a fraction f of the hubs. When nodes are removed, remaining nodes with a certain degree k will lose some links when some of their neighbors vanish. We can calculate the fraction $\tilde{f}$ of links that disappear as a consequence of removing a fraction f of the hubs. We have

$$\tilde{f} = \frac{\int_{\tilde{K}_{max}}^{K_{max}} P(k)dk}{\langle k \rangle} = \frac{1}{\langle k \rangle}(\gamma - 1)K_{min}^{\gamma-1} \int_{\tilde{K}_{max}}^{K_{max}} k^{1-\gamma} dk = -\frac{1}{\langle k \rangle}\frac{\gamma - 1}{2 - \gamma}K_{min}f^{\frac{2-\gamma}{1-\gamma}}.$$

For the mth moment we have

$$\langle k^m \rangle = -\frac{(\gamma - 1)}{(m - \gamma + 1)}K_{min}^m$$

and

$$\langle k \rangle = -\frac{(\gamma - 1)}{(2 - \gamma)}K_{min}.$$

We conclude that

$$\tilde{f} = f^{\frac{2-\gamma}{1-\gamma}}.$$

This equation can be solved numerically to obtain $\tilde{K}$ as a function of m and γ, and $f_c(m, \gamma)$ can then be determined. The critical fraction f_c is very small for all γ, of the order of a few percent. A graph of $f_c(\gamma)$ exhibits a maximum at $\gamma \approx 2.25$, which represents the highest robustness against attacks. When γ approaches 2, $\tilde{f} \to 1$, which can be interpreted as when only a small fraction of the hubs is removed, the network is broken up. This follows from that for $\gamma = 2$ hubs dominate the network.

Just like in the robustness problem, we can find a threshold f_c at which an initial giant cluster is destroyed. Actually, with corrections for the changes in K_{max} and $P(k)$, the attack problem can be reduced to the robustness problem. Therefore, replacing K_{max} by $\tilde{K}_{max}$ and f by $\tilde{f}$, we have

$$\tilde{f} = 1 - \frac{1}{\kappa' - 1}$$

and

$$\kappa' = \frac{\langle \tilde{k}^2 \rangle}{\langle \tilde{k} \rangle} = \frac{\langle k^2 \rangle}{(1 - f_c)\langle k \rangle} = \frac{\kappa}{1 - f_c}.$$

We recall that κ is defined by Eq. (3.18) and the expression for f_c becomes

$$f_c^{\frac{2-\gamma}{1-\gamma}} = 2 + \frac{2 - \gamma}{3 - \gamma}K_{min}\left(f_c^{\frac{3-\gamma}{1-\gamma}} - 1\right).$$

We have the following facts regarding robustness and attach tolerance.

- The threshold f_c for random failures decreases monotonically with γ, whereas for attacks it can have a non-monotonic behavior, increasing for small γ and decreasing for large γ.
- For attacks f_c is always smaller than f_c for random failures.
- For large γ a scale-free network behaves similar to a random network. Since a random network does not have hubs, it follows that the attack and random failure thresholds converge for large γ. In the limit $\gamma \to \infty$, we get

$$f_c \to 1 - \frac{1}{(k_{\min} - 1)}.$$

We conclude that the higher average degree a network has, the more robust it is to failures and attacks.

3.9 Fault Propagation

Suppose we have a random network with N nodes where initially all nodes are functional. At the introduction of a fault in a node, we are interested in how this fault propagates throughout the network. To model fault propagation, we define constants f_i associated with each node that reflects the fraction (or probability) of neighbors of i that fail as a result of a fault at i if exceeding a critical spreading rate, that is $f_i \geq \phi$. This mechanism models how a fault introduced in a randomly selected node may cascade throughout the network. Figs. 3.13–3.15 show the result of fault propagation where the propagation rates f_i are uniformly distributed.

In contrast to random networks, a fault spreads easier in scale-free networks. In fact, any spreading rate leads to corruption of the whole network. That is, in scale-free networks the critical spreading rate approaches zero.

3.10 Improving Robustness

An interesting design question is how to increase the robustness of a network with respect to both random failures and targeted attacks without increasing the cost. We have shown that the robustness of a network is given by the size of the threshold f_c – the fraction of the nodes we must remove for the network to fragmentize. To enhance the robustness of a network we must increase f_c. Since f_c only depends on $\langle k \rangle$ and $\langle k^2 \rangle$, a degree distribution maximizing f_c has to maximize $\langle k^2 \rangle$ if we wish to keep the cost fixed. The general solution is a bimodal degree distribution, leading to a network with only two kinds of nodes with degrees $k_{\min}$ and $k_{\max}$.

Finally, we note that there is a fundamental difference between transportation networks and evolving networks. Cost optimal transportation networks can have widely different topologies, strongly dependent on the cost structure. Such topologies result from global optimization

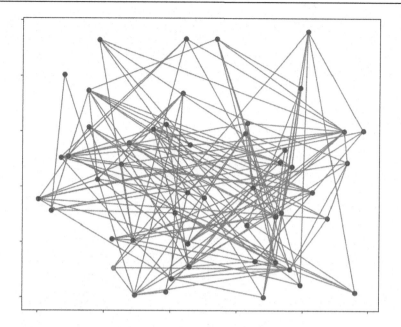

Figure 3.13: Fault propagation in a random network with random uniform propagation rates – only one node affected.

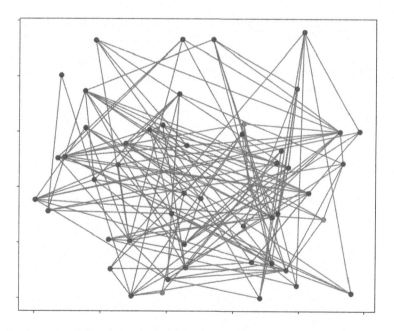

Figure 3.14: Fault propagation in a random network with random uniform propagation rates – a few nodes affected.

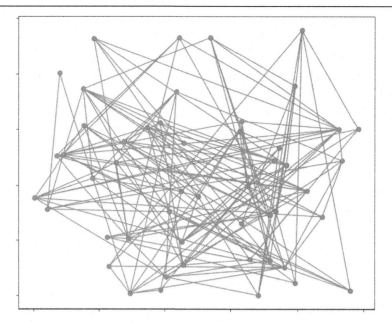

Figure 3.15: Fault propagation in a random network with random uniform propagation rates – the whole network affected.

processes, whereas evolving networks are mainly governed by local processes, on node level. However, realistic transportation networks are shaped by a certain degree of evolution as well as it is expanded. On the other hand, in scale-free models the nodes are assumed to possess some knowledge about the degrees of other nodes in the network.

CHAPTER 4

Self-Similarity, Fractality, and Chaos

The chapter discusses the origin, analysis and implications of data that behaves differently from "regular" data. Such data includes self-similar and fractal data. Self-similarity and fractality are two closely related concepts. Self-similar data shows approximately similar behavior on different scales. This may, for example, be the relative variation of traffic as the result of aggregation. Fractals have the property that they scale differently from "regular" geometric objects. In a square or circle, for example, the area is proportional to the square of the longest side or diameter, whereas a fractal may scale differently. The way fractals scale is known as its fractal dimension.

Chaos theory describes the behavior of dynamical systems that show apparent randomness in its evolution. Chaos in communication networks often results from complex feedback, retransmissions, and self-similar and fractal processes.

4.1 Self-Similarity: Causes and Implications

Network traffic is very heterogeneous, exhibiting variation on various time scales. It has been widely accepted that many traffic types are self-similar (long range–dependent), which may lead to starvation of traffic with shorter memory, just as prioritization of short-memory traffic may throttle long range–dependent traffic. Inasmuch long range–dependent traffic may cause congestion where resources are allocated statically, it might be possible to improve the network performance by allocating resources dynamically based on the characteristics of the traffic. In fact, the long-range dependence of traffic means that the load it induces can be predicted using its autocorrelation structure. The predicted load may serve as a control variable for dynamic traffic aggregation, which is performed on much longer time scales than, for example, routing table lookups.

Even if the control is assumed to be much slower than the traffic processing speed, the analysis of offered traffic and network load should be parsimonious and fast enough to capture variations on, say, time scales of seconds or fractions of a second.

Firstly, it is instructive to look at the scaling properties of different processes (see, for example, [21]). We define the time-aggregated traffic as the average of a time block of size m, so that

$$X_t^{(m)} = \frac{1}{m}\left(X_{tm-m+1} + \ldots + X_{tm}\right). \tag{4.1}$$

5G Networks. https://doi.org/10.1016/B978-0-12-812707-0.00009-7

When studying the sample mean $\bar{X}$ of a traffic process X, a standard result in statistics is that the variance of $\bar{X}$ decreases linearly with sample size. That is, if $X_1, X_2, \ldots, X_n$ represent instantaneous traffic with mean $\mu = \mathbf{E}(X_i)$ and variance $\sigma^2 = \text{Var}(X_i) = \mathbf{E}((X_i - \mu)^2)$, then the variance of $\bar{X} = n^{-1} \sum_{i=1}^{n} X_i$ equals

$$\text{Var}(\bar{X}) = \sigma^2 n^{-1}. \tag{4.2}$$

For the sample mean $\bar{X}$, we have for large samples

$$\mu \in [\bar{X} \pm z_{\alpha/2} s \cdot n^{-1/2}], \tag{4.3}$$

where $z_{\alpha/2}$ is the upper $(1 - \alpha/2)$ quantile of the standard normal distribution and $s^2 = (n - 1)^{-1} \sum_{i=1}^{n} (X_i - \bar{X})^2$ is the sample variance estimating σ^2.

The condition under which Eqs. (4.2) and (4.3) hold are:

(1)   the process mean $\lambda = \mathbf{E}(X_i)$ exists and is finite,
(2)   the process variance $\sigma^2 = \text{Var}(X_i)$ exists and is finite,
(3)   the observations $X_1, X_2, \ldots, X_n$ are uncorrelated, that is,

$$\rho(i, j) = 0, \quad \text{for} \quad i \neq j.$$

We assume that conditions (1) and (2) always hold, but not necessarily condition (3).

The autocorrelation of a process X estimates the inherent memory in a process, and is defined as

$$\rho(i, j) = \gamma(i, j)/\sigma^2, \tag{4.4}$$

where

$$\gamma(i, j) = \mathbf{E}((X_i - \lambda)(X_j - \lambda)).$$

The maximum likelihood estimator with $h = j - i$ is

$$\gamma(h) = \sum_{i=1}^{n-h} ((X_i - \bar{X})(X_{i+h} - \bar{X}))/((n - h)\sigma^2),$$

where X_i and X_j are observations (number of packets) at times i and j, respectively, on some time scale. The quantity λ is the average arrival rate of the process.

Smooth Traffic

Smooth traffic is traditionally modeled by a Poisson process. The Poisson process, also called Markovian or *memoryless*, aggregates in time so that the variance to the mean decreases fast. The average magnitude tends to a well-defined limit (that is, the aggregate variance tends to zero) as m increases. This produces a smoothing effect in time, which is very advantageous from a performance point of view. Thus, only one parameter – the intensity – is required to characterize a Poisson process, and the same parameter determines the server capacities necessary for a given performance.

In contrast, self-similar traffic does not exhibit such convenient behavior [4][21]. Time aggregates do not smoothen such processes rapidly, and large variations can be experienced even on large time scales. This phenomenon motivates an adaptive traffic aggregation that allocates resources in proportion to some measure of the load on a particular time scale.

The Poisson process

The traditional traffic arrival model is the Poisson process, which can be derived in a straightforward manner. We note that the Poisson process is a discrete process (for example, the number of packets) in continuous time. Fixing a time t and looking ahead a short time interval $t + h$, a packet may or may not arrive in the interval $(t, t + h]$. If h is small enough and the packet arrivals are independent, then the probability of a packet arrival in this interval can be assumed to be approximately proportional to the length of the interval, h. The probability of two or more arrivals in this interval can be considered negligible. We define a Poisson process as follows.

Definition 4.1.1. *A Poisson process with intensity λ is a process $X = \{N(t) : t \geq 0\}$ taking values in $S = \{0, 1, 2, \ldots\}$ such that*

(1)

$$N(0) = 0 \text{ and if } s < t \text{ then } N(s) \leq N(t),$$

(2)

$$\mathbf{P}(N(t + h) = n + m \mid N(t) = n) = \begin{cases} \lambda h + o(h) & \text{if } m = 1, \\ o(h) & \text{if } m > 1, \\ 1 - \lambda h + o(h) & \text{if } m = 0, \end{cases}$$

(3) if $s < t$ then the number $N(t) - N(s)$ of arrivals in the interval $(s, t]$ is independent of the times of arrivals during $[0, s]$.

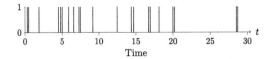

Figure 4.1: A Poisson call arrival process.

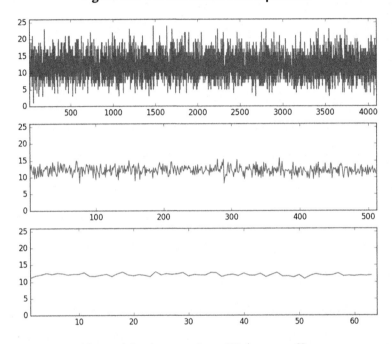

Figure 4.2: Aggregation of Poisson traffic.

An example of packets arriving according to a Poisson process is illustrated in Fig. 4.1, showing arrivals in continuous time. In discrete time, we are concerned with the arrivals in a time interval $(t, t + h]$, which can be described by the counting process related to the Poisson process, where $N(t)$ represents the number of arrivals of a process $X(t)$ up to time t.

Theorem 4.1.1. *The counting process $N(t)$ has the Poisson distribution with parameter λt, that is,*

$$\mathbf{P}(N(t) = j) = \frac{(\lambda t)^j}{j!} e^{\lambda t}, \quad j = 0, 1, \ldots.$$

By substituting t for h, the length of a discrete-time interval, $N(h)$ is the number of arrivals in each interval. This is the way Poisson traffic is generated in our simulations. We note that the Poisson process is one of the simplest continuous-time Markov processes, which means that it is memoryless, which follows from condition (3) in the definition. Fig. 4.2 shows the scaling of a Poisson process, where the time interval becomes successively longer in Eq. (4.1) with a scaling factor $m = 8$ in two consecutive steps.

Bursty Traffic

It has been shown that video sources exhibit various degrees of burstiness and autocorrelation [22][23]. The autocorrelation can be interpreted as *short-range memory*. This is partly explained by the nature of most video image sequences and their coding. In its original form, a movie consists of *frames* that change in rapid succession. Within a movie, it is common that scenes do not change with very high frequency. To save bandwidth, video coding therefore utilizes this fact and *changes* to a reference frame are encoded. When a scene changes, more data are needed to set up a new scene, whereas within the same scene, only updates with lower bandwidth need to be transmitted. The size of these updates also depend on the type of scene.

Modeling of video traffic is complicated by a number of possible encodings, compressions, and source types, such as movies or video conferences. We simply refer to bursty traffic as video traffic, without any claim that the chosen model (Markovian additive process) is suitable for all video sources. We also note that interactive multimedia, such as video conferencing, is more sensitive to delay, whereas streaming video is more sensitive to delay variation (jitter) [24]. We are using delay as the optimization target, so our results may be considered more relevant for the former type of video source.

The Markovian Additive Process

The Markovian Additive Process (MAP) has been suggested as a model for traffic types exhibiting burstiness and autocorrelation [25][26][27]. In its simplest form, the process switches between two states – active and silent – controlled by a Markov chain. The parameters are set so that the probability of remaining in either state is high if the process was in this state immediately before. This generates a bursty arrival process with autocorrelation (memory), contrary to the arrivals following a Poisson process.

To define the MAP, we assume that there are N independent traffic sources which are controlled by the same Markov chain. The chain jumps between the active and the silent states, represented by the states $S = \{0, 1\}$, with zero representing the silent state and one the active state. The Markov chain is defined by the transition probabilities between the silent and the active states as

$$a = \mathbf{P}(X_t = 1 | X_{t-1} = 0),$$
$$d = \mathbf{P}(X_t = 0 | X_{t-1} = 1).$$

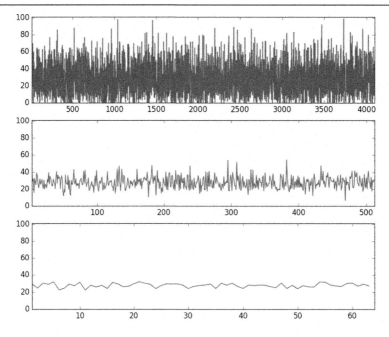

Figure 4.3: Aggregation of MAP traffic.

The Markov chain can now be expressed as

$$M = \begin{pmatrix} 1-a & a \\ d & 1-d \end{pmatrix},$$

where the steady-state probabilities are $\pi_0 = \frac{d}{a+d}$ and $\pi_1 = \frac{a}{a+d}$, respectively. The smaller the parameters a and d are, the burstier is the traffic generated by the model. As defined above, the MAP is a discrete-time model, which makes simulation straightforward. We note that the length of a source remaining in a state is geometrically distributed. Thus, the mean duration of an active period is $\sum_{n=1}^{\infty} dn(1-d)^{n-1} = 1/d$ time units, and the mean duration of a silence period is $1/a$ time units.

When fed to a queue with service capacity s, we require a stability criterion to have a limited stationary queue length, that is,

$$\frac{a}{a+d} < s/N.$$

Fig. 4.3 shows the scaling of an MAP as in Eq. (4.1) with a scaling factor $m = 8$ in two steps. Although bursty, the MAP is still Markovian, which means that even if packet arrivals are correlated, the change of states of the source is memoryless. The aggregation of an MAP is shown in Fig. 4.3. It scales similarly to the Poisson process (Fig. 4.2), but converges slightly slower to its process mean.

Long Range–Dependent Traffic

It may happen that the autocorrelation (4.4) decays very slowly, so that

$$\sum_{k=-\infty}^{\infty} \rho(k) = \infty.$$

This is the case when

$$\rho(k) \approx C_1 |k|^{-\alpha} \text{ as the time lag } k \to \infty, \tag{4.5}$$

where $\alpha \in (0, 1)$ and $C_1 > 0$ is a constant, that is, the autocorrelation decays according to a power law distribution. A process for which (4.5) holds is known as a long range–dependent process, or a long-memory process. The aggregation of such processes is such that

$$\text{Var}(X^{(m)}) = \sigma^2 m^{-\alpha},$$

with $0 < \alpha < 1$, that is, the aggregate converges to the sample mean slower than for a short-memory process. In contrast, for Markovian and short-memory processes, the autocorrelation is bounded as

$$|\rho(k)| \leq b \cdot a^k,$$

where $0 < b < \infty$ and $0 < a < 1$ are positive constants and k is the time lag. It follows that the sum of autocorrelations

$$\sum_{k=-\infty}^{\infty} \rho(k) = C_2 < \infty$$

is finite. Long range–dependent traffic shows significant dependence in time on long time scales. This type of traffic can be generated by superimposing a large number of short-memory processes or by certain flow control, such as the bandwidth of Transmission Control Protocol (TCP) controlled traffic [28]. It should be noted that long-range dependence is difficult to ascertain, since it requires determination of how correlations converge to zero and therefore needs investigation of traffic measured during long time intervals.

This type of processes have the interesting property that they can be predicted with better accuracy than short-memory processes. The stronger dependence of an observation X_t and past values X_{t-1}, X_{t-2}, ..., the more certain a future value X_{t+h} is likely to be close to past values X_t. This fact is used by Tuan and Park [29] for predictive congestion control.

Long-range dependence is closely related to the concept of self-similarity, where the latter describes the scaling properties of a process. The degree of dependence of the increments $X(t) = Y(t) - Y(t - 1)$ of a process is specified by the *Hurst parameter H*, where

$$H \in (0, 1).$$

If $H = \frac{1}{2}$, the process is independent (or memoryless). When $H > \frac{1}{2}$, the process is positively correlated, and when $H < \frac{1}{2}$ it is negatively correlated. We will assume that the process is positively correlated, so that $\frac{1}{2} < H < 1$. This follows from the nature of internet protocols as well as, for example, human browsing behavior. A cumulative process in continuous time $Y(t)$ is self-similar with Hurst parameter H if

$$Y(at) = |a|^H Y(t),$$

for all $a > 0$ and $t \geq 0$. Long-range dependence and self-similarity are not equivalent in general. However, when $\frac{1}{2} < H < 1$, self-similarity implies long-range dependence and vice versa. In this case, the autocorrelation function expressed in the Hurst parameter H is

$$\rho(k) \approx H(2H - 1)k^{2H-2},$$

and $\rho(k)$ behaves asymptotically as (4.5) with $\alpha = 2 - 2H$.

Fractional Brownian motion

A parsimonious model for self-similar, long range–dependent traffic is the *fractional Brownian motion*, $B_H(t)$, defined on a time interval $(0, T)$. The fractional Brownian motion is a process such that

(1) $B_H(t)$ is Gaussian on $t \in (0, T)$,
(2) the process starts from zero, that is, $B_H(0) = 0$ almost surely,
(3) $B_H(t)$ has stationary increments,
(4) the expectation is $\mathbf{E}(B(t) - B(s)) = 0$ for any $s, t \in (0, T)$,
(5) the autocovariance of $B_H(t)$ is

$$\mathbf{E}(B_H(t)B_H(s)) = \frac{1}{2}\left(|t|^{2H} + |s|^{2H} - |t - s|^{2H}\right) \quad \text{for any } s, t \in (0, T). \tag{4.6}$$

The process is self-similar and long range–dependent for $H > \frac{1}{2}$. Fig. 4.4 shows the aggregation of a fractional Brownian motion subject to a scaling factor of $m = 8$. The aggregate

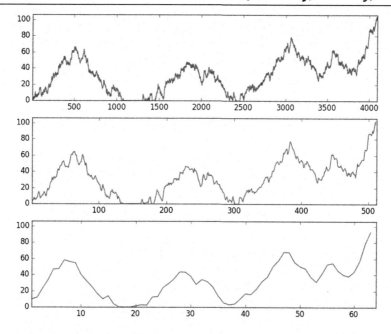

Figure 4.4: Aggregation of fractional Brownian motion traffic.

does not converge to any well-defined mean value, but a noticeable variability persists for the aggregate traffic.

4.2 Stochastic Processes

One of the main tools for network analysis is the theory of stochastic processes. By imposing some type of constraints on a probability governing the data, rich theories have developed for several classes of processes. Examples include *stationary stochastic processes*, *Markov processes*, and *queues*. A more recent class of processes are *self-similar processes*.

Basic Definitions

The following basic definitions and results can be found in for example [30] and [31].

A general *stochastic process* consists of three components:

(1) A *state space X*, the set of all states of the system, normally some topological space.
(2) An *index set T*, usually time. The index set T should have two basic properties:
    (a) T should be linearly ordered. In case the T represents time, this condition gives a meaning to concepts like "the future" and "the past".

(b) T should have some sort of additive structure, giving meaning to "moving forward" or "backward". However, T need not be closed under addition (which, in the case of time, would imply an "infinite future" or "infinite past").

T can be chosen to be discrete or continuous.

(3) A probability measure $\mathbf{P}$ on the space of all possible maps from T into X.

Assuming that the index set T is one-dimensional, the stochastic process is formally defined as follows.

Definition 4.2.1 (Stochastic process). A *stochastic process* is a family of random variables $\{X_t, t \in T\}$ defined on a probability space $(\Omega, \mathcal{F}, \mathbf{P})$, where Ω is the sample space and $\mathcal{F}$ its σ-algebra.

Some of the main statistics related to stochastic processes are

$$
\begin{aligned}
\mu &= \mathbf{E}(X_n), \quad \text{(mean)} \\
\sigma^2 &= \text{Var}(X_n) = \mathbf{E}[(X_n - \mu)^2], \quad \text{(variance)} \\
\gamma(k, n) &= \mathbf{E}((X_n - \mu)(X_{n+k} - \mu)), \quad \text{(autocovariance function)} \\
\rho(k, n) &= \frac{\gamma(k, n)}{\sigma^2}, \quad \text{(autocorrelation function)} \\
S(\nu) &= \sum_{k=-\infty}^{\infty} \rho(k) e^{-ik\nu}, \quad \text{(spectral density)},
\end{aligned}
$$

where ν represents the frequency in radians.

In analysis on telecommunication networks, mostly discrete stochastic processes are used, where $T \in \mathbb{Z}$ or $T \in \mathbb{N}$, since measurements on such networks are done by sampling at discrete-time intervals. It is usually assumed that the stochastic process is *stationary*.

Definition 4.2.2 (Strict stationarity). A *discrete stochastic process* $\{X_t\}$ *is said to be* strictly stationary *if the joint distributions*

$$
(X_{t_1}, X_{t_2}, \ldots, X_{t_k}) \quad and
$$
$$
(X_{t_1+h}, X_{t_2+h}, \ldots, X_{t_k+h})
$$

are the same for all $k \in \mathbb{Z}^+$ *and all* $t_1, t_2, \ldots, t_k, h \in \mathbb{Z}$.

The strict stationary definition is often difficult to apply, and the following definition, referred to as *weak stationarity* (second-order stationarity, or stationarity in the wide sense) is used instead.

Definition 4.2.3 (Weak stationarity). *A real-valued discrete stochastic process $\{X_t\}$ is said to be* weakly stationary *if*

(1) $\mathbf{E}(X_t)^2 < \infty$ *for all $t \in \mathbb{Z}$,*
(2) $\mathbf{E}(X_t) = m$ *for all $t \in \mathbb{Z}$, and*
(3) $\gamma(r, s) = \gamma(r + t, s + t)$ *for all $r, s, t \in \mathbb{Z}$.*

A naturally occurring discrete stochastic process in telecommunication engineering is the *counting process.*

Definition 4.2.4 (Counting process). *Consider a stochastic process $\{X_n, n \in \mathbb{Z}^+\}$ representing the number X_n units of some quantity arriving during the nth time interval with duration Δt, so that*

$$X_n = N(n\Delta t) - N((n - 1)\Delta t].$$

The number of units arriving up to time T represents a counting process,

$$N(t) = \int_0^T \mathrm{d}N(z),$$

where $\mathrm{d}N(z)$ is a point process *that represents arrivals of units; X_n is referred to as the* increment process.

Two fundamental classes of stochastic processes are *Gaussian processes* and *Markov processes.*

Definition 4.2.5 (Gaussian process). *A real-valued continuous time process $X(t)$ is a* Gaussian process *if each finite-dimensional vector $(X(t_1), \ldots, X(t_n))$ has the multivariate normal distribution*

$$\mathrm{N}(\mu(\mathbf{t}), \mathbf{V}(\mathbf{t}))$$

for some mean vector μ and some covariance matrix $\mathbf{V}$.

Definition 4.2.6 (Markov process). *A real-valued continuous-time process $X(t)$ is a* Markov process *if*

$$\mathbf{P}(X(t_n) \leq x | X(t_1) = x_1, \ldots, X(t_{n-1}) = x_{n-1}) = \mathbf{P}(X(t_n) \leq x | X(t_{n-1}) = x_{n-1})$$

for all $x, x_1, \ldots, x_{n-1}$ and all increasing sequences $t_1 < t_2 < \ldots < t_n$.

A Brownian motion (or Wiener process) is a continuous-time process describing the erratic movements of a particle in a suspension, colliding with other particles and thereby changing momentum and direction.

Definition 4.2.7 (Wiener process). *A Wiener process is a Gaussian process with independent increments for which:*

(1) $W(0) = 0$ with probability 1;
(2) $\mathbf{E}(W(t)) = 0$;
(3) $\mathrm{Var}(W(t) - W(s)) = t - s$ for all $0 \leq s \leq t$.

Self-Similar and Long Range–Dependent Processes

Traditional *short range–dependent* models have been shown not to be adequate for modeling certain processes, such as the packet arrivals observed in most packet data networks ([32][33]). Two related concepts that such processes show are *long-range dependence* and *self-similarity*. A self-similar process shows structural similarities across many different time scales. This arises for example in ethernet traffic, where there is no natural length of a *burst* of data [32], where a burst is a transmission period with high data signaling rate. A comprehensive discussion on self-similar processes can be found in [4]. See also [34] for a summary of self-similar process characteristics and models.

Let $X_t, t \in \mathbb{Z}$ be a discrete stochastic process where X_t represents the traffic volume at time t. Related to X_t is the accumulated process up to time t,

$$Y(t) = \sum_{\tau=0}^{t} X_\tau.$$

Definition 4.2.8 (Aggregated process). *The aggregated process $X^{(m)}$ of X at aggregation level m is*

$$X_t^{(m)} = \frac{1}{m} \sum_{\tau=m(t-1)+1}^{mt} X_\tau.$$

Thus, X_t is partitioned into nonoverlapping blocks of size m and their values are averaged.

For Poisson-distributed traffic, aggregate traffic becomes smoother (less "bursty") as the number of sources increases. This is a consequence of the finite variance of the Poisson process and the weak law of large numbers. Contrary to this property, the burstiness of Local Area Network (LAN) traffic tends to *increase* as the number of active sources increases [32]. Let $\gamma^{(m)}(k)$ denote the autocovariance function of $X^{(m)}$.

Definition 4.2.9 (Second-order self-similarity). *The process X_n is called second-order self-similar with* Hurst parameter H ($\frac{1}{2} < H < 1$) *if*

$$\gamma(k) = \frac{\sigma^2}{2}\left((k+1)^{2H} - 2k^{2H} + (k-1)^{2H}\right),$$

for all $k \geq 1$; X_n is called asymptotically second-order self-similar *if*

$$\lim_{m \to \infty} \gamma^{(m)}(k) = \frac{\sigma^2}{2} \left((k+1)^{2H} - 2k^{2H} + (k-1)^{2H} \right).$$

Definition 4.2.10 (H-ss). The process $Y(t)$ is called *self-similar* with *Hurst parameter H* $(0 < H < 1)$, denoted H-ss, if for all $a > 0$ and $t \geq 0$

$$Y(t) \stackrel{d}{=} a^{-H} Y(at),$$

where $\stackrel{d}{=}$ denotes equality in distribution, unless $Y(t) \equiv 0$, $Y(t)$ cannot be stationary due to the normalization factor a^{-H}.

If $Y(t)$ is H-ss and has stationary increments, $Y(t)$ is called H-sssi. Then

$$
\begin{aligned}
\mathbf{E}(Y(t)) &= 0, \\
\mathbf{E}(Y^2(t)) &= \sigma^2 |t|^{2H}, \\
\gamma(k) &= \frac{\sigma^2}{2} \left(|t|^{2H} - |t-s|^{2H} + |s|^{2H} \right).
\end{aligned}
$$

The increment process X_t has mean 0 and autocovariance

$$\gamma(k) = \frac{\sigma^2}{2} \left((k+1)^{2H} - 2k^{2H} + (k-1)^{2H} \right).$$

If $Y(t)$ is an H-sssi process, then its increment process X_t satisfies

$$X_t \stackrel{d}{=} m^{1-H} X_t^{(m)}.$$

Depending on whether a discrete-time process X_t satisfies the relation for all $m \geq 0$ or only in the limit as $m \to \infty$, X_t is said to be exactly self-similar or asymptotically self-similar. In the Gaussian case, this definition coincides with second-order self-similarity. If $\frac{1}{2} < H < 1$, then

$$\mathrm{Var}(X^{(m)}) = \sigma^2 m^{-\beta},$$

with $0 < \beta < 1$ and $H = 1 - \beta/2$.

There is a dependency structure in the time series that causes $\mathrm{Var}(X^{(m)})$ to converge to zero slower than m^{-1}. Let $\rho(k) = \gamma(k)/\sigma^2$ denote the autocorrelation function. For $0 < H < 1$, $H \neq \frac{1}{2}$,

$$\rho(k) \sim H(2H-1)k^{2H-2}, \quad k \to \infty.$$

If $\frac{1}{2} < H < 1$, $\rho(k)$ asymptotically behaves as $ck^{-\beta}$ for $0 < \beta < 1$, $c > 0$, $\beta = 2 - 2H$, and

$$\sum_{k=-\infty}^{\infty} \rho(k) = \infty.$$

Thus, the autocorrelation function decays hyperbolically. The corresponding stationary process X_t is called *long range–dependent*. The process X_t is called *short range–dependent* if the autocorrelation function is summable.

An equivalent definition of long-range dependence is to require that the spectral density

$$S(\nu) = (2\pi)^{-1} \sum_{k=-\infty}^{\infty} \rho(k)e^{ik\nu}$$

satisfies

$$S(\nu) \sim c|\nu|^{-\alpha}, \quad \nu \to 0,$$

where $c > 0$ and $0 < \alpha = 2H - 1 < 1$. Thus $S(\nu)$ diverges around the origin, implying even larger contributions by low-frequency components.

If $H = \frac{1}{2}$, then $\rho(k) = 0$ and X_t is trivially short range–dependent (uncorrelated). When $0 < H < \frac{1}{2}$, then

$$\sum_{k=-\infty}^{\infty} \rho(k) = 0;$$

$H = 1$ leads to $\rho(k) = 1$ for all $k \geq 1$, and values of $H > 1$ are prohibited due to the stationarity condition on X_t. For asymptotic self-similarity ($\frac{1}{2} < H < 1$), self-similarity is equivalent to long-range dependence. For practical applications only $\frac{1}{2} < H < 1$ needs to be considered.

Definition 4.2.11 (Heavy-tailed distribution). A random variable Z has a *heavy-tailed distribution* if

$$\mathbf{P}(Z > x) \sim cx^{-\alpha}, \quad x \to \infty,$$

where $0 < \alpha < 2$ is called the tail index (shape parameter) and c is a positive constant (or a slowly varying function). The tail of the distribution decays hyperbolically.

If $0 < \alpha < 2$, the process has infinite variance and finite mean; if $0 < \alpha \leq 1$, the process has infinite variance and unbounded mean. In the networking context the case $1 < \alpha < 2$ is of primary interest. The most common heavy-tailed distribution is the *Pareto distribution*.

Definition 4.2.12 (Pareto distribution). *A stochastic variable X has the* Pareto distribution *with parameters* $\kappa, \alpha > 0$, *if it has the complementary cumulative probability function* $\bar{F} = 1 - F$ *given by*

$$\bar{F}(x) = \left(\frac{\kappa}{\kappa + x} \right)^{\alpha}.$$

One of the most common models for self-similar phenomena is *fractional Brownian motion*.

Definition 4.2.13 (Fractional Brownian motion). *The process* $Y(t), t \in \mathbb{R}$ *is called* fractional Brownian motion *with parameter* $H, 0 < H < 1$, *if* $Y(t)$ *is Gaussian and H-sssi.*

Definition 4.2.14 (Fractional Gaussian noise). *The process* $X_t, t \in \mathbb{Z}^+$ *is called* fractional Gaussian noise (fGn) *with parameter* H, *if* X_t *is the increment process of a fractional Brownian motion with parameter* H.

Fractional Brownian motion reduces to Brownian motion and fractional Gaussian noise to Gaussian noise when $H = \frac{1}{2}$. Thus X_t becomes completely uncorrelated. For each H, $0 < H < 1$, there is a unique Gaussian process that is the stationary increment of an H-sssi process. Fractional Brownian motion is the corresponding unique Gaussian H-sssi process. The main difference between fractional Brownian motion and regular Brownian motion is that while the increments in Brownian motion are independent, they are dependent in fractional Brownian motion.

A commonly used structural model for bursty traffic sources is the *on/off* model, in which a source is assumed to transmit data with constant rate during the *on* period and is silent during the *off* period. The *on* and *off* periods are independent and identically distributed stochastic variables. Consider N independent *on/off* traffic sources $X_i(t), i \in [1, N]$.

Let

$$S_N(t) = \sum_{i=1}^{N} X_i(t)$$

denote the aggregate traffic at time t. The cumulative process $Y_N(Tt)$ is defined as

$$Y_N(Tt) = \int_0^{Tt} \left(\sum_{i=1}^{N} X_i(s) \right) ds,$$

where $T > 0$ is a scale factor. Thus $Y_N(Tt)$ measures the total traffic up to time Tt. Let τ_{on} be the random variable describing the duration of the *on* periods and let τ_{off} be the random variable associated with the durations of the *off* periods. Furthermore, let τ_{on} be distributed according to a heavy-tailed distribution, so that

$$\mathbf{P}(\tau_{on} > x) \sim cx^{-\alpha}, \quad x \to \infty,$$

where $1 < \alpha < 2$ and $c > 0$ is a constant; τ_{off} can be either heavy tailed or light tailed, but with finite variance. It turns out that $Y_N(Tt)$ behaves asymptotically as fractional Brownian motion.

Theorem 4.2.1 (*On/off* model and fractional Brownian motion). *The process $Y_N(Tt)$ behaves statistically as*

$$\frac{\mathbf{E}(\tau_{on})}{\mathbf{E}(\tau_{on}) + \mathbf{E}(\tau_{off})} NTt + cN^{1/2}T^H B_H(t)$$

for large T and N, where $H = (3 - \alpha)/2$, $B_H(t)$ is fractional Brownian motion with parameter H, and $c > 0$ is a quantity depending only on the distributions of τ_{on} and τ_{off}.

The process $Y_N(Tt)$ is long range–dependent ($\frac{1}{2} < H < 1$) if and only if $1 < \alpha < 2$, that is, if τ_{on}'s distribution is heavy-tailed.

If neither τ_{on} nor τ_{off} is heavy-tailed, then $Y_N(Tt)$ is short range–dependent. If the *off* period is heavy-tailed but the *on* period is not, the process is long range–dependent. Thus, heavy-tailedness causes long range–dependence. An infinite aggregation of short range–dependent sources (for example heterogeneous *on/off* sources with exponential *on/off* times) can produce long-range dependence. Finite aggregations of short-range sources cannot induce long-range dependence.

4.3 Detection and Estimation

Estimation of traffic characteristics and model parameters is a challenging task. In particular, long-range dependence is difficult to establish with some precision, since it would require long data series, and long-range dependence is then often clouded by nonstationary effects.

Detection of Poisson Characteristics

There are several methods to investigate whether a time series can be described by a Poisson process. A Poisson distribution can be fitted to the data and the result tested for goodness of fit with the χ^2 statistic, defined as

$$\chi^2 = \sum_{i=1}^{N} \frac{(Y_i - Np_i)^2}{Np_i}, \tag{4.7}$$

where Y_i is the observed frequency of data point i and p_i is the theoretical probability density of data point i (see for example [35]). Alternatively, if it can be shown that the interarrival times are exponentially distributed and the increments are independent, then the process can be concluded to be Poisson-distributed. The latter approach is used in [36].

Detection and Estimation of Long-Range Dependence and Self-Similarity

Estimation of H is a difficult task. Furthermore, the expressions provided for H generally require time series of infinite length [4]. Several methods are available to estimate long-range dependence and self-similarity in a time series. The most popular are variance-time analysis, rescaled-range analysis, periodogram-based analysis, the Whittle estimator, and wavelet-based analysis [21][37].

The *variance-time analysis* is a graphical method based on the property of the slowly decaying variance of long range–dependent processes under aggregation. For such a process,

$$\text{Var}(X^{(m)}) \sim cm^{-\hat{b}}, \quad \text{as } m \to \infty,$$

where $\hat{b} \in (0, 1)$, whereas for short range–dependent processes the variance of the aggregated time series

$$\text{Var}(X^{(m)}) \sim cm^{-1}, \quad \text{as } m \to \infty.$$

By computing the variances for the aggregated time series $\{X^{(m)}\}$ and plotting the variance against m in a log-log diagram, $\hat{b}$ can be estimated. By performing a least square fit analysis with the data, a numerical value of $\hat{b}$ is achieved. Fig. 4.5 shows an example of variance-time analysis of self-similar traffic data.

The *rescaled-range (R/S) analysis* is a normalized, dimensionless measure to characterize variability. For a given set of observations $X = \{X_n, n \in \mathbb{Z}^+\}$ with sample mean $\bar{X}(n)$, sample variance $S^2(n)$, and range $R(n)$, the rescaled adjusted range statistic is given by

$$\frac{R(n)}{S(n)} = \frac{\max(0, \Delta_1, \Delta_2, \ldots, \Delta_n) - \min(0, \Delta_1, \Delta_2, \ldots, \Delta_n)}{S(n)},$$

where

$$\Delta_k = \sum_{i=1}^{} k X_i - k\bar{X},$$

for $k = 1, 2, \ldots, n$. For many natural phenomena,

$$\mathbf{E}\left(\frac{R(n)}{S(n)}\right) \sim cn^H,$$

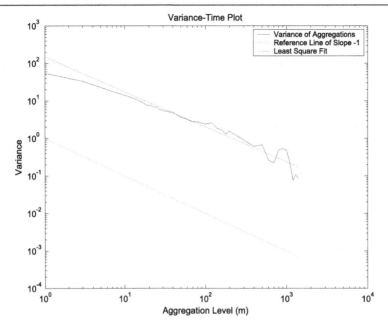

Figure 4.5: Variance-time analysis of self-similar traffic.

as $n \to \infty$. Thus, an estimate of H is given by plotting $\log(R(n)/S(n))$ versus $\log(n)$. The R/S method only provides an estimation of the level of self-similarity in a time series. The method can be used to test whether a time series is self-similar or not, and if so, give a rough estimation of H. An example of R/S analysis of a trace of self-similar traffic is shown in Fig. 4.6.

4.4 Wavelet Analysis

In analysis of communication networks, different characteristics appear on different *scales*. *Wavelets* are mathematical tools for analyzing time series or images by decomposing them with respect to different scales. A wavelet is a "small wave", which essentially grows and decays in a limited time period. A comprehensive exposition on wavelet analysis of time series can be found in [132]. See also [4] for wavelet analysis of long range–dependent processes.

A wavelet transform is a representation of a signal – in this case packet arrivals – that allows decomposition into fluctuations on different scales. The discrete wavelet transform (DWT) is defined on discrete-time instances that correspond to the sampling rate in the system. A mul-

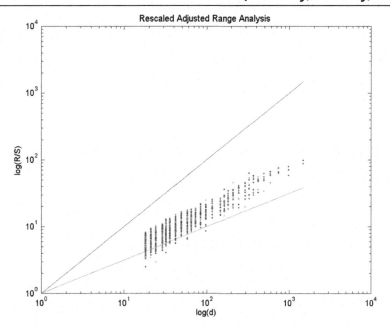

Figure 4.6: Rescaled-range analysis of self-similar traffic.

tiresolution analysis is such a decomposition which is aligned in time. Due to its construction, it is a suitable method to analyze self-similar data.

In continuous time, a wavelet is a transform that satisfies

$$\int_{-\infty}^{\infty} \psi(u) du \;\; = \;\; 0,$$

$$\int_{-\infty}^{\infty} \psi^2(u) du \;\; = \;\; 1,$$

for some kernel (or filter) $\psi(u)$, together with other regularity conditions. The kernel can be chosen in many ways and this generates different types of wavelet transforms.

The discrete wavelet transform can be regarded as an approximation to the continuous wavelet transform. We consider the real-valued wavelet filter $\mathcal{W}$ with filter coefficients $\{h_l : l = 0, \ldots, L - 1\}$ where the width L of the filter is an even integer, so that $h_0 \neq 0$, $h_{L-1} \neq 0$. Define $h_l = 0$ for $l < 0$ and $l \geq L$. A wavelet filter must satisfy

$$\sum_{l=0}^{L-1} h_l \;\; = \;\; 0,$$

$$\sum_{l=0}^{L-1} h_l^2 = 1,$$

$$\sum_{l=0}^{L-1} h_l h_{l+2n} = \sum_{l=-\infty}^{\infty} h_l h_{l+2n} = 0, \quad n \neq 0.$$

An interesting and important fact is that the wavelet coefficients are approximately uncorrelated. The DWT therefore uncorrelates even highly correlated series.

Let $\mathbf{X}$ be a time series represented by a column vector and let $\mathbf{W} = \mathcal{W}\mathbf{X}$, where $\mathcal{W}$ is the discrete wavelet filter. Define the wavelet *details* as $\mathcal{D}_j = \mathcal{W}_j^T \mathbf{W}_j, \quad j = 1, \ldots, J$, and $\mathcal{S}_J = \mathcal{V}_J^T \mathbf{V}_J$, where $\mathcal{V}$ is the scaling filter. Then

$$\mathbf{X} = \sum_{j=1}^{J} \mathcal{D}_j + \mathcal{S}_J$$

is called a *multiresolution analysis* of $\mathbf{X}$. The multiresolution forms an additive decomposition where each component can be associated with a particular scale $\lambda_j = 2^j$.

The *Daubechies D(4) wavelet filter* is based on the filter coefficients

$$h_0 = \frac{1 - \sqrt{3}}{4\sqrt{2}}, \tag{4.8}$$

$$h_1 = \frac{-3 + \sqrt{3}}{4\sqrt{2}}, \tag{4.9}$$

$$h_2 = \frac{3 + \sqrt{3}}{4\sqrt{2}}, \tag{4.10}$$

$$h_3 = \frac{-1 - \sqrt{3}}{4\sqrt{2}}. \tag{4.11}$$

Let $\mathcal{T}$ be the time shift operator defined by

$$\mathcal{T}\mathbf{X} = [X_{N-1}, X_0, X_1, \ldots, X_{N-2}].$$

Let $\mathcal{W}_i$ denote the ith row in the wavelet filter. Then the rows in the wavelet transformation matrix $\mathcal{W}$ are related by $\mathcal{W}_{i+1} = \mathcal{T}^2 \mathcal{W}_i$ and

$$\mathcal{W} = \begin{pmatrix} h_1 & h_0 & 0 & 0 & \cdots & 0 & 0 & h_3 & h_2 \\ h_3 & h_2 & h_1 & h_0 & \cdots & 0 & 0 & 0 & 0 \\ 0 & 0 & h_3 & h_2 & \cdots & 0 & 0 & 0 & 0 \\ 0 & 0 & 0 & 0 & \cdots & 0 & 0 & 0 & 0 \\ \vdots & \vdots & \vdots & \vdots & \ddots & \vdots & \vdots & \vdots & \vdots \\ 0 & 0 & 0 & 0 & \cdots & 0 & 0 & 0 & 0 \\ 0 & 0 & 0 & 0 & \cdots & 0 & 0 & 0 & 0 \\ 0 & 0 & 0 & 0 & \cdots & h_1 & h_0 & 0 & 0 \\ 0 & 0 & 0 & 0 & \cdots & h_3 & h_2 & h_1 & h_0 \end{pmatrix}.$$

The *pyramid algorithm* provides an efficient algorithm to compute the discrete wavelet transform of a time series. The pyramid algorithm can be expressed in linear matrix operations as follows.

Step 1: Define the $N \times \frac{N}{2}$ matrices

$$\mathcal{W}_1 = \begin{pmatrix} h_0 & h_1 & h_2 & h_3 & 0 & 0 & \cdots & 0 & 0 & 0 & 0 \\ 0 & 0 & h_0 & h_1 & h_2 & h_3 & \cdots & 0 & 0 & 0 & 0 \\ \vdots & \vdots & \vdots & \vdots & \vdots & \vdots & \ddots & \vdots & \vdots & \vdots & \vdots \\ 0 & 0 & 0 & 0 & 0 & 0 & \cdots & h_0 & h_1 & h_2 & h_3 \\ h_2 & h_3 & 0 & 0 & 0 & 0 & \cdots & 0 & 0 & h_0 & h_1 \end{pmatrix}$$

$$\mathcal{V}_1 = \begin{pmatrix} g_0 & g_1 & g_2 & g_3 & 0 & 0 & \cdots & 0 & 0 & 0 & 0 \\ 0 & 0 & g_0 & g_1 & g_2 & g_3 & \cdots & 0 & 0 & 0 & 0 \\ \vdots & \vdots & \vdots & \vdots & \vdots & \vdots & \ddots & \vdots & \vdots & \vdots & \vdots \\ 0 & 0 & 0 & 0 & 0 & 0 & \cdots & g_0 & g_1 & g_2 & g_3 \\ g_2 & g_3 & 0 & 0 & 0 & 0 & \cdots & 0 & 0 & g_0 & g_1 \end{pmatrix}$$

where h_i and g_i are found from (4.8)–(4.11) and $g_l = (-1)^{l+1} h_{L-1-l}$.

Step 2: Multiply the time series vector $\mathbf{X}$ with $\mathcal{W}_1$ and $\mathcal{V}_1$, respectively, which yields the first order wavelet coefficients and scaling coefficients:

$$\begin{aligned} \mathbf{W}_1 &= \mathcal{W}_1 \mathbf{X} \\ \mathbf{V}_1 &= \mathcal{V}_1 \mathbf{X}. \end{aligned}$$

Divide N by 2, goto step 1 and apply the filters to the data vector $\mathbf{V}_1$.

Step j: Let $N := N_j = N/2^j$, goto step 1 and apply the filters to the data vector $\mathbf{V}_{j-1}$. Repeat until $N = 2$.

The wavelet transform of $\mathbf{X}$ is given by

$$\mathbf{W} = \begin{pmatrix} \mathbf{W}_1 \\ \vdots \\ \mathbf{W}_J \\ \mathbf{V}_J \end{pmatrix}.$$

Unfortunately, the discrete wavelet transform requires that the number of data points is a power of 2. This requirement is relaxed for the maximum-overlap discrete wavelet transform (MODWT), which is well defined for any sample size N. The MODWT is also suitable for multiresolution analysis. Define the *MODWT wavelet filter* $\{\tilde{h}_l\} : \tilde{h}_l \equiv h_l/\sqrt{2}$ and the *MODWT scaling filter* $\{\tilde{g}_l\} : \tilde{g}_l \equiv g_l/\sqrt{2}$ so that

$$\sum_{l=0}^{L-1} \tilde{h}_l = 0,$$

$$\sum_{l=0}^{L-1} \tilde{h}_l^2 = \frac{1}{2},$$

$$\sum_{l=-\infty}^{\infty} \tilde{h}_l \tilde{h}_{l+2n} = 0,$$

for all nonzero integers n. This gives the first level MODWT ($J_0 = 1$)

$$\tilde{W}_{1,t} = \sum_{l=0}^{L-1} \tilde{h}_l X_{t-l \mod N},$$

$$\tilde{V}_{1,t} = \sum_{l=0}^{L-1} \tilde{g}_l X_{t-l \mod N},$$

for $t = 0, \ldots, N-1$. Repeating this operation on $\tilde{V}_{1,t}$ gives the details on successively longer scales.

In the following, we illustrate multiresolution analyses of Markovian, self-similar, and aggregate traffic. We use a Daubechies MODWT of width $L = 4$. The figures show, from the top, fluctuations on levels $J = 1, 2, \ldots 5$, the *details*, and the *smoothness* on level $J = 5$.

Multiresolution analyses up to level $J_0 = 5$ of Poisson and MAP traffic are shown in Figs. 4.7 and 4.8, respectively. We note the rapidly decreasing amplitudes on large scales. This indi-

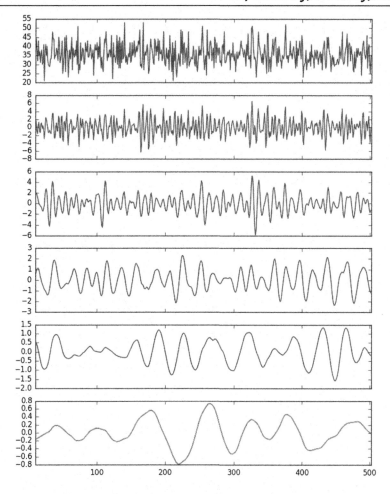

Figure 4.7: Multiresolution analysis of Poisson traffic.

cates that the traffic is fairly regular and does not cause long periods of saturation in a queue, provided that the intensity is less than the processing capacity ($\lambda < s$).

For the fractional Brownian motion, the multiresolution analysis shows large variations on both short and long scales (see Fig. 4.9). Even if the average load is lower than the processing capacity of the queue, there are relatively long periods of time where the workload in a queue builds up and causes congestion.

Comparing the multiresolution analyses of the total aggregate traffic shown in Fig. 4.10 and some of the Bellcore used in [28] and shown in Fig. 4.11 reveals similar behavior on both short and long scales. Notably, the amplitude on longer scales does not decrease significantly, but remains between 10–20% of the first detail.

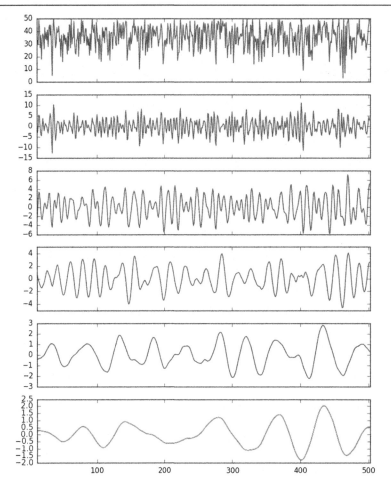

Figure 4.8: Multiresolution analysis of MAP traffic.

4.5 Fractal Maps

With large volumes of data exhibiting self-similar, fractal behavior, it is desirable to have a way to monitor processes online and to analyze such data within a single framework. This is the idea behind fractal maps, proposed by Ruschin-Rimini et al. [38]. The authors describe the use of fractal maps and their application in statistical process control (SPC).

A fractal map is based on a transformation that preserves the correlation between integer data points, and therefore no information is lost. An inverse transform can easily be constructed to recover the original data from a transformed data set.

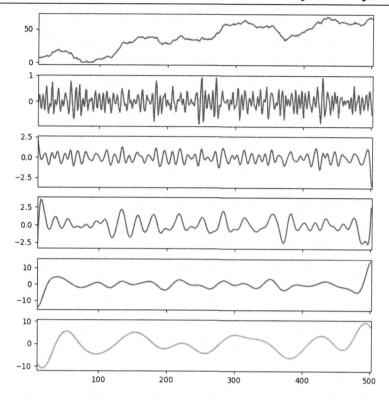

Figure 4.9: Multiresolution analysis of fractional Brownian motion traffic.

After transforming the data, measures of the fractal dimension can be applied to derive properties of and to compare maps.

In SPC, online data are represented in the fractal map, which can be used for pattern detection and root cause analysis. The fractal dimension measures properties of the mapped data, complementing other statistical measures.

Data analytics in many fields aim at representation and prediction of processes. For data with strong correlations, this is often accomplished by modeling the data generating processes. When data correlation is complex this is a difficult task, and therefore a nonparametric, model-free method is preferable, not imposing any assumptions on the data. In particular, this is important for state-dependent nonlinear data, such as traffic and other load-level data. We consider variables taking values from a finite discrete set. Such a representation is suitable for any categorical or binned data. Continuous values can be discretized to fit into this data type.

Ruschin-Rimini et al. [38] list the main differences between traditional methods of SPC and requirements in a high-volume data environment:

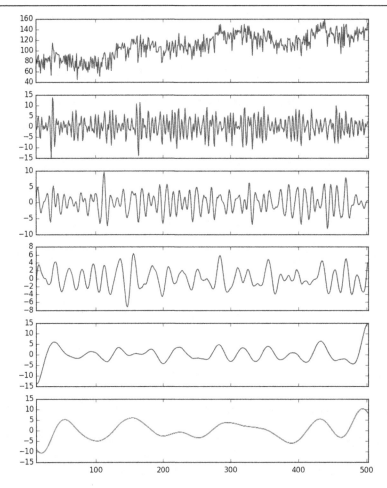

Figure 4.10: Multiresolution analysis of the total simulated traffic.

- (1) Traditional methods are often based on various model assumptions, whereas data-rich environments with many different data sources often require a model-free approach.
- (2) Many methods are unsuitable to represent nonlinear dynamics of complex systems with feedback control.
- (3) In addition to pattern detection, there is often a need to identify probable causes and relationships between variables, often hidden in traditional methods.

Fractal maps are suitable for representation and monitoring of both univariate and multivariate data, particularly in data-intensive environments. They capture complex dependence structures and can be used to visualize large data sets, all without any model assumptions.

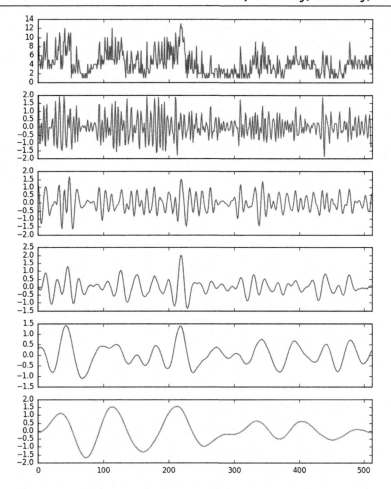

Figure 4.11: Multiresolution analysis of Bellcore traffic.

The Iterated Function System

The iterated function system (IFS) is used as an iterative contractive mapping from the process values to vectors in $\mathcal{R}^2$. Formally, having m categories, an IFS consists of a finite set of contraction mappings $w_i : X \rightarrow X$ having contractivity factors s_i, $i = 1, 2, \ldots, m$, defined on a complete metric space (X, d). By a contraction we mean the following.

Definition 4.5.1. *A mapping $w_i(x)$ is called contractive in (X, d), if $d(w_i(\mathbf{y}), w_i(\mathbf{z})) \geq s_i \times d(\mathbf{y}, \mathbf{z}) \forall \mathbf{y}, \mathbf{z} \in X$ for some contractivity factor $0 < s_i < 1$, where $\mathbf{y} = (y_1, y_2, \ldots, y_D)$ and $\mathbf{z} = (z_1, z_2, \ldots, z_D)$ are vectors in $\mathcal{R}^D$.*

The transformation converts a sequence into a self-similar, or fractal, shape. It is made up of several smaller copies of itself, which in turn are also made up of copies of itself, etc. This is the origin of its self-similar nature. Notably, the mapping has the following two properties.

(1) The transformation gives a unique representation of a sequence in that each point on the map contains the history up to that point, and the map captures all subsequences in the sequence.
(2) The original sequence can be fully reconstructed from the map.

We use an iterated function system with a circular transformation, which gives a geometrically efficient representation in $\mathcal{R}^2$. In practice, different colors can be used to increase the visibility.

Suppose we have a process representing m categories. Then, we define the transformation as

$$w_i \left\{ \begin{pmatrix} x_1 \\ x_2 \end{pmatrix} \right\} = \begin{pmatrix} \alpha & 0 \\ 0 & \alpha \end{pmatrix} \begin{pmatrix} x_1 \\ x_2 \end{pmatrix} + \begin{pmatrix} \beta_i \\ \delta_i \end{pmatrix}, \tag{4.12}$$

for $i = 1, 2, \ldots, m$, where

$$\beta_i = \cos \left(\frac{2\pi}{m} i \right), \text{ for } i = 1, 2, \ldots, m, \tag{4.13}$$

$$\delta_i = \sin \left(\frac{2\pi}{m} i \right), \text{ for } i = 1, 2, \ldots, m, \tag{4.14}$$

where we require that α satisfies

$$\frac{\alpha}{1 - \alpha} <\sim \left(\frac{\pi}{m} \right),$$

to guarantee that every point on the map has a unique address.

Given a sequence with m different symbols of length N, the transformation assigns a two-dimensional vector to each data point as follows.

(1) Associate each category with a contractive mapping $w_i(\mathbf{x})$, $i \in \{1, 2, \ldots, m\}$.
(2) Transform a sequence of length N, with values $i \in \{0, 1, \ldots, m-1\}$, using the contractive mappings $\{w_{i(n+1)}(\mathbf{x}n), n = 1, 2, \ldots, N\}$. The expression $w_i(\mathbf{x}_n)$ implies that the vector $\mathbf{x}_n$ is transformed by the contractive mapping associated with the data point in position $n + 1$ of value i. The shift in index is due to the initiation, which does not correspond to any data point; we start the transformation process by choosing an arbitrary initial point $\mathbf{x}_{(0)}$ in $\mathcal{R}^2$.
(3) Recursively apply each of the N contractive mappings $w_i(x_0), w_i(x_1), \ldots, w_i(x_{N-1})$ in sequential order, obtaining $x(n) = w_i(x_{(n-1)})$, $n = 1, 2, \ldots, N$, and $i \in \{1, 2, \ldots, m\}$. The transformation defines a mapping onto a sequence of N points in $\mathcal{R}^2$.

We illustrate how this works with an example from Ruschin-Rimini et al. [38].

Example 4.5.1. *Suppose we are given a sequence with* $m = 9$ *different values (symbols), starting with* $0, 3, 6 \ldots$.

Each of the m process symbols is associated with a contractive mapping. Following the notation used, we associate variable 1 with contractive mapping w_1, *variable 2 with contractive mapping* $w_2, \ldots$, *and so on, and finally variable 0 with contractive mapping* w_9. *Note that we cannot use the mapping* w_0, *since it is degenerate, with loss of information as a result. We consecutively apply the mappings defined according to Eq. (4.12), with* $\alpha = 0.08$. *In the following discussion, we take the initial radius to be* $r_0 = 1$. *We have*

$$w_1\left(\begin{pmatrix} x_1 \\ x_2 \end{pmatrix}\right) = \begin{pmatrix} 0.08 & 0 \\ 0 & 0.08 \end{pmatrix} \begin{pmatrix} x_1 \\ x_2 \end{pmatrix} + \begin{pmatrix} \cos(2\pi/9) \\ \sin(2\pi/9) \end{pmatrix},$$

$$w_2\left(\begin{pmatrix} x_1 \\ x_2 \end{pmatrix}\right) = \begin{pmatrix} 0.08 & 0 \\ 0 & 0.08 \end{pmatrix} \begin{pmatrix} x_1 \\ x_2 \end{pmatrix} + \begin{pmatrix} \cos(2 \cdot 2\pi/9) \\ \sin(2 \cdot 2\pi/9) \end{pmatrix},$$

$$\vdots$$

$$w_9\left(\begin{pmatrix} x_1 \\ x_2 \end{pmatrix}\right) = \begin{pmatrix} 0.08 & 0 \\ 0 & 0.08 \end{pmatrix} \begin{pmatrix} x_1 \\ x_2 \end{pmatrix} + \begin{pmatrix} \cos(9 \cdot 2\pi/9) \\ \sin(9 \cdot 2\pi/9) \end{pmatrix}.$$

Next, the sequence of symbols 0, 3, 6 is represented by vectors given by the consecutive application of contractive mappings, $\{w_9, w_3, w_6\}$. *The initial vector* $\mathbf{x}_{(0)}$ *is arbitrarily chosen as*

$$\mathbf{x}_{(0)} = \begin{pmatrix} 0 \\ 0 \end{pmatrix}.$$

The vector $\mathbf{x}_{(1)}$, *given* $\mathbf{x}_{(0)}$ *and the first symbol, 0, is given by*

$$\mathbf{x}_{(1)} = w_9(\mathbf{x}_{(0)})$$

$$= \begin{pmatrix} 0.08 & 0 \\ 0 & 0.08 \end{pmatrix} \begin{pmatrix} 0 \\ 0 \end{pmatrix} + \begin{pmatrix} \cos(9 \cdot 2\pi/9) \\ \sin(9 \cdot 2\pi/9) \end{pmatrix} = \begin{pmatrix} 1 \\ 0 \end{pmatrix}.$$

Similarly, $\mathbf{x}_{(2)}$ *is computed from* $\mathbf{x}_{(1)}$ *and the symbol, 3, as*

$$\mathbf{x}_{(2)} = w_3(\mathbf{x}_{(1)})$$

$$= \begin{pmatrix} 0.08 & 0 \\ 0 & 0.08 \end{pmatrix} \begin{pmatrix} 1 \\ 0 \end{pmatrix} + \begin{pmatrix} \cos(3 \cdot 2\pi/9) \\ \sin(3 \cdot 2\pi/9) \end{pmatrix} = \begin{pmatrix} -0.42 \\ 0.87 \end{pmatrix}.$$

The construction of a fractal map of the sequence $(0, 3, 6, 1, 4, 7, 5, 3, 2, 0, 7, 2, 9, 3)$ is shown in Fig. 4.12. The data points are mapped onto contracting circles whose center point is the

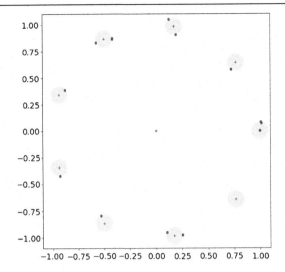

Figure 4.12: Fractal map of the sequence (0, 3, 6, 1, 4, 7, 5, 3, 2, 0, 7, 2, 9, 3).

vector corresponding to the immediately preceding data point. Here, the initial point is the origin, which together with the first data point creates the first nontrivial vector, shown as the center point in the circle at $(1.00, 0.00)$. Moving counter-clockwise, the following circles represent the symbols $1, 2, \ldots, 8$. Next, the vector $(1.00, 0.00)$ is contracted and mapped onto the circle representing symbol 3. In this circle, the point is drawn at a zero angle and radius $r = \alpha$, defining the next position vector (or address). Note that the circle representing symbol 8 is empty; the symbol does not appear in the subsequence.

The construction of a fractal map of the uniformly distributed sequence of $m = 9$ symbols and 15 series of $N = 1000$ points is shown in Fig. 4.13. Each cluster here represent one of the symbols 0–8. By addition of a large number of points under the IFS, the map assumes a self-similar structure. We refer to this figure as the map view at level 1.

By zooming in on one of the clusters, say the one for symbol 3, we get a very similar image, as shown in Fig. 4.14. This similarity illustrates the self-similar nature of the map. The clusters now represent subsequences of length 2, that is, $03, 13, 23, 33, 43, 53, 63, 73, 83$. The density of the clusters indicates how frequently they appear in the data.

An example of the next level is shown in Fig. 4.15, representing subsequences of length 3, that is, zooming in on subsequence 23. Specific subsequences become rarer, but the overall pattern persists.

As we drill down in the map, missing or rare subsequences of various lengths become discernable. This is the main idea behind SPC. The ability to zoom in on subsequences provides a visual tool for fault and root cause analyses.

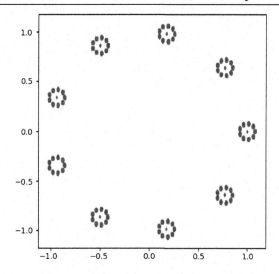

Figure 4.13: Fractal map of sequences with $m = 9$ – view at level 1.

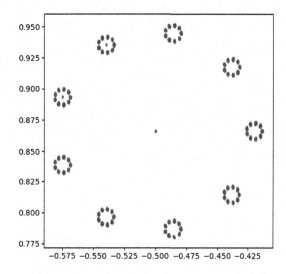

Figure 4.14: Fractal map of sequences with $m = 9$ – view at level 2.

The Fractal Dimension

A useful metric defined on a fractal is the measure on how the detail in the fractal changes with scale, known as its fractal dimension, D. Another view is, when projected onto a grid, how many elements the fractal covers as the number of elements increases. The fractal dimension need not be an integer.

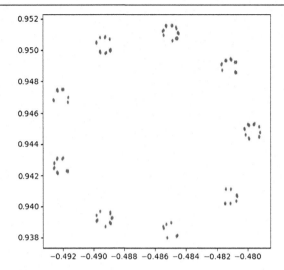

Figure 4.15: Fractal map of sequences with $m = 9$ – view at level 3.

We define three statistics defined on a fractal measuring its fractal dimension, the box counting dimension, the information dimension, and the correlation dimension.

Box counting dimension

To compute the box counting dimension, we divide the fractal space into hypercubes of side length r. Letting $N(r)$ be the number of hypercubes occupied by points of the fractal, the statistic is calculated as

$$D_{\text{bc}} = - \lim_{r \to 0} \frac{\log N(r)}{\log r}. \tag{4.15}$$

In practice, the circular element (hyperball) is very convenient for box counting with the used transformation, since, for any point, we just need to calculate its distance from the center of the hyperball to check whether it is included in it or not, which is given by its radius r.

It is convenient to compute the statistic per level k. Each level consists of m circles of radius α for $k = 1$, m^2 circles of radius α^2 for $k = 2$, and so on. To compute the statistic at level k, we count the number of circles containing at least one point, out of a total of m^k circles, and use Eq. (4.15) with $r = \alpha^k$.

As we have seen, each circle at level k having radius α^k represents certain k-length subsequences of the original data. The box counting dimension changes whenever a new point falls within an empty circle. This happens only when a new subsequence appears in the data. It is therefore reasonable to assume that the box counting dimension would be a good statistic to detect outliers. However, the box counting dimension does not depend on the number of points in each box, and so it is not sensitive to changes in the distribution of the data.

Information dimension

Again, we divide the fractal space into hypercubes of side length r and let $p_i(r)$ be the frequency with which points fall into the ith hypercube. The information dimension is calculated as

$$D_{\text{inf}} = \lim_{r \to 0} \frac{\sum_i p_i(r) \log p_i(r)}{\log r}.$$

The information dimension at level k is closely related to the entropy of subsequences of length k.

Set $r = \alpha_k$ to be the radius of the kth level and let $p_i(\alpha_k)$ be the frequency of data points in circle i. Then we have m^k such circles, and each represents a subsequence of length k in the data. The information dimension therefore captures changes in the distribution of subsequences, and the more "randomly" these are spread out, the larger the value of D_{inf}. We can compute the statistic as

$$D_{\text{inf}} = \lim_{\alpha^k \to 0} \frac{\sum_{i=1}^{m^k} p_i(\alpha^k) \log p_i(\alpha^k)}{\log \alpha^k}. \tag{4.16}$$

Note that the numerator $\sum_{i=1}^{m^k} p_i(\alpha^k) \log p_i(\alpha^k)$ is the Shannon entropy of all k-length subsequences in the data.

Correlation dimension

The correlation dimension is based on counting points that are close, that is, whose distance is smaller than some constant $\epsilon > 0$. We have

$$D_{\text{cor}} = -\lim_{\varepsilon \to 0} \frac{\log C(\varepsilon)}{\log \varepsilon}, \tag{4.17}$$

$$C(\varepsilon)) = \lim_{N \to \infty} N^{-2} \times \{\text{number of pairs } (x_i, y_j) : |x_i - x_j| < \epsilon, \tag{4.18}$$

$$i \neq j, j = 1, \dots N.$$

The correlation dimension measures the probability of correlated pairs of points (x_i, x_j), that is, subsequences, as set by the constant ϵ. It can be implemented from Eqs. (4.17)–(4.18). We select ϵ as the radius of a circle at level k, $\epsilon = \alpha k$. The statistic increases with the correlation between subsequences of length k, so it can be used to find changes in the correlation between subsequences.

The D_{inf} statistics defined in Eq. (4.16) is often most useful for process monitoring, but it is fruitful to exploit both D_{bc} and D_{cor} for the purpose of root cause analysis when an out-of-control signal is triggered. Such signals are compared to apriori set control limits for the fractal dimension statistics that can be determined numerically by using the in-control data.

All three fractal dimensions can be used for pattern detection. Even when the information dimension is used as the main investigative statistic, we may still wish to compute the box counting dimension and the correlation dimension whenever an anomaly is detected.

(1)   Fractal mapping: Select an IFS as data transformation and apply this to historical data to generate a fractal of training data.
(2)   Selection of a fractal statistic: We choose a fractal dimension to use as monitoring statistic. The information dimension is often a good choice, and it can be used with theoretically and numerically derived control limits.
(3)   Online process monitoring: Used online, each process sample is transformed and mapped as points in the fractal. Next, the selected fractal dimension is recalculated for the data sample, together with numerical control signals used to detect deviations.
(4)   Visual root cause analysis: Visual inspection of the fractal together with various fractal dimension statistics can be used to identify causes of detected deviations.

Control Limits

To investigate fractal dimension statistics and derive control limits, we use some results for the estimator of the Shannon entropy $\hat{H}$.

Let p_i be the actual frequencies for category i in a sample of size N, corresponding to the absolute observations N_i. The entropy is then given by $H = -\sum_{i=1}^{m} p_i \log p_i$ while $\hat{H} = -\sum_{i=1}^{m} (N_i/N) \log(N_i/N)$ is its maximum likelihood estimator. Let $\tilde{H} = \sqrt{N}(H - \hat{H})$. It can be shown that in the limit, $\tilde{H}$ has a normal distribution with mean zero, $\mathbf{E}(\tilde{H}) = 0$, and variance

$$\sigma^2(\tilde{H}) = \sum_{i=1}^{n} p_i (\log p_i + H)^2. \tag{4.19}$$

Miller and Madow [39] show that when $p_i = 1/n$ for every i, then $(2N/\log \ell)(H - \tilde{H})$ approaches a chi-square distribution with $(n - 1)$ degrees of freedom. The fractal map has $(m^k - 1)$ degrees of freedom for m symbols at level k.

Consequently, the monitoring statistic can be assumed to be normally distributed even in the case of a uniform distribution when $p_i = 1/m^k$.

Since the chi-square distribution approaches a normal distribution when the number of degrees of freedom is large, we approximate the statistic with a normal variate. For small samples, we can use the following expression, which contains a correction term for bias:

$$H = \mathbf{E}(\hat{H}) + (\log \ell) \left(\frac{n-1}{2N} - \frac{1}{12N^2} + \frac{1}{12N^2} \sum_{i=1}^{n} \frac{1}{p(i)} \right) + O\left(\frac{1}{N^3} \right). \tag{4.20}$$

Therefore, as the information dimension approximately is normally and independently distributed for long sequences, we can derive the control limits as follows:

(1) Calculate the information dimension of the fractal based on the given $\hat{D}_1 = \hat{H}(\alpha^k)/\log \alpha^k$.
(2) For small samples, we use Eq. (4.20) to estimate the entropy, compensating for bias.
(3) Use Eq. (4.19) to estimate the variance of $\tilde{H}(\alpha^k)$, dividing by $N(\log \alpha^k)^2$ to derive the variance $\sigma^2(\hat{D}_1)$ of the fractal dimension.
(4) Now, set the control limits as $\mathbf{E}(\hat{D}_1) \pm z_{\alpha/2} \cdot \sigma(\hat{D}_1)$.

The variance estimator $\sigma^2(\hat{D}_1)$ can be written

$$\sigma^2(\hat{D}_1) = \frac{1}{N(\log \alpha^k)^2} \sum_{i=1}^{n} p_i (\log p_i + H)^2,$$

which has an upper limit of $\sigma^2(\tilde{H})_{\max}$ when the data is deterministic and the entropy H is zero. Thus, we have

$$\sigma^2(\hat{D}_1)_{\max} = \frac{(\log m^k)^2}{N(\log \alpha^k)^2},$$

since $H = 0$.

The lower bound of $\sigma^2(\tilde{H})$ is attained when the data is uniformly distributed, so that $\sigma^2(\tilde{H})_{\min} = \sum_{i=1}^{n} p_i (\log p_i + H)^2 = 0$ and $\sigma^2(\hat{D}_1)_{\min} = 0$.

Online Process Monitoring

In a framework for online process monitoring, we can proceed as follows. In the first step, the fractal corresponding to some reference data is generated, and the monitoring statistics and the corresponding control limits are computed.

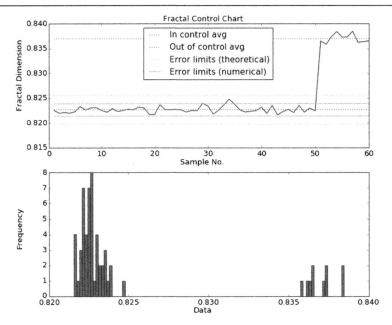

Figure 4.16: Example of a control chart, showing the development of the estimated information dimension and its empirical distribution.

During online monitoring, each data sample is transformed and mapped onto the fractal initially created. The fractal dimensions are recomputed for each new data sample, and deviations from the reference data are indicated by the fractal dimension estimates and the previously determined control limits.

The results can be shown in a line graph, showing the development of the statistic, and a histogram of its values. An example is shown in Fig. 4.16. It shows the information dimension before and after a slight change in the data sequence generation and the histogram of its values.

We summarize these steps for pattern detection in online data as follows:

(1) Set level $k = 1$.
(2) By inspection, find a circle of high density, possibly using predefined density limits.
(3) Drill down the fractal in the area of the relevant circle. Set $k \rightarrow k + 1$.
(4) Repeat steps (2) and (3) until the relevant circle on level k shows points that are approximately uniformly distributed between the m circles. At this level, the dominating pattern ends.
(5) Find the address of the identified subsequence represented by the selected circles, which can be used to recover the pattern.
(6) Repeat steps (1) to (5) for all circles indicating dominating patterns.

CHAPTER 5

Optimization Techniques

In this book we study a large number of optimization problems, many of which are hard to solve. In network optimization, a common challenge is avoiding getting stuck at local minima when the global minimum is sought. At the same time, the search space is restricted by boundary conditions that, whenever exceeded, contain only infeasible solutions of no interest to us.

Metaheuristics is a class of optimization methods which can be characterized as advanced search methods that by construction try to avoid getting trapped in local minima. Many of these methods are inspired by and mimic biological evolution or systems.

5.1 Optimization Problems in 5G

Due to the $\mathcal{NP}$-hardness of many network optimization problems, we have to resort to approximations, heuristics, or randomization. Metaheuristics combines randomization with a set of heuristic rules that makes up a general framework for solving hard problems.

There is a plethora of metaheuristic methods having slightly different properties. The best choice of method to use on a given problem, however, is in general hard to tell. Usually, it is advisable to use the method which makes modeling the easiest. Correctly implemented, different methods are likely to show similar performance.

Many network optimization problems can be translated into conventional combinatorial problems. These include clustering, facility location, bin packing, the traveling salesman's problem, and other design problems. There is vast literature on heuristics and approximations, a selection of which is described in this book. There is no generally applicable, efficient method to solve such problems, however. From a modeling perspective, it is often a good idea to formulate the problem at hand as an integer program. A common approach is to relax the inter restriction to obtain an ordinary linear program. We can find an approximate solution to such a program by *branch-and-bound*, which would give us an idea about the nature of the integer solution.

Oftentimes, the problem itself gives a hint to the most appropriate method to use. Firstly, we need to have a fairly clear view of what we want to optimize – the target function. This can be a straightforward cost we wish to minimize, several more or less dependent cost items, or a combination of cost and benefit. For more complex optimization targets, a utility function of some kind can be set up. It is important to analyze how this function behaves for extreme values. It is also beneficial to look at the variables. Sometimes a change of variables is warranted for, a restriction or a relaxation of their value sets.

Optimization problems in 5G range from access networks and data centers to antenna arrays and the internet of things (IoT). We may choose to optimize with respect to equipment or operational cost, energy consumption, or performance metrics including resilience and coverage. In this chapter we outline some of the main optimization principles applied to specific case studies in later chapters.

5.2 Mixed-Integer Programs

Most network optimization problems can be expressed as integer programs (IP) or mixed integer programs (MIP) having some variables taking real values and others taking integer values.

We illustrate the formulation and the approximate solution of an MIP by a concrete example – the knapsack problem. Suppose we are given n items, where each item j is associated with a value $c'_j > 0$ and a size a_j. We also have a knapsack, or bin, of size b, and we wish to choose $m < n$ items so that the total value is maximized. We assume that all the individual items fit into the bin, so that $a_j \leq b$ for all j, and that the total size of the items exceeds the size of the bin, $\sum_j a_j > b$. Otherwise the problem is trivial – just place all the items in the bin.

Note that the restriction $c'_j > 0$ also is necessary for the problem to be well defined; otherwise we would increase the value by not putting the item in the bin. Similarly, we need to have $0 < a_j \leq b$ for the problem to be well defined. We often assume that some or all constants in the constraints of the problem are integers. This is not a severe restriction, since any real number can be approximated arbitrarily close by a rational number, and the problem can then be rescaled to have integer constraints. Usually, we formulate the problem as a minimization problem. To do so, we simply set $c_j = -c'_j$. Then we have

$$v(K) = \min \quad \sum_{j=1}^{n} c_j x_j, \tag{5.1}$$

$$\sum_{j=1}^{n} a_j x_j \geq b, \tag{5.2}$$

$$x_j \quad \in \{0, 1\}, j = 1, \dots, n, \tag{5.3}$$

where $v(K)$ is the problem instance with K representing the data given and x_j are the variables that define the selection, taking value 1 if item j is chosen and 0 otherwise. This is the variable that makes the problem $\mathcal{NP}$-hard. Indeed, if we could choose fractional items, we could find the optimal solution by sorting items by decreasing value c'_j and for each item put as much as fits into the bin until it is full. This is the result we obtain when we relax the problem by letting $0 \leq x \leq 1$ in (5.3) and solving the corresponding linear program (5.1)–(5.2).

Dynamic Programming

Dynamic programming is a powerful and easily implemented method for solving the integer knapsack problem. For dynamic programming to work, the flows and capacities must be integers. However, the costs may be real numbers. This is not much of a limitation, as noted above, since any rational approximation of a real number can be used by multiplying by an appropriate factor to yield integer values. Dynamic programming solves the problem (5.1)–(5.3) by using the recursive relation

$$F_j(y) = \min\{F_{j-1}(y), F_j(y - a_j) + c_j\}, \tag{5.4}$$
$$F_0(y) = \infty,$$
$$F_j(y) = 0, \quad \text{for } y \leq 0. \tag{5.5}$$

The equation gives $F_j(y)$, which is the minimum cost using only the first j link types on edges with flow y, that is,

$$F_j(y) = \min \sum_{i=1}^{j} c_j x_j, \quad j < n,$$

with the condition that

$$\sum_{i=1}^{j} a_i x_i \geq y, \quad y < b.$$

Eqs. (5.4)–(5.5) work by first deciding how to best cover all flow values using only one line type. Then when a second line type is considered, it looks at all possible ways of dividing the flow between the two line types. When a third line type is added, Eq. (5.4) is simply choosing the best amount of flow to cover with the third line type, leaving the rest of the flow to be covered optimally among the first two line types (it had decided those questions optimally after the first two iterations of the recursion). The term $F_{j-1}(y)$ means "Don't take any more of the ith line type", while the term $F_j(y - a_j) + c_j$ means "Take at least one more of the ith line type" in the final decision. Since all the previous decisions have been made optimally, the only decision to make in Eq. (5.4) is whether one more instance of line type i is necessary to cover the flow optimally.

If the unit costs are the same for all edges in the network, only one instance of the capacity assignment problem needs to be solved, with b being the maximum flow value of any edge in the network. If the local tariffs add a fixed charge per line (based on the line type) in addition to the distance cost, then that fixed cost must be divided by the length of the line before being added to the cost per unit distance. The new unit cost would be computed as

$$c_k(i, j) = c_k^1 + \frac{c_k^2}{d_{ij}}. \tag{5.6}$$

In Eq. (5.6), $c_k(i, j)$ is the unit cost of line type k on edge (i, j), c_k^1 is the cost per unit distance of line type k, c_k^2 is the fixed cost for line type k, and d_{ij} is the distance from node i to node j. When both cost per unit distance and fixed costs appear in our cost function, the capacity assignment problem must be recalculated for every edge in the topology (since unit cost is now a function of the distance of each edge). Still, the solution given by the dynamic programming method outlined above would be optimum.

Example 5.2.1. *A manufacturer of microwave transmission equipment provides, say, microwave equipment in capacities 2, 4, 8, and 17 Mbps, corresponding to 1, 2, 4, and 8 E1 links. Choosing equipment in a cost-effective manner constitutes an integer knapsack problem. We assume that all links in the network have a distance such that one hop is required per link. Thus, the unit cost is the cost of the equipment and installation, which is assumed to be the same for all links.*

Suppose the link costs and capacities in some arbitrary units are given as per the following table,

Variable	x_1	x_2	x_3	x_4
Capacity	2	4	8	17
Cost	2	3	5	9
Density	1	0.75	0.63	0.53

Let $v_k(y)$ be the value of the knapsack subproblem defined for the first k variables and for the right-hand side $b = y$,

$$v_k(y) = \max \left\{ \sum_{j=1}^{k} c_j x_j \mid \sum_{j=1}^{k} a_j x \le y, x_j \ge 0, x_j \in \mathbb{Z}, j = 1, \ldots, k \right\}. \tag{5.7}$$

If $k \ge 2$, then for $y = 0, 1, \ldots, b$, we may write (5.7) in the form

$$v_k(y) = \max_{x_k=0,1,\ldots,\lfloor y/a_k \rfloor} c_k x_k +$$

$$+ \max \left\{ \sum_{j=1}^{k-1} c_j x_j \mid \sum_{j=1}^{k-1} a_j x_j \le y - a_k x_k, x_j \ge 0, x_j \in \mathbb{Z}, j = 1, \ldots, k - 1 \right\}.$$

The expression in the brackets equals $v_{k-1}(y - a_k x_k)$, so we may write (5.7) as

$$v_k(y) = \max_{x_k = 0, 1, \ldots, \lfloor y/a_k \rfloor} \{c_k x_k + v_{k-1}(y - a_k x_k)\}. \tag{5.8}$$

Putting $v_0(y) = 0$ for $y = 0, 1, \ldots, b$ we extend (5.8) for the case $k = 1$. The relation (5.8) expresses the so-called *dynamic programming principle of optimality*, which says that regardless of the number of the kth item chosen, the remaining space, $y - a_k x_k$, must be allocated optimally over the first $k - 1$ items. In other words, looking for an optimal decision at the n-stage process, we have to take an optimal decision at each stage of the process.

If, for a given y and k, there is an optimal solution to (5.8) with $x_k = 0$, then $v_k(y) = v_{k-1}(y)$. On the other hand, if $x_k > 0$, then in an optimal solution to (5.8) one item of the kth type is combined with an optimal knapsack of size $y - a_k$ over the first k items. Thus we have

$$v_k(y) = \max\{v_{k-1}(y), c_k + v_k(y - a_k)\} \tag{5.9}$$

for $k = 1, \ldots, n$ and $y = 0, 1, \ldots, b$. Then obviously $v(\mathbb{Z}) = v_n(b)$.

The computation of $v_k(y)$ requires by (5.9) comparison of two numbers. Thus the computational complexity of the dynamic programming is $O(nb)$. Dynamic programming is not a polynomial algorithm for solving K, since the length of data is K is $O(n \log(n))$.

Branch-and-Bound

The branch-and-bound approach can easily be described by the example above. In essence, the problem is solved as a linear program, probably giving a noninteger number as the optimum value. We then choose the integer upper and lower bounds of one of these variables and solve two new linear programs with this restriction. In the presented example, the optimal value is choosing a variable with the lowest density. We then choose the upper bounding integer (the "ceiling") and the lower bounding integer (the "floor"), which then must be feasible, and solve the problem again for these two different cases. See Fig. 5.1.

5.3 Rounding

Denote an integer program by Π and its relaxation by Π_L. The relaxation can be solved by standard methods, such as the simplex method, and its solution $\hat{x}$ has a cost $\hat{c}$ that by necessity is a lower bound of the cost c_{OPT} of the integer solution (since the integers are a subset of the set of real numbers), and the solution itself can give us some idea of the integer solution. These facts are explored to construct an approximate integer solution by a method known as *rounding*.

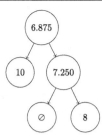

Figure 5.1: When using the branch-and-bound strategy we solve a number of linear problems. We identify the solution with the lowest cost satisfying the integer constraints (5.2)–(5.3).

Rounding was introduced as a technique by Raghavan and Thompson [40] and generalized by Lin and Vitter [41]. It consists of three basic steps. Given a combinatorial problem:

(1) Formulate the problem as an integer program Π.
(2) Relax the integer restriction on the IP to obtain a linear program relaxation Π_L; find an optimal *fractional solution* $\hat{\mathbf{x}}$.
(3) Round the fractional solution $\hat{\mathbf{x}}$ of the LP to an integer solution $\mathbf{x}'$ of the IP.

The first step may present challenges in itself. Some integer program formulations are presented throughout the text; for other problems, searching formulations in the literature is often a good idea to avoid pitfalls.

In the second step, the restriction to integer-valued variables is relaxed. The optimal fractional solution to the so-obtained linear program can typically be computed in polynomial time using any standard linear programming algorithm.

In the third step, the fractional solution is converted into an integer solution, that is, a feasible solution to the original problem (unless the solution in step two consists of integer values, which is unlikely). This step is known as rounding of the fractional solution. We know that the thus obtained integer solution should have a cost not much larger than the cost of the fractional solution. This, in turn, ensures that the cost of the integer solution is not much larger than the cost of the optimal integer solution.

Even if the rounding can be done both by deterministic and randomized methods, we will describe only randomization. We use the following claim.

Proposition 5.3.1. *Given a fractional solution $\hat{\mathbf{x}}$ of the LP relaxation to an IP. Then, with positive probability, the randomized rounding process produces an integer solution $\mathbf{x}'$ that approximates the optimal solution $\mathbf{x}$ according to some desired criterion.*

Assuming that the solution takes component values $x_i \in \{0, 1\}$ and using the values in the fractional solution $\hat{x}_i$, it seems reasonable to assume that a small fractional value of $\hat{x}_i$ should round to zero with a high probability and one with a small probability. We therefore select a value 1 with a probability proportional to the fractional value. This principle leads us to the most intuitive rounding scheme. For each $i = 1, 2, \ldots, d$ solution component, let $u_i \sim \mathcal{U}(0, 1)$ be a uniformly distributed random variable and $\hat{x}_i$ the fractional solution. Then, if $u_i < \min\{\hat{x}_i, 1\}$, set $x'_i = 1$ and $x'_i = 0$ otherwise.

Raghavan and Thompson [40] analyze a multicommodity flow problem using this technique, and we return to this type of problems in Chapter 11. It is also used in the capacitated facility location problem in Chapter 8.

There is, however, a problem with this basic rounding scheme: for small variable values $\hat{x}_i$, the probability that the variable will be covered (represented) is very low. We therefore modify the scheme as follows.

Let $\lambda \geq 1$ be a scaling parameter with which we scale up the probabilities. For each solution component i, let $u_i \sim \mathcal{U}(0, 1)$ be a uniformly distributed random variable and $\hat{x}_i$ the fractional solution. Then, if $u_i < \min\{\lambda \hat{x}_i, 1\}$, set $x'_i = 1$ and $x'_i = 0$ otherwise. The drawback is that by increasing λ, the cost is also scaled up by λ. The strategy is to choose λ as small as possible, while giving all components nonnegligible probability of being covered.

Lin and Vitter [41] generalize this idea to a wider class of problems. They consider integer programs of the form

$$\min \mathbf{c}^T \mathbf{x} \tag{5.10}$$
$$A\mathbf{x} = 1, \tag{5.11}$$
$$B\mathbf{x} \leq \mathbf{b}, \tag{5.12}$$
$$\mathbf{x} \in \mathcal{P}, \tag{5.13}$$
$$x_i \in \{0, 1\}, \quad \text{for all } i \in \mathcal{I}, \tag{5.14}$$

where $\mathcal{I}$ is the set of indices for the vector $\mathbf{x}$, $\mathbf{c}$ and $\mathbf{b}$ are nonnegative rational vectors, A is a 0-1 matrix, B is a nonnegative rational matrix, and $\mathcal{P}$ is a convex set that corresponds to other linear constraints. The linear program relaxation of the above program is obtained by removing restriction (5.14) and allow the x_is to take rational values in $[0, 1]$.

The ϵ-approximation of an $\mathcal{NP}$-hard problem is an approximation to the optimal solution which lies within a factor of $\epsilon > 0$ of additional cost. Finding an ϵ-approximation is for many problems of the type above $\mathcal{NP}$-hard as well.

Let Π be an integer program of the given type with linear program relaxation Π_L. The filtering/rounding algorithm consists of three phases.

Phase 1: Solve Π_L by any method for solving linear programs; denote the fractional solution by $\hat{\mathbf{x}}$.

Phase 2:   Filtering. Given $\epsilon > 0$ and the fractional solution $\hat{\mathbf{x}}$, we transform Π into an integer program $\bar{\Pi}(\epsilon, \hat{\mathbf{x}})$ of minimizing L, subject to

$$Ax = 1, \tag{5.15}$$

$$Bx \leq L\mathbf{b}, \tag{5.16}$$

$$\mathbf{x} \in \mathcal{Z}, \tag{5.17}$$

$$x_i \in \{0, 1\}, \quad \text{for all } i \in \mathcal{I} - \mathcal{Z}, \tag{5.18}$$

where $\mathcal{Z} \subset \mathcal{I}$ depends on ϵ and $\hat{\mathbf{x}}$. Besides setting a subset of variables to 0, another effect of $\mathcal{Z}$ is that, for each $i \in \mathcal{Z}$, column i of A and B can be considered to be zeroed out. This transformation is said to be valid if any feasible solution $\bar{x}$ for $\bar{\Pi}(\epsilon, \hat{\mathbf{x}})$ satisfies

$$\mathbf{c}^{\mathsf{T}}\bar{\mathbf{x}} \leq (1 + \epsilon)\mathbf{c}^{\mathsf{T}}\hat{\mathbf{x}} \leq (1 + \epsilon)\mathbf{c}^{\mathsf{T}}\mathbf{x}^*, \tag{5.19}$$

where $\mathbf{x}^*$ is the optimal solution for Π. Let $\bar{\Pi}_L(\epsilon, \hat{\mathbf{x}})$ be the linear program relaxation of $\bar{\Pi}(\epsilon, \hat{\mathbf{x}})$.

Phase 3:   Rounding. Solve $\bar{\Pi}(\epsilon, \hat{\mathbf{x}})$, that is, minimize the packing constraint violation, which is represented by the variable L. We solve $\bar{\Pi}(\epsilon, \hat{\mathbf{x}})$ by first converting $\hat{\mathbf{x}}$ into a fractional solution $\tilde{\mathbf{x}}$ for $\bar{\Pi}_L(\epsilon, \hat{\mathbf{x}})$. We can then transform $\tilde{\mathbf{x}}$ into a good integer solution for $\bar{\Pi}(\epsilon, \hat{\mathbf{x}})$ by deploying various techniques.

We will refer to $\bar{\Pi}(\epsilon, \hat{\mathbf{x}})$ as the filtered program of Π with respect to ϵ and $\hat{\mathbf{x}}$.

Phase 2 can be combined with Phase 3. In Phase 3, we can derive from $\hat{\mathbf{x}}$ an upper bound for the optimal solution of the relaxed filtered program. Therefore, any provably good rounding algorithms for transforming a solution for $\bar{\Pi}_L(\epsilon, \hat{\mathbf{x}})$ into an integer solution for $\bar{\Pi}(\epsilon, \hat{\mathbf{x}})$ will also provide performance guarantees for the packing constraint violation.

Lin and Vitter apply the method to the k-median problem, for which they describe the following rationalized random sampling technique for rounding.

(1)   Solve the linear program relaxation of the k-median problem by linear programming techniques; denote the fractional solution by $\hat{\mathbf{y}}, \hat{\mathbf{x}}$.

(2)   Given $\epsilon > 0$ and $0 < \delta < 1$, we select $(1 + 1/\epsilon)s \ln(n/\delta)$ vertices randomly, where vertex j has relative weight $\hat{y}_j/k$. Let U be the set of vertices that the sampling algorithm selects. Then the solution for the filtered program is obtained by setting $y_j = 1$ for each $j \in U$ and $y_j = 0$ for each $j \in V - U$. The values for x_{ij} are given by the proof of Lemma 5.3.2.

Lemma 5.3.2. *Given a solution $\hat{\mathbf{y}} = (\hat{y}_1, \ldots, \hat{y}_n)$ for the fractional k-median problem, we can determine the optimal fractional values for x_{ij}.*

Proof. Each vertex i is assigned to its nearest fractional medians at vertices $j_1, j_2, \ldots$ such that their total sum of weights $y_{j_1} + y_{j_2} + \cdots$ reaches 1. In other words, we sort the values c_{ij}, $j \in V$, so that $c_{ij_1(i)} \leq c_{ij_2(i)} \leq \ldots c_{ij_n(i)}$, and let p be such that $\sum_{l=1}^{p-1} \hat{y}_{j_l(i)} \leq 1 \leq \sum_{l=1}^{p} \hat{y}_{j_l(i)}$. Then set $\hat{x}_{ij} = \hat{y}_j$ for $j = j_1(i), \ldots, j_{p-1}(i)$, $\hat{x}_{ij_p(i)} = 1 - \sum_{l=1}^{p-1} \hat{y}_{j_l(i)}$, and $\hat{x}_{ij} = 0$ otherwise. $\square$

5.4 Simulated Annealing

We can think of optimization as a virtual hill climbing. To reach a global extreme point, we need to cross many smaller, local ones. This is one of the contributing factors to the hardness of many design problems; we cannot just look at whether the slope is uphill or downhill at our current position. However, we do not know in which direction we will find the global optimum. To improve our search for a global extreme, we therefore allow a move in the "wrong" direction determined locally with a certain probability. This improves our chances to reach the global peak. This is also the idea behind *simulated annealing*, inspired by the cooling process in metallurgy.

First, we define a local search strategy. Starting from an initial solution (or candidate) c, we evaluate it by computing some function $f(c)$ to find its value (as included in the target function). The search is performed randomly, since no analytic search method is assumed available. A new candidate is generated by modifying c slightly, called a mutant m. We say that we are searching the neighborhood of c. By evaluation of m through $f(m)$ we can determine which of c and m is better. Throughout the search, we keep record of the best solution found so far. A natural strategy is choosing $c = m$ whenever $f(m) \geq f(c)$ and keeping c unaltered otherwise, and repeating the search. However, it may happen that c is in a local maximum with $f(m) < f(c)$ in a neighborhood of c. It may therefore be necessary to move in a direction where $f(m) < f(c)$ to be able to move further uphill. We therefore change strategies choosing $c = m$ if $f(m) \geq f(c)$ and also with a certain probability if $f(m) \leq f(c)$. The probability also depends on how much worse the mutant is than the candidate, so that worse points are chosen with lower probability. The probability of choosing an inferior point also decreases with time.

To implement such a control of the search, we introduce two parameters: the initial temperature T and the cooling rate r, $0 < r < 1$. We determine the probability of choosing a worse point as candidate using a *test function*, such as

$$p = \exp(f(m) - f(c))/T. \tag{5.20}$$

Next, we generate a uniformly distributed random number $u \in [0, 1]$ and choose $c = m$ if $u < p$. Note that when $f(m) \geq f(c)$, $p \geq 1$, so then the mutant is always chosen. After each iteration, we set $T \leftarrow rT$. As T decreases, the exponent increases.

Algorithm 5.4.1 (Phase I – assignment).

Set $p_j = 0, \forall j \in \mathcal{D}$.

STEP 1:$|\mathcal{D}|$ **while** there are unassigned clients: All unassigned clients j linearly increase
their prices p_j until they reach a facility i at $p_j \geq c_{ij}$,
      Then they start contributing to the facility's opening cost f_i by $p_j - c_{ij}$;   When the
facility's opening cost f_i is covered, the prices remain unchanged;   If an unassigned
client j reaches an open facility i, it is assigned to it without contributing to its opening
cost f_i; **end**

Output the set of coefficients C.

5.5 *Genetic Algorithms*

This section will not treat genetic algorithms in detail, but rather illustrate the principles by
a relatively simple example. The literature on genetic algorithms is vast and this section is
intended to serve only as an introduction of its application to network design. Genetic algorithms belong to a class of numerical methods known as metaheuristics. It mimics an evolutionary process where the "survival of the fittest" has a higher probability to reproduce, just
like favorable genes promote better biological specimens. The algorithm has lent much of its
vocabulary from biological evolution, such as "population", "chromosome", "generation", and
"reproduction".

A genetic algorithm is based on a number of subroutines which require some care in its programming. Since it is a randomized algorithm, the quality of random number generators
should be considered. Based on an initial population coded as binary strings, the procedure
performs three steps in its search for an optimal solution. The steps are *reproduction, recombination*, and *mutation*, which represent the selection processes, but also allow random changes
to the population in order to cover as large a search space as possible.

The first task is to find a proper binary coding that can represent any network topology. The
main reason for this is that the recombination and mutation processes are harder to implement
for nonbinary representations. Given a number of binary represented strings, we need to evaluate their "fitness", representing the quality of a particular configuration, and which typically
is a number reciprocal of its cost or proportional to some performance measure.

Genetic algorithms work with a population of individuals, each representing a possible solution to a given problem. Each individual is assigned a fitness value according to how good
a solution to the problem it is. Highly fit individuals are given opportunities to reproduce by

crossbreeding with other individuals in the population. This produces new individuals as "off-spring", which share some features taken from each "parent". The least fit members are less likely to get selected for reproduction and "die out".

The reproduction (or crossover) operator combines randomly selected parts of two parent chromosomes to form an offspring chromosome. In addition, a random mutation is applied to the resulting configuration. The idea is that having two good solutions, combining them may lead to an even better one. Mutations are applied to allow for cases that may not have been represented in earlier generations. We describe a genetic algorithm for network design by considering the implementation of the operators.

A whole new population of possible solutions is thus produced by selecting the best individuals from the current "generation" and mating them to produce a new set of individuals. Thus new generations contain a higher proportion of the characteristics possessed by the good members of the previous generations. In this way, over many generations, good characteristics are spread throughout the population. By favoring the mating of the more fit individuals, the most promising areas of the search space are explored. If the genetic algorithm has been designed well, the population will converge to an optimal solution to the problem.

The evaluation function, or objective function, provides a measure of performance with respect to a particular set of parameters. The fitness function transforms that measure of performance into an allocation of reproductive opportunities. The evaluation of a string representing a set of parameters is independent of the evaluation of any other string. The fitness of that string, however, is always defined with respect to the other members of the current population. In the genetic algorithm, fitness is defined by f_i/f_A, where f_i is the evaluation associated with string i and f_A is the average evaluation of all the strings in the population A.

Fitness can also be assigned based on a string's rank in the population or by sampling methods, such as tournament selection. The execution of the genetic algorithm is a two-stage process. It starts with the current population. Selection is applied to the current population to create an intermediate population. Then recombination and mutation are applied to the intermediate population. The process of going from the current population to the next population constitutes one generation in the execution of a genetic algorithm.

In the first generation, the current population is also the initial population. After calculating f_i/f_A for all the strings in the current population, the selection is performed. The probability that strings in the current population are copied (that is, duplicated) and placed in the intermediate generation is proportional to their fitness value.

Individuals are chosen using stochastic sampling with replacement to fill the intermediate population. A selection process that will more closely match the expected fitness values is *remainder stochastic sampling*. For each string i where f_i/f_A is greater than 1.0, the integer

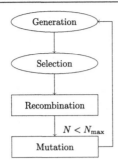

Figure 5.2: Diagram of a genetic algorithm for topological design.

portion of this number indicates how many copies of that string are directly placed in the intermediate population. All strings (including those with $f_i/f_A < 1.0$) then place additional copies in the intermediate population with a probability corresponding to the fractional portion of f_i/f_A. For example, a string with $f_i/f_A = 1.36$ places one copy in the intermediate population and then receives a 0.36 chance of placing a second copy.

Remainder stochastic sampling is most efficiently implemented using a method known as stochastic universal sampling. Assume that the population is laid out in random order as in a pie graph, where each individual is assigned space on the pie graph in proportion to fitness. An outer roulette wheel is placed around the pie with N equally spaced pointers. A single spin of the roulette wheel will now simultaneously pick all N members of the intermediate population.

The first step in the implementation of any genetic algorithm is to generate an initial population. We can use so-called *splicers* for this purpose. The following algorithm due to Aldous and Broder (see [52]) is a simple and easy to use algorithm for generation of k-splicers.

Algorithm 5.5.1. STEP 1: Start a random walk at some arbitrary vertex of the complete graph.

STEP 2: When the walk visits a vertex for the first time, include into the tree the edge used to reach that vertex.

STEP 3: When all the vertices have been visited, stop. We have a spanning tree which is uniformly random regardless of the initial vertex.

It is necessary to decide the size of the initial population and the number of generations (that is, the number of iterations) N that the algorithm will be run. Usually a trial-and-error approach is the only way to determine these algorithm parameters. The overall schematic optimization process is shown in Fig. 5.2.

Binary representation

In order to implement the functions of a genetic algorithm, we need to have a binary representation of each candidate. Since the edges are the most important part of a network topology, it is advantageous to find a mapping from each node pair (i, j) to a number identifying the corresponding edge e_{ij}. Suppose that we have an enumeration of the vertices in the graph. For any node pair (i, j) let l denote the largest number and s the smallest number, so that $l = \max\{i, j\}$ and $s = \min\{i, j\}$. Then let the string position be indexed by $l \cdot (l - 1)/2 + s$. If an edge is present, we put a 1 in that position; if not, a 0.

Fitness function

The fitness function may include both feasibility constraints and cost evaluation. We may require, for example, that each candidate topology is connected.

The fitness function determines the probability assigned to each candidate that it will be chosen during the reproduction phase. Since we are interested in the lowest cost, we can use a number reciprocal to the cost, so that a lower cost yields higher probability. For example, we can assign the fitness value

$$C_{\max}/C_i, \quad \text{for each candidate } i, \tag{5.21}$$

where $C_{\max}$ is the cost of the most expensive candidate in the population. To obtain probabilities, we then normalize the numbers so that they sum to one.

Reproduction

Based on the fitness (or rather the normalized fitness), a selection of all candidates in the current population is made, proportional to their respective fitness. A common method is the so-called roulette wheel selection. An easy implementation of this is to generate a random number u uniformly in the interval $[0, 1]$. Next we sum the individual fitness values for the candidates, where edges are selected in the order of the binary representation of the topology, until the sum just but exceeds u. This is repeated as many times as there are candidates in the population. For each new generation, the fitness function for its members needs to recomputed.

Recombination (crossover)

The next step is to select two candidates, each with equal probability, and a random number c between zero and the length b of the binary representation. A new candidate is formed by combining the c first binaries/binary digits from the first candidate and the $b - c$ last binaries/binary digits from the second candidate to form an "offspring". This is repeated for as many times as the size of the population.

Mutation

For each new candidate created by the recombination step, we "scan" the binary string and generate a uniform random number for each allele of the chromosome. If the number is less than a small probability p, we change the current allele from its current value to its complement (that is, from 1 to 0 or vice versa). This step increases the search space. This is the final step in the creation of a new generation, after which the algorithm returns to the reproduction phase.

5.6 Swarm Algorithms

Swarm algorithms mimic the collective behavior and interaction between elements of certain organisms. Many biological systems exhibit sophisticated self-organizing traits, which give a collective of organisms the ability to quickly adapt to changing conditions. There exists a long list of swarm algorithms, of which we describe a few. The possibly subjective choice is based on their successful application to some of the practical problems we discuss in later chapters. Swarm algorithms have in common that a population of organisms performs searches in parallel.

Ant Colony Optimization

Ant colony optimization (ACO) is a metaheuristic technique that models the foraging behavior of ant colonies, first proposed by Dorigo [42]; see also [43][44][45]. An ant communicates with other ants by means of chemicals known as pheromones, which help other members of the colony to find short paths to food sources. The algorithm is based on population search, that is, a number of organisms each performing a search subject to communication of their success with other organisms. Initially, ants explore their surroundings randomly. When moving about, they leave a trail of a chemical substance, pheromone, that can be detected by other ants. At any point, an ant is likely to choose a direction marked by a high pheromone concentration. When an ant finds a food source, it brings some of it back to the nest, leaving a trail of pheromone that typically depends on the quality and quantity of the food, thus guiding other ants to the source. It has been shown that the indirect communication between ants using pheromone trails leads them to find the shortest paths between their colony and food sources.

To formulate a model of the ant colony, we consider a simple graph $G = (V, E)$, where V contains two nodes, v_s, representing the nest, and v_d, representing the food source. The set E consists of links (making up two paths), e_1 and e_2, between v_s and v_d, with the assigned lengths l_1 and l_2 such that $l_2 > l_1$. Thus, e_1 represents the shortest path between v_s and v_d, and

e_2 represents a longer path. We let τ_i denote the pheromone concentration deposited on the two edges e_i, $i = 1, 2$. An ant, originally at v_s, chooses the path $i = 1, 2$ with a probability given by

$$p_i = \frac{\tau_i}{\tau_1 + \tau_2}, \quad i = 1, 2, \tag{5.22}$$

to reach v_d. When returning from v_d to v_s, the ant uses the same path as when moving from v_s to v_d and deposits pheromone on the used edge so that

$$\tau_i \leftarrow \tau_i + \frac{Q}{l_i}, \tag{5.23}$$

where the constant $Q > 0$ is a model parameter. The deposited pheromone concentration is therefore inversely proportional to the length of the path. During the iterations, we let the pheromone deposited in earlier steps evaporate. The concentrations are updated so that

$$\tau_i \leftarrow (1 - \rho)\tau_i, \quad i = 1, 2, \tag{5.24}$$

where the model parameter $\rho \in (0, 1]$ controls the rate of pheromone evaporation. The ACO metaheuristics can be regarded as a framework that needs to be properly adapted to the problem it is applied to. The obvious first step is a thorough understanding of the nature of a solution and of how to produce a new candidate solution from two other candidates.

The optimization is controlled by the set of pheromone values $\mathcal{T}$, referred to as the pheromone model, that defines the probabilistic search process. It is used to generate candidate solutions from candidates previously found, weighted by assigned pheromone levels $\tau_i \in \mathcal{T}$ in a manner akin to trails leading to food sources. The ACO framework operates iteratively, using two general steps:

(1) Candidate solutions are generated based on the pheromone model.
(2) The candidate solutions are evaluated and the results are used to modify the pheromone levels to bias future sampling toward candidates of high quality.

The pheromone model thereby successively concentrates the search to subspaces where there is a high probability of finding good candidates, but with nonzero probability of searching other parts of the search space as well. We formalize the ACO into a master algorithm, following [44].

Algorithm 5.6.1 (Ant colony optimization).

Given a set of candidate solutions $\mathcal{S} = \{s_1, s_2, \ldots, s_m\}$ and a pheromone model $\mathcal{T}$.

STEP 1:

>   **while** convergence conditions not met **do**
>   Perform `AntBasedSolutionConstruction`,   Perform `PheromoneUpdate`, **end**

Output optimal solution candidate s^*. •

We describe the steps of the algorithm in some more detail.

Ant-based solution construction

The construction of a candidate solution starts with an empty set $S = \{\varnothing\}$. A solution candidate is formed by adding feasible components $\mathcal{F} \subseteq \mathcal{C}$ to build a solution, with respect to the current neighborhood. This specification is clearly problem-dependent. In ACO, this construction is probabilistic and governed by the transition probabilities,

$$\mathbf{P}(c_i|s) = \frac{\tau_i^\alpha \cdot \eta(c_i)^\beta}{\sum_{c_j \in \mathcal{F}(s)} \tau_j^\alpha \cdot \eta(c_j)^\beta}, \tag{5.25}$$

where η is an optional weighting function. This function may assign a value $\eta(c_j)$ to each feasible solution component $c_j \in \mathcal{F}(s)$, known as heuristic information. The exponents α and β are positive parameters whose values determine the relation between pheromone information and heuristic information.

Pheromone update

The pheromone model controls the search pattern and is updated in each iteration. Different ACO variants mainly differ in how the pheromone levels are updated. The update consists of two steps, evaporation and pheromone adjustment on solution components, following the evaluation of new solutions. Evaporation prevents a too fast convergence to a region of suboptimal solutions. When a new solution is tried out by chance, its pheromone levels may cause ants to "stick to" it, with too fast decrease of the probability of moving in other directions as a result. The entire update can be expressed as

$$\tau_i \leftarrow (1 - \rho) \cdot \tau_i + \rho \cdot \sum_{s \in \mathcal{S}_{upd} c_i \in s} w_s F(s), \tag{5.26}$$

where $i = 1, \ldots, n$ and $\mathcal{S}_{upd}$ denotes the set of candidate solutions used for the update. The constant $\rho \in (0, 1]$ is the evaporation rate and $F : S \mapsto \mathbb{R}^+$ is a *quality function* such that $f(s) < f(s') \Rightarrow F(s) \geq F(s'), \forall s = s' \in S$. The quality function is such that if a solution s evaluates to be better than a solution s', then the quality of s is at least as high as the quality of s'. The update is based on the current solution sets and the best solution found since the start of the iteration s_{bs}.

ACO can be used to solve assignment problems as follows. We let the ants represent the jobs and the nodes define the transition probabilities subject to feasibility. That is, the transition probability is zero unless a job having certain parameters can be accommodated by the node. The transition probability is based on the chosen policy. The transition probability is updated and normalized in each cycle.

The assignment of a job, that is, the node position of an ant, is controlled by a roulette-wheel selection based on transition probabilities. After, assigning a job, the available resources are updated, results are stored and the cycle is completed.

Particle Swarm Optimization

Particle swarm optimization (PSO), proposed by Kennedy and Eberhart [47][48], is another optimization method mimicking social interaction, where small entities – the particles – are spread out in the search space and used to evaluate the solutions they represent. In each iteration, the particles determine its movement by combining its current and best points, information about the success of one or more other particles in the swarm, with some random perturbations. In an iteration, every particle is moved accordingly. The idea is that the swarm as a whole eventually moves to a near-optimal solution.

An individual particle consists of three D-dimensional vectors, representing points in the D-dimensional search space. These vectors are the position of the current solution $\mathbf{x}_i$, the position of the previous best solution $\mathbf{p}_i$, and the velocity vector $\mathbf{v}_i$.

The current position $\mathbf{x}_i$ and the velocity $\mathbf{v}_i$ of a particle determine its next step. At each step, the solution is evaluated; its value is compared to the value of the best solution encountered so far. If the current value is better than the previously found best solution, the current position is stored in $\mathbf{p}_i$.

The governing part of the algorithm is the computation of the velocity $\mathbf{v}_i$, which can be interpreted as the step size and direction of the search. This ties particles together in the swarm, as it is determined on the interaction between particles. The search is therefore the outcome of the collective behavior of the particles.

The particle swarm itself is organized in some sort of communication structure, or topology, which can be viewed in terms of a graph. In this structure, any particle i is connected to neighboring particles by bidirectional edges. A particle j connected to i is said to be in the neighborhood of i, and the bidirectionality of the edges imply that then i is in the neighborhood of j as well. The velocity of particle i is determined by the direction to best solution points found by any of its neighbors, say g, and its position is denoted by $\mathbf{p}_g$. We have

$$
\begin{cases}
\mathbf{v}_i \leftarrow \mathbf{v}_i + \mathbf{U}(0, \phi_1) \times (\mathbf{p}_i - \mathbf{x}_i) + \mathbf{U}(0, \phi_2) \times (\mathbf{v}_g - \mathbf{x}_i), \\
\mathbf{x}_i \leftarrow \mathbf{x}_i + \mathbf{v}_i
\end{cases}
\tag{5.27}
$$

where ϕ_1 and ϕ_2 are the maximum strengths of the forces pulling the particle in either direction and $\mathbf{U}(\cdot)$ is a vector containing uniformly distributed random numbers.

In the particle swarm optimization process, the velocity of each particle is iteratively adjusted so that the particle stochastically oscillates around $\mathbf{p}_i$ and $\mathbf{p}_g$ locations. The (original) process for implementing PSO is shown in Algorithm 5.6.2.

Algorithm 5.6.2 (Particle swarm optimization).

Initialize a set of particles with random positions and velocities in the D dimensions of the search space.

STEP 1:$|N|$

>   **while** convergence criteria not met **do**
>   >   **for** each particle **do**
>   >   >   Evaluate the objective function to find the value v_i of each solution.
>   >   >   Compare the value v_i with the value of the best solution found by particle i so far.
>   >   >   If the current value is better, set the best value equal to the current value,
>   >   >   and $\mathbf{p}_i$ equal to the current location $\mathbf{p}_i$.
>   >   >   Identify the neighbor of i having found the best solution so far, and assign
>   >   >   its index to the variable g.
>   >   >   Change the velocity and position of the particle according to Eq. (5.27).
>   >   **end**
>   **end**

Output best found solution $\mathbf{p}_{\text{best}}$. •

Parameters

The PSO method needs to be initiated with a few parameters. The first parameter that needs to be set is the number of particles in the swarm. The more particles, the more search points and the slower the algorithm. A suitable number is usually found empirically and is selected based on the number of dimensions of the problem and its perceived difficulty. Common values range between 20 and 50.

The parameters ϕ_1 and ϕ_2 in Eq. (5.27) determine the strengths of the random forces in the directions of the individual best solution $\mathbf{p}_i$ and neighborhood best solution $\mathbf{p}_g$. In analogy with physics, these are often referred to as acceleration coefficients. The particle can be envisaged as being suspended by two springs of random stiffness and end points fixed at each solution. Parameter values chosen without due care can cause the particle velocity to oscillate wildly or even become unbounded.

One way of preventing such a situation is to impose minimum and maximum values $[-V_{\text{max}}, +V_{\text{max}}]$ that the velocity $\mathbf{v}_i$ of any particle i can assume.

Different magnitudes of the acceleration coefficients create two search strategies: exploration and exploitation. The former returns a rougher approximation but scans a larger area, whereas the latter searches a smaller area more thoroughly. The use of a maximum value V_{max} for the acceleration has influence on the trade off between exploration and exploitation and on the convergence of the algorithm.

To attain better control of the search and stabilize the particle movement, the following modification of the PSO was proposed by Shi and Eberhart [49]:

$$\mathbf{v}_i \leftarrow \omega\mathbf{v}_i + \mathbf{U}(0, \phi_1) \times (\mathbf{p}_i - \mathbf{x}_i)\mathbf{U}(0, \phi_2) \times (\mathbf{p}_g - \mathbf{x}_i), \tag{5.28}$$

$$\mathbf{x}_i \leftarrow \mathbf{x}_i + \mathbf{v}_i, \tag{5.29}$$

where ω is called "inertia weight". Note that if we interpret $\mathbf{U}(0, \phi_1) \times (\mathbf{p}_i - \mathbf{x}_i) + \mathbf{U}(0, \phi_2) \times (\mathbf{p}_g - \mathbf{x}_i)$ as the external force, $\mathbf{f}_i$, acting on a particle, the change in the velocity of the particle can be written as $\Delta\mathbf{v}_i = \mathbf{f}_i - (1 - \omega)\mathbf{v}_i$. The constant $1 - \omega$ can therefore be interpreted as a coefficient of friction.

As a rule of thumb, good results are achieved by an initially high value $\omega \approx 0.9$, which gives a system where particles perform exploration, and gradually decreasing ω to ≈ 0.4 to perform exploitation. Rather than a deterministic strategy for ω, randomly assigning its value has also been proposed, such as $\omega = U(0.5, 1)$.

PSO is used for example to find optimal solutions in antenna arrays [81][82].

Firefly Algorithm

The firefly optimization (FFO) algorithm is one of many biologically inspired swarm optimization algorithms, proposed by Xin-She Yang [50] (see also [51]). The firefly algorithm is in many respects similar to the particle swarm algorithm and can be shown to reduce to the latter with some choices of functions.

The firefly algorithm is based on the following principles.

(1) All fireflies are unisex: all individual insects can attract each other equally.
(2) The attractiveness between fireflies is proportional to their emitted brightness. A less bright firefly will move towards a brighter one, but the light intensity (brightness) decreases with increasing mutual distance.
(3) If no one firefly is brighter than a particular firefly, it moves randomly.
(4) The brightness of a firefly is related to the target function of the optimization problem.

For firefly i, the communication with other insects is based on the emitted light intensity I_0 and the intensity I_i perceived by another firefly j at a distance r_{ij} away, which are related by

$$I_i = I_0 e^{-\gamma r_{ij}}, \tag{5.30}$$

where γ is a light absorption coefficient. The constant γ can take any value, but in practice $\gamma = 1$.

We define the *attractiveness* β_{ij} a firefly i exerts on another firefly j at a distance r_{ij} away as

$$\beta_{ij} = \beta_0 e^{-\gamma r_{ij}^2}, \tag{5.31}$$

where β_0 is the attractiveness at $r_{ij} = 0$. The distance r_{ij} between any two fireflies i and j at points $\mathbf{x}_i$ and $\mathbf{x}_j$ is usually the Euclidean distance

$$r_{ij} = ||\mathbf{x}_i - \mathbf{x}_j||_2. \tag{5.32}$$

In step t, a firefly i moves towards another (brighter) firefly j by means of

$$\mathbf{x}_i^{(t+1)} = \mathbf{x}_i^{(t)} + \beta_{ij}(\mathbf{x}_j^{(t)} - \mathbf{x}_i^{(t)}) + \alpha^{(t)}\epsilon_{ij}^{(t)}, \tag{5.33}$$

where ϵ_{ij} is a random parameter generated by a uniform, Gaussian, or some other distribution, and α is a parameter controlling the step size. We summarize these steps of the algorithm in pseudocode.

Algorithm 5.6.3 (Firefly optimization).
Given a set of m candidate solutions $\mathcal{S} = \{s_1, s_2, \ldots, s_m\}$ at points $\mathbf{x}_i$ and computed light intensities l_i for $i = 1, 2, \ldots, m$. Let N be the maximum number of generations and/or define convergence criteria.

STEP 1: N

       while convergence conditions not met **do**
         for $i = 1$ to m **do**
         for $j = 1$ to m **do**
          if $I_i > I_j$, move i towards j by Eq. (5.33)
         end
         end
         end

Output optimal solution candidate s^*.   ●

The FFO algorithm thus requires Nm evaluations of cost. The algorithm can be seen as a generalization of the PSO algorithm. It has been applied to antenna systems as well (see Chapter 9). We model both a circular array with binary excitation levels ($e_i = \{0, 1\}$) and a hexagonal array with arbitrary exitation levels $0 \leq e_i \leq 1$.

CHAPTER 6

Clustering

Clustering is frequently used in data analysis and machine learning, for example in pattern recognition. It is also closely related to network design and so plays a prominent role throughout this text. Clustering techniques are used to find clusters meeting some objectives, such as finding the largest or smallest cluster sizes, or division of data into a predefined number of clusters k in an optimal way in some sense. To determine the properties of the clustering itself (as opposed to the data), cluster quality measures are used.

We can use different techniques to group general data in an efficient way according to some given criteria. Such a grouping often reveals underlying structures and dependencies in the data that may not be immediately obvious. We will refer to a single observation as a data point, or more generally, a data object, that possesses various properties that can be defined independently of other data objects.

Since the relationships between data points can be represented by a graph, many clustering techniques are closely related to graph theory. More formally, the minimum k-clustering problem where a finite data set D is given together with a distance function $d : D \times D \to \mathbb{N}$, satisfying the triangle inequality [53]. The goal is to partition D into k clusters $C_1, C_2 \ldots, C_k$, where $C_i \cap C_j = \oslash$ for $i \neq j$, so that the maximum intercluster distance is minimized (that is, the maximum distance between two points assigned to the same cluster). This problem is approximable within a factor of 2, but not approximable within $(2 - \epsilon)$ for any $\epsilon > 0$.

A related problem is the minimum k-center problem, where a complete graph is given with a distance function $d : V \times V \to \mathbb{N}$ and the goal is to construct a set of centers $C \subseteq V$ of fixed order $|C| = k$ such that the maximum distance from a vertex to the nearest center is minimized. Essentially, this is not a graph problem as the data set is simply a set of data and their distances – the edges play no role here. If the distance function satisfies the triangle inequality, the minimum k-center problem can be approximated within a factor of 2, but it is not approximable within $(2 - \epsilon)$ for any $\epsilon > 0$. Without the assumption of the distance satisfying the triangle inequality, the problem is harder.

A capacitated version, where the triangle inequality does hold, but the number of vertices "served" by a single center vertex is bounded from above by a constant, is approximable within a factor of 5. A center serves a vertex if it is the closest center to that vertex. Another capacitated version where the maximum distance is bounded by a constant and the task is to choose a minimum-order set of centers is approximable within a factor $\log(c) + 1$, where c is the capacity of each center. The problem is also referred to as the *facility location* problem.

A weighted version of the k-center problem, where the distance of a vertex to a center is multiplied by the weight of the vertex and the maximum of this product is to be minimized, is approximable within a factor of 2, but it can not be approximated within $(2 - \epsilon)$ for any $\epsilon > 0$. If it is not the maximum distance that is of interest, but the sum of the distances to the nearest center is minimized instead while keeping the order of the center set fixed, the problem is called the minimum k-median problem.

Unfortunately, no single definition of a cluster in graphs is universally accepted, and the variants used in the literature are numerous. In the setting of graphs, each cluster should intuitively be connected: there should be at least one, preferably several paths connecting each pair of vertices within a cluster. If a vertex u cannot be reached from a vertex v, they should not be grouped in the same cluster. Furthermore, the paths should be internal to the cluster: in addition to the vertex set C being connected in G, the subgraph induced by C should be connected in itself, meaning that it is not sufficient for two vertices v and u in C to be connected by a path that passes through vertices in $V\ C$ but they also need to be connected by a path that only visits vertices included in C.

As a consequence, when clustering a disconnected graph with known components, the clustering should usually be conducted on each component separately, unless some global restriction on the resulting clusters is imposed. In some applications one may wish to obtain clusters of similar order and/or density, in which case the clusters computed in one component also influence the clustering of other components. We classify the edges incident on $v \in C$ into two groups: internal edges, which connect v to other vertices also in C, and external edges, which connect to vertices that are not included in the cluster C. We have

$$
\begin{aligned}
\deg_{\text{int}}(v, C) &= |\Gamma(v) \cup C|, \\
\deg_{\text{ext}}(v, C) &= |\Gamma(v) \cup V\ C|, \\
\deg(v) &= \deg_{\text{int}}(v, C) + \deg_{\text{ext}}(v, C).
\end{aligned}
$$

Clearly, $\deg_{\text{ext}}(v) = 0$ implies that C containing v could be a good cluster, as v has no connections outside it. Similarly, if $\deg_{\text{int}}(v) = 0$, v should not be included in C as it is not connected to any of the other vertices included.

It is generally agreed upon that a subset of vertices forms a good cluster if the induced subgraph is dense, but there are relatively few connections from the included vertices to vertices in the rest of the graph.

A measure that helps evaluate the sparsity of connections from the cluster to the rest of the graph is the cut size $c(C_i, V \ C_i)$. The smaller the cut size, the better "isolated" the cluster. Determining when a cluster is dense is naturally done by computing the graph density. We refer to the density of the subgraph induced by the cluster as the internal or intra-cluster density, i.e.,

$$\delta_{int}(C_i) = \frac{|\{\{u, v\}|u, v \in C_i\}|}{|C_i|(|C_i| - 1)}.$$

The intercluster density of a given clustering of a graph G into k clusters $C_1, C_2, \ldots, C_k$ is the average of the intercluster densities of the included clusters, i.e.,

$$\delta_{int} = (G|C_1, \ldots, C_k) = \frac{1}{k} \sum_{i=1}^{k} \delta_{int}(C_i).$$

The external or intercluster density of a clustering is defined as the ratio of intercluster edges to the maximum number of intercluster edges possible, which is effectively the cut sizes of the clusters with edges having weight 1. We have

$$\delta_{ext}(G|C_1, \ldots, C_k) = \frac{|\{\{u, v\}|u \in C_i, v \in C_j, i \neq j\}|}{n(n-1) - \sum_{l=1}^{k}(|C_l|(|C_l| - 1))}.$$

The internal density of a good clustering should be notably higher than the density of the graph $\delta(G)$, and the intercluster density of the clustering should be lower than the graph density. The loosest possible definition of a graph cluster is that of a connected component, and the strictest definition is that each cluster should be a maximal clique.

There are two main approaches for identifying a good cluster: one may either compute some values for the vertices and then classify the vertices into clusters based on the values obtained, or compute a fitness measure over the set of possible clusters and then choose among the set of cluster candidates those that optimize the measure used. The measures described below belong to the second category.

Firstly, we need to define what we mean by *similarity* of data points in this context. Consider a number T of points distributed in the two-dimensional plane. In case the connection costs are assumed to be proportional to the length of the links connecting the terminals with a concentrator, the intuition behind clustering of these terminals is to group terminals that are geometrically close into the same cluster. It is convenient to form a *similarity matrix* for the terminals, whose entries are the reciprocals of the mutual distances between them. Since we do not allow self-loops in an undirected graph, the diagonal elements are set to zero. The choice of similarity is dependent both on application and, to some degree, on the selected method.

6.1 Applications of Clustering

Data analysis often studies information that can be categorized along different dimensions. Such dimensions may include independent variables, such as time, location, price, or various degrees of association. Association is any possible dependence between two data objects, such as delay or time lag, distance, price difference, or other relational variables.

Due to its close relationship to graph theory, clustering is also used in network design, such as C-RAN and network capacity and resilience planning. In the former case, the goal is to maximize the network performance subject to strict delay and resilience constraints. In the latter case, subnetworks with high capacity or resilience properties can be analyzed.

A cluster can be said to have a center of gravity, identified as the point closest to all points based on some distance measure. The clusters' centers of gravity can be used to assign concentrator or switching facilities in a manner that minimizes the cost of facilities and transportation of traffic.

6.2 Complexity

Clustering is $\mathcal{NP}$-hard. The number of possible clusterings of m data points into k clusters is bounded above by k^m, which is exponential in m. For example, if m is the number of links in a network, so that $m = n(n-1)/2$, and the number of clusters is two, representing present links and absent links, the upper limit is $2^{n(n-1)/2}$.

Due to the $\mathcal{NP}$-hardness of the problems involved in network design, decomposition (also known as divide-and-conquer) methods are efficient to reduce problem complexity. The idea behind decomposition is to scale down the problem instances to a level which can be solved with reasonable effort. Since complexity for many graph problems increases exponentially with the order of the graph, decomposition methods yield a large reduction in effort if performed properly. Many of the algorithms presented in the text therefore use decomposition, such as approximations, local search, and randomization algorithms.

6.3 Cluster Properties and Quality Measures

Clustering is the process of grouping data objects together based on some measure of *similarity* between these objects. In network design, this similarity is usually related to cost and can therefore be translated into geographical distance between terminals and concentrators and the amount of traffic they may carry. Roughly speaking, the more traffic a cluster may carry per area unit, the denser it is and the higher the quality it possesses. At the same time, the distance to other clusters should be large, so that the number of concentration devices is minimized. It is therefore desirable to measure the cluster quality in an efficient way.

Even if the quality of clusters can be considered as a rather subjective matter that is very dependent on the application, there are some general measures that may be used for the evaluation of a decomposition. In general terms a cluster should be dense within a cluster and sparse between clusters. It is illuminating to use the analogy with graphs to illustrate cluster quality. The data points are represented by the vertices V in a graph $G = (V, E)$, and the similarity between the data points is represented by the lengths of the edges E connecting the vertices. Note that the edge lengths need not be restricted to a two- or three-dimensional space, and so it may not be possible to visually depict the graph in a low-dimensional space.

We use the following terminology to describe clusters. Let $G = (V, E)$ be a connected, undirected graph with $|V| = n$, $|E| = m$ and let $\mathcal{C} = (C_1, \ldots, C_k)$ be a partition of V. We call $\mathcal{C}$ a clustering of G and C_i clusters; $\mathcal{C}$ is called trivial if either $k = 1$ or all clusters C_i contain only one element. We often identify a cluster C_i with the induced subgraph of G, that is, the graph $G[C_i] = (C_i, E(C_i))$, where $E(C_i) = \{\{v, w\} \in E : v, w \in C_i\}$. Then $E(\mathcal{C}) = \cup_{i=1}^{k} E(C_i)$ is the set of intracluster edges, denoted $m(\mathcal{C})$, and $E \setminus E(\mathcal{C})$ the set of intercluster edges, denoted $\bar{m}(\mathcal{C})$. A clustering $\mathcal{C} = (C, V \setminus C)$ is the cut of G, and $\bar{m}(\mathcal{C})$ is the size of the cut.

The clustering problem can formally be stated as follows. Given an undirected graph $G = (V, E)$, a density measure $\delta(\cdot)$ defined over vertex subsets $S \subseteq V$, a positive integer $k \leq n$, and a rational number $\eta \in [0, 1]$. Is there a subset $S \subseteq V$ such that $|S| = k$ and the density $\delta(S) \geq \eta$?

Note that simple maximization of any density measure without fixing k would result in choosing any clique. This fact shows that computing the density measure is $\mathcal{NP}$-complete as well, since for $\eta = 1$ it coincides with the maximum clique problem.

Vertex Similarity

Central to clustering is the distance (in some sense) between points, or rather its reciprocal – their similarity. A distance measure $\text{dist}(d_i, d_j)$ between two points d_i and d_j is usually required to fulfill the following criteria:

(1) $\text{dist}(d_i, d_i) = 0$,
(2) $\text{dist}(d_i, d_j) = \text{dist}(d_j, d_i)$ (symmetry),
(3) $\text{dist}(d_i, d_j) \leq \text{dist}(d_i, d_k) + \text{dist}(d_k, d_j)$ (triangle inequality).

For points in n-dimensional Euclidean space, possible distance measures between two data points $d_i = (d_{i,1}, d_{i,2}, \ldots, d_{i,n})$ and $d_j = (d_{j,1}, d_{j,2}, \ldots, d_{j,n})$ include the Euclidean distance

(L_2 norm)

$$\text{dist}(d_i, d_j) = \sum_{k=1}^{n} \sqrt{(d_{i,k} - d_{j,k})^2},$$

the Manhattan distance (L_1 norm),

$$\text{dist}(d_i, d_j) = \sum_{k=1}^{n} |d_{i,k} - d_{j,k}|,$$

and the L_∞ norm,

$$\text{dist}(d_i, d_j) = \max_{k \in [1,n]} |d_{i,k} - d_{j,k}|.$$

Possibly the most straightforward manner of determining whether two vertices are similar using only the adjacency information is to study the overlap of their neighborhoods in $G = (V, E)$: a straightforward way is to compute the intersection and the union of the two sets

$$\omega(u, v) = \frac{|\Gamma(u) \cap \Gamma(v)|}{|\Gamma(u) \cup \Gamma(v)|},$$

arriving at the Jaccard similarity. The measure takes values in $[0, 1]$; it is zero when there are no common neighbors and one when the neighbors are identical. Another measure is the Pearson correlation of columns (or rows) in a modified adjacency matrix $C = A_G + I$ (the modification simply forces all reflective edges to be present). The Pearson correlation is defined for two vertices v_i and v_j corresponding to the columns i and j of C as

$$\frac{n \left(\sum_{k=1}^{n} (c_{i,k} c_{j,k}) \right) - \deg(v_i) \deg(v_j)}{\sqrt{\deg(v_i) \deg(v_j)(n - \deg(v_i))(n - \deg(v_j))}}.$$

This value can be used as an edge weight $\omega(v_i, v_j)$ to construct a symmetric similarity matrix.

In a graph, closeness can be seen as the degree of connectivity, that is, the number of edge-disjoint paths that exist between each pair of vertices. With this metric, vertices belong to the same cluster if they are highly connected to each other.

It is, however, not necessary that two vertices u and v belonging to the same cluster are connected by a direct edge if they are connected by a short path. Therefore, a similarity matrix can be based on the distance between each vertex pair, where a short distance implies a high degree of similarity. We can use a threshold k of the path length, so that similar vertices must be at distance at most k from each other. Such a subgraph is called a k-clique.

If we require that the induced subgraph be a k-clique, this implies that the k-shortest paths connecting the cluster members must be restricted to intracluster edges only. The threshold k should be compared with the diameter, the maximum distance between any two nodes in the graph. A threshold close to the diameter may lead to too large clusters, whereas too small values of the threshold k may force splitting of natural clusters.

Expansion

Starting the reasoning from the opposite end, a well-performing clustering algorithm assigns similar points to the same cluster and dissimilar points to different clusters. Expressing the clustering as a graph, points within the same cluster induce low-cost edges, and points that are farther apart induce high-cost edges. We therefore interpret the clustering problem as a node partitioning problem on an edge-weighted complete graph. In the graph, the edge weight a_{uv} then represents the similarity of vertices u and v. Associated with the graph is an $n \times n$ symmetric matrix A with entries a_{uv}. We shall assume that the a_{uv} are nonnegative.

The quality of a clustering can be described by the size (weight) of a cut relative to the sizes of the clusters it creates. The *expansion* measures the relative cut size of a partitioned graph. The expansion of a graph is the minimum ratio of the total weight of edges of a cut to the number of vertices in the smaller part separated by the cut. The expansion of a cut $(S, \bar{S})$ is defined as

$$\varphi(S) = \frac{\sum_{i \in S, j \notin S} a_{ij}}{\min(|S|, |\bar{S}|)}.$$

We say that the minimum expansion of a graph is the minimum expansion over all the cuts of the graph. A measure of the quality of a cluster is the expansion of the subgraph corresponding to this cluster. The expansion of a clustering is the minimum expansion of one of the clusters. Expansion gives equal importance to all vertices of the given graph, which may lead to a rather strong requirement, particularly for outliers.

Coverage

The coverage of a graph clustering $\mathcal{C}$ is defined as

$$\text{coverage}(\mathcal{C}) = \frac{m(\mathcal{C})}{m},$$

where $m(\cdot)$ is the number of intracluster edges.

Intuitively, the larger the value of the coverage, the better the quality of a clustering $\mathcal{C}$. Notice that a minimum cut has maximum coverage. However, in general a minimum cut is not considered to be a good clustering of a graph.

Performance

The *performance* of a clustering $\mathcal{C}$ is based on a count of the number of "correctly assigned pairs of nodes" in a graph. It computes the fraction of intracluster edges together with nonadjacent pairs of nodes in different clusters of the set of all pairs of nodes, i.e.,

$$\text{performance}(\mathcal{C}) = \frac{m(\mathcal{C}) + \sum_{\{v,w\} \notin E, v \in C_i, w \in C_j, i \neq j} 1}{\frac{1}{2}n(n-1)}.$$

Alternatively, the performance can be computed as

$$1 - \text{performance}(\mathcal{C}) = \frac{2m(1 - coverage(\mathcal{C})) + \sum_{i=1}^{k} |C_i|(|C_i| - 1)}{n(n-1)}.$$

Conductance

The *conductance* of a cut compares the size of a cut and the number of edges in either of the two cut-separated subgraphs. The conductance $\phi(G)$ of a graph G is the minimum conductance over all cuts of G. The conductance actually allows defining two measures – the quality of an individual cluster (and therefore of the overall clustering) and the weight of the intercluster edges, providing a measure of the cost of the clustering. The quality of a clustering is given by two parameters: α, the minimum conductance of the clusters, and ε, the ratio of the weight of intercluster edges to the total weight of all edges. The objective is to find an (α, ε) clustering that maximizes α and minimizes ε. The conductance of a cut $(S, \bar{S})$ in G is denoted by

$$\phi(S) = \frac{\sum_{i \in S, j \notin S} a_{ij}}{\min(a(S), a(\bar{S}))},$$

where $a(S) = a(S, V) = \sum_{i \in S} \sum_{j \in V} a_{ij}$. The conductance of a graph is the minimum conductance over all the cuts in the graph, i.e.,

$$\phi(G) = \min_{S \subseteq V} \phi(S).$$

In order to quantify the quality of a clustering we generalize the definition of conductance further. Take a cluster $C \subseteq V$ and a cut $(S, C \backslash S)$ within C, where $S \subseteq C$. Then we say that the conductance of S in C is

$$\phi(S, C) = \frac{\sum_{i \in S, j \in C \backslash S} a_{ij}}{\min(a(S), a(C \backslash S))}.$$

The conductance of a cluster $\phi(C)$ will then be the smallest conductance of a cut within the cluster. The conductance of a clustering is the minimum conductance of its clusters. We then obtain the following optimization problem. Given a graph and an integer k, find a k-clustering with the maximum conductance.

There is still a problem with the above clustering measure. The graph might consist mostly of clusters of high quality and a few points that create clusters of poor quality, and this leads to the notion that any clustering has a poor overall quality. One way to handle this is avoiding to restrict the number of clusters, but this could lead to many singletons or very small clusters. Rather than simply relaxing the number of clusters, we introduce a measure of the clustering quality using two criteria – the minimum quality of the clusters, α, and the fraction of the total weight of edges that are not internal to the clusters, ε.

Definition 6.3.1 ((α, ε)-partition). We call a partition $\{C_1, C_2, \ldots, C_l\}$ of V an (α, ε)-partition if

(1) the conductance of each C_i is at least α;
(2) the total weight of intercluster edges is at most an ε fraction of the total edge weight. •

Associated with this bicriterion is the following optimization problem (relaxing the number of clusters). Given a value of α, find an (α, ε)-partition that minimizes ε. Alternatively, given a value of ε, find an (α, ε)-partition that maximizes α. There is a monotonic function f that represents the optimal (α, ε) pairings. For example, for each α there is a minimum value of ε, equal to $f(\alpha)$ such that an (α, ε)-partition exists.

In addition to direct density measures, conductance also measures connectivity with the rest of the graph to identify high-quality clusters. Measures of the "independence" of a subgraph of the vertices of the graph have been defined based on cut sizes. For any proper nonempty subset $S \subset V$ in a graph $G = (V, E)$, the conductance is defined as

$$\phi(S) = \frac{c(S, V\ C)}{\min\{\deg(S), \deg(V\ S)\}}.$$

The internal and external degrees of a cluster C are defined as

$$\deg_{int}(C) = |\{\{u, v\} \in E | u, v \in C\}|,$$
$$\deg_{ext}(C) = |\{\{u, v\} \in E | u \in C, v \in V\ C\}|.$$

Note that the external degree is in fact the size of the cut $(C, V\ C)$. The relative density is

$$\rho(C) = \frac{\deg_{int}(C)}{\deg_{int}(C) + \deg_{ext}(C)}$$
$$= \frac{\sum_{v \in C} \deg_{int}(v, C)}{\sum_{v \in C} \deg_{int}(v, C) + 2\deg_{ext}(v, C)}.$$

For cluster candidates with only one vertex (and any other candidate that is an independent set), we set $\rho(C) = 0$.

The computational challenge lies in identifying subgraphs within the input graph that reach a certain value of a measure, whether of density or independence, as the number of possible subgraphs is exponential. Consequently, finding the subgraph that optimizes the measure (that is, a subgraph of a given order k that reaches the maximum value of a measure in the graph) is computationally hard. However, as the computation of the measure for a known subgraph is polynomial, we may use these measures to evaluate whether or not a given subgraph is a good cluster.

For a clustering $C = (C_i, \ldots, C_k)$ of a graph G, the intracluster conductance $\alpha(C)$ is the minimum conductance value over all induced subgraphs $G[C_i]$, while the intercluster conductance $\delta(C)$ is the maximum conductance value over all induced cuts $(C_i, V \; C_i)$. For a formal definition of the different notions of conductance, let us first consider a cut $C = (C, V \; C)$ of G and define conductance $\phi(C)$ and $\phi(G)$ as follows:

$$
\phi(C) \;=\; \begin{cases} 1, & C \in \{\emptyset, V\}, \\ 0, & C \notin \{\emptyset, V\} \text{ and } \bar{m}(C) = \emptyset, \\ \dfrac{\bar{m}(C)}{\min(\sum_{v \in C} \deg(v), \sum_{v \in V \; C} \deg(v))}, & \text{otherwise}, \end{cases}
$$

$$
\phi(G) \;=\; \min_{C \subseteq V} \phi(C).
$$

Then a cut has low conductance if its size is small relative to the density of either side of the cut. Such a cut can be considered as a bottleneck. Minimizing the conductance over all cuts of a graph and finding the according cut is $\mathcal{NP}$-hard, but it can be approximated with polylogarithmic approximation guarantee in general, and constant guarantee for special cases.

Based on the notion of conductance, we can now define intracluster conductance $\alpha(C)$ and intercluster conductance $\delta(C)$. We have

$$
\alpha(C) \;=\; \min_{i \in \{1, \ldots, k\}} \phi(G[C_i]),
$$

$$
\delta(C) \;=\; 1 - \max_{i \in \{1, \ldots, k\}} \phi(C_i).
$$

In a clustering with small intracluster conductance there is supposed to be at least one cluster containing a bottleneck, that is, the clustering is possibly too coarse in this case. On the other hand, a clustering with small intercluster conductance is supposed to contain at least one cluster that has relatively strong connections outside, that is, the clustering is possibly too fine. To see that a clustering with maximum intracluster conductance can be found in polynomial time, consider first $m = 0$. Then $\alpha(C) = 0$ for every nontrivial clustering C, since it contains at least one cluster C_j with $\phi(G[C_j]) = 0$. If $m \neq 0$, consider and edge $\{u, v\} \in E$ and the clustering $\mathcal{C}$ with $C_1 = \{u, v\}$ and $|C_i| = 1$ for $i \geq 2$. Then $\alpha(\mathcal{C}) = 1$, which is a maximum.

Intracluster conductance may exhibit some artificial behavior for clusterings with many small clusters. This justifies the restriction to clusterings satisfying certain additional constraints on the size or number of clusters. However, under these constraints, maximizing intracluster conductance becomes an $\mathcal{NP}$-hard problem. Finding a clustering with maximum intercluster conductance is $\mathcal{NP}$-hard as well, because it is at least as hard as finding a cut with minimum conductance. Although finding an exact solution is $\mathcal{NP}$-hard, the algorithm presented in [54] is shown to have simultaneous polylogarithmic approximation guarantees for the two parameters in the bicriterion measure.

6.4 Heuristic Clustering Methods

Heuristic methods do not deliver a result with any guarantee, but they are often built on straightforward principles and therefore easy to modify. We discuss the k-nearest neighbor and the k-means algorithms, which are used not only in clustering, but also as subroutines in many other algorithms.

k-Nearest Neighbor

A conceptually simple and powerful method that can be used as a clustering technique is the k-nearest neighbor. In graph theory and other combinatorial problems, the nearest neighbor is a much used principle in search applications and greedy algorithms.

The k-nearest neighbor simply takes a parameter k, the size of the neighborhood, and a data point p and determines the k data points that in some sense are closest to p. A straightforward way of implementing this is to create a vector $\mathbf{p}$ of points $p_i \neq p$ of size k. We can then successively replace a data point p_i in the vector by p_j whenever the distance $d(p_j, p) < d(p_i, p)$.

k-Means and k-Median

A popular algorithm for clustering data with respect to a distance function is the k-means algorithm. The basic idea is to assign a set of points in some metric space into k clusters by iteration, successively improving the location of the k cluster centers and assigning each point to the cluster having the closest center. The centers are often chosen to minimize the sum-of-squares of the distances within each cluster. This is the metric used in the k-means algorithm. If the median is used instead, we have the k-median algorithm. Collectively, these are known as centroids.

The method starts with $k > 1$ initial cluster centers and then the data points are assigned greedily to the cluster centers. Next, the algorithm switches between recomputing the center positions, that is, the centroids, from the data points, and assigning the data points to the new locations. These steps are repeated until the algorithm converges.

The choice of k may or may not be given by the problem. Usually, we can form some idea of the magnitude of k, for example from expected cluster sizes. Otherwise, we can either perform trial-and-error to find a suitable k, or resort to methods that determine k as well. Such methods are discussed in Chapter 8.

The next step is to estimate initial positions of cluster centers. Again, this can be more or less obvious from the problem itself and the locations can be estimated by inspection. Alternatively, the k center points randomly from the data points. Another heuristic approach is to select the two mutually most distant data points as the two first center points, and subsequent center points as the data points with the largest distance to already chosen center points. The heuristic guarantees a good spread of the center points, but it usually not particularly accurate.

The k-means algorithm uses the Euclidean distance to compute the centroids,

$$m_i = \frac{1}{|C_i|} \sum_{x_j \in C_i} x_j,$$

for $i = 1, \ldots, k$ and clusters C_i.

Algorithm 6.4.1 (Forgy).

Given a data set and k initial centroid estimates $m_1^{(0)}, m_2^{(0)}, \ldots, m_k^{(0)}$. Let t denote the current iteration.

STEP 1:n

> **Assignment:**
> Construct clusters C_i as $C_i^t = \{x_p : ||x_p - m_i^{(t)}||^2 \le ||x_p - m_j^{(t)}||^2$ for all $1 \le j \le k$.
> The data points are assigned to exactly one cluster C_i, and ties are broken arbitrarily.
> **Update:**
> Calculate new centroid locations as
>
> $$m_i^{(t+1)} = \frac{1}{|C_i|} \sum_{x_j \in C_i^{(t)}} x_j.$$

> The algorithm has converged when the assignments no longer change.

Output $C_1, C_2, \ldots, C_k$.

6.5 Spectral Clustering

Spectral clustering is a general technique of partitioning the rows of an $n \times n$-matrix A according to their components in the first k eigenvectors (or more generally, singular vectors) of the matrix. The matrix contains the pairwise similarities of data points or nodes of a graph.

Let the rows of A contain points in a high-dimensional space. In essence, the data may have n dimensions. From linear algebra we know that the subspace defined by the k first eigenvectors of A, the eigenvectors corresponding to the k smallest eigenvalues, defines the subspace of rank k that best approximates A. The spectral algorithm projects all the points onto this subspace. Each eigenvector then defines a cluster. To obtain a clustering each point is projected onto the eigenvector that is closest to it in angle.

In a similarity matrix, diagonal elements are zero since we do not allow self-loops. We then form the *Laplacian L*, a matrix with sum of the row elements (also known as the volume) of A on the diagonals. The eigenvectors of the Laplacian L are then used for clustering. Given a matrix A, the spectral algorithm for clustering the rows of L can be summarized as follows.

Algorithm 6.5.1 (Spectral clustering).

Given an $n \times n$ similarity matrix A and its Laplacian L.

STEP 1:   Find the top k right singular vectors $\mathbf{v}_1, \mathbf{v}_2, \ldots, \mathbf{v}_k$ of L.
STEP 2:   Let C be the matrix whose jth column is given by $A\mathbf{u}_j$.
STEP 3:   Place row i in cluster j if C_{ij} is the largest entry in the ith row of C.

Output clustering $C_1, C_2, \ldots, C_k$.   •

We discuss the steps of the algorithm in some more detail.

Similarity Matrices

There are several ways to construct a similarity matrix for a given set of data points $x_1, \ldots, x_n$ with pairwise distances d_{ij} or similarities $s_{ij} = 1/d_{ij}$ (that is, a short distance means strong similarity). This matrix serves as a model of the local neighborhood relationships between the data points, and this neighborhood can be defined using different principles, which may be suggested by the problem at hand. When the relationship between nodes cannot easily be expressed by a single distance measure, the similarity matrix can be defined in alternative ways.

The ϵ-neighborhood

Let points belong to the same neighborhood whose pairwise distances are smaller than a threshold ϵ. Since the distances between all points in the neighborhood are roughly ϵ, the entries a_{ij} in a neighborhood are often simply set to the same value, for example $s_{ij} = 1$, and entries representing nodes not in the same neighborhood by zero.

k-Nearest neighbors

We let node v_i be in the same neighborhood as v_j if v_j is among the k-nearest neighbors of v_i. This relation, however, is not symmetric, and we therefore need to handle cases where v_j is in the neighborhood of v_i, but v_i is not in the neighborhood of v_j. The first way to do this is to simply ignore this asymmetry, so that v_i and v_j are in the same neighborhood whenever v_i is a k-nearest neighbor of v_j or v_j is a k-nearest neighbor of v_i. Alternatively, we may choose to let v_i and v_j be in the same neighborhood whenever both v_i is a k-nearest neighbor of v_j and v_j is a k-nearest neighbor of v_i. The entries a_{ij} in the similarity matrix between nodes belonging to the same neighborhood can then be set to their pairwise similarity s_{ij} and to zero otherwise.

Laplacians

Spectral clustering is based on Laplacian matrices, which know different variants. Although we discuss a particular form of matrices, it is beneficial to express the problem in terms of a graph.

Let therefore G be an undirected weighted graph, with weight matrix W with entries $w_{ij} = w_{ji} \geq 0$. Also, let D be the diagonal matrix containing the weighted degrees of the nodes (that is, the sum $d_i = \sum_{j \neq i} w_{ij}$). The (unnormalized) Laplacian is then defined as

$$L = D - W,$$

and has the following important properties.

Proposition 6.5.2. *For the matrix L the following is true:*

(1) *For every vector $f \in \mathbb{R}^n$,*

$$f^{\mathsf{T}} L f = \sum_{i,j=1}^{n} w_{ij}(f_i - f_j)^2.$$

(2) *L is symmetric and positive definite.*
(3) *The smallest eigenvalue of L is 0, corresponding to the eigenvector $\mathbf{1}$.*
(4) *L has n nonnegative, real eigenvalues $0 = \lambda_1 \leq \lambda_2 \leq \ldots \leq \lambda_n$.*

Proof. (1) From the definition of d_i, we have

$$f^{\mathsf{T}}Lf = f^{\mathsf{T}}Df - f^{\mathsf{T}}Wf = \sum_{i=1}^{n} d_i f_i^2 - \sum_{i,j=1}^{n} f_i f_j w_{ij}$$

$$= \frac{1}{2}\left(\sum_{i=1}^{n} d_i f_i^2 - 2\sum_{i,j=1}^{n} f_i f_j w_{ij} + \sum_{j=1}^{n} d_j f_j^2\right) = \frac{1}{2}\sum_{i,j=1}^{n} w_{ij}(f_i - f_j)^2.$$

(2) The symmetry of L follows from the symmetry of W and D. The positive semidefiniteness is a direct consequence of (1), which shows that $f^{\mathsf{T}}Lf \geq 0$ for all $f \in \mathbb{R}^n$.

(3) This is obvious.

(4) Follows directly from (1)–(3). $\qquad\square$

We have the following result, tying together the connectivity of a graph and the spectrum of the associated Laplacian. For a proof see, for example, [55].

Proposition 6.5.3. *Let G be an undirected graph with nonnegative weights and L its (unnormalized) Laplacian. Then the multiplicity k of the eigenvalue 0 of L equals the number of connected components $G_1, \ldots, G_k$ in the graph. The eigenspace of eigenvalue 0 is spanned by the indicator vectors $\mathbf{1}_{A_1}, \ldots, \mathbf{1}_{A_k}$ of those components.*

As the use of the qualifier "unnormalized" suggests, the Laplacian can also be normalized. The symmetric normalized and the random walk Laplacians are as follows:

$$L_{\text{sym}} = D^{-1/2}LD^{-1/2} = I - D^{-1/2}WD^{-1/2},$$
$$L_{\text{rw}} = D^{-1}L = I - D^{-1}W.$$

Similar properties are valid for these Laplacians as for the unnormalized version, but the details are omitted here. The eigenvalues λ_i and eigenvectors $\mathbf{v}_i$ of an $n \times n$-matrix satisfy the equation

$$(A - \lambda_i I)\mathbf{v}_i = 0, \quad \mathbf{v}_i \neq \mathbf{0}.$$

Eigenvectors

Most computational packages include routines for eigenvalues and eigenvectors. Alternatively, an easy to use method is the power method. Let $\mathbf{z}_0$ be an arbitrary initial vector (possibly random) and calculate

$$\mathbf{z}_{s+1} = A\mathbf{z}_s/\|A\mathbf{z}_s\|, \quad s = 0, 1, 2, \ldots.$$

Assuming that A has n linearly independent vectors and a unique (dominant) eigenvalue of maximum magnitude and z_0 has a nonzero component in the direction of an eigenvector of the dominant eigenvalue, z_s converges to an eigenvector corresponding to this eigenvalue. The dominant eigenvalue is the limiting value of the sequence

$$\mu_s = \frac{z_k^T A z_k}{z_k^T z_k}.$$

When the first eigenpair has been found, the eigenpair corresponding to the second-most dominant eigenvector can be found by modifying $A_0 \triangleq A$, i.e.,

$$A_{i+1} = A_i - \lambda_i v_i v_i^T.$$

We here use the convention of ordering the eigenvalues in increasing order. The first k eigenvectors are therefore the eigenvectors corresponding to the k smallest eigenvalues.

Projection

For mapping of data points we can use the k-means algorithm. Selecting the eigenvectors corresponding to the k smallest nonzero eigenvalues, the initial centroid can be taken as the first k elements of the k eigenvectors. The eigenvectors contain the approximate projection (restricted to k dimensions) of the data onto the k centroids. We compute the distance to the centroid for all rows in the eigenvectors and update the centroid coordinates. Iterating until the assignment does not change, we obtain the clustering.

Example 6.5.1. *Spectral clustering is applied to a data set consisting of 400 geographical objects. The eigenvalues of the Laplacian are computed with the QR-method and the corresponding eigenvectors using the power method with deflation (see [59]), and clusters are formed using Forgy's k-means algorithm. The clustering in Fig. 6.1 shows well-formed clusters. The colors are reused for different clusters.*

6.6 Iterative Improvement

Once an approximate minimum cut, generating a bisection, that is, two clusters of approximately equal size, has been found, this approximation can be improved by an iterative improvement algorithm.

Let $G = (V, E)$ be a graph. The algorithm attempts to find a partition of V into two disjoint subsets A and B of equal size, such that the sum T of the weights of the edges between nodes in A and B is minimized. Let I_a be the internal cost of a, that is, the sum of the costs of edges

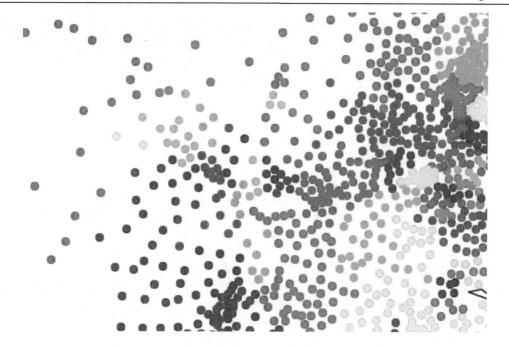

Figure 6.1: Spectral clustering of geographical data.

between a and other nodes in A, and let E_a be the external cost of a, that is, the sum of the costs of edges between a and nodes in B. Furthermore, let

$$D_a = E_a - I_a$$

be the difference between the external and internal costs of a. If a and b are interchanged, then the reduction in cost is

$$T_{\text{old}} - T_{\text{new}} = D_a + D_b - 2c_{ab},$$

where c_{ab} is the cost of the possible edge between a and b. The algorithm attempts to find an optimal series of interchange operations between elements of A and B which maximizes $T_{\text{old}} - T_{\text{new}}$ and then executes the operations, producing a partition of the graph to A and B.

Simulated annealing and genetic algorithms have also been used to partition graphs, but some results show that they provide clusters of inferior quality and require much greater computational resources than when using the spectral partitioning algorithm [56].

Uniform Graph Partitioning

Iterative improvement is particularly efficient when data consist of two clusters of nearly equal size.

Definition 6.6.1. *Given a symmetric cost matrix c_{ij} defined on the edges of a complete undirected graph $G = (V, E)$ with $|V| = 2n$ vertices, a partition $V = A \cup B$ such that $|A| = |B|$ is called a* uniform *partition. The* uniform graph partitioning *problem is that of finding a uniform partition $V = A \cup B$ such that the cost*

$$C(A, B) = \sum_{i \in A, j \in B} c_{ij}$$

is minimal over all uniform partitions.

The problem can be thought of as dividing a load into two pieces of equal size so that the weight of the connection between the two pieces is as small as possible. This represents common situations in engineering, such as VLSI design, parallel computing, or load balancing.

Suppose (A^*, B^*) is an optimal uniform partition and we are considering some partition (A, B). Let X be those elements of A that are not in A^* – the "misplaced" elements – and let Y be similarly defined for B. Then $|X| = |Y|$ and

$$\begin{aligned} A^* &= (A - X) \cup Y, \\ B^* &= (B - Y) \cup X. \end{aligned}$$

That is, we can obtain the optimal uniform partition by interchanging the elements in set X with those in Y.

Definition 6.6.2. *Given a uniform partition A, B and elements $a \in A$ and $b \in B$, the operation of forming*

$$\begin{aligned} A' &= (A - \{a\}) \cup \{b\}, \\ B' &= (B - \{b\}) \cup \{a\} \end{aligned}$$

is called a swap.

We next consider how to determine the effect that a swap has on the cost of a partition (A, B). We define the *external cost* $E(a)$ associated with an element $a \in A$ by

$$E(a) = \sum_{i \in B} d_{ai}$$

and the *internal cost* $I(a)$ by

$$I(a) = \sum_{j \in A} d_{aj}$$

(and similarly for elements of B). Let

$$D(v) = E(v) - I(v)$$

be the difference between external and internal cost for all $v \in V$.

Lemma 6.6.1. *The swap of a and b results in a reduction of cost (gain) of*

$$g(a, b) = D(a) + D(b) - 2d_{ab}.$$

The swap neighborhood N_s for the uniform graph partitioning problem is

$$N_s(A, B),$$

that is, all uniform partitions A', B' that can be obtained from the uniform partition A, B by a single swap [46].

CHAPTER 7

Bayesian Analysis

Bayesian analysis is an important technique in machine learning. Being a large and diverse field, in this chapter we are merely scratching the surface. Some techniques discussed in other chapters – such as clustering or metaheuristic optimization – are also often used in machine learning. Here, we will turn the matter around somewhat and formulate some methods based on Bayesian statistics which in turn can be used for clustering.

Bayesian methods are based on the idea that any statistical inference is based not only on our observed data, but also on a certain prior belief of the nature of these data. Bayes' famous theorem can be written

$$\mathbf{P}(A|B) = \frac{\mathbf{P}(B|A)\,\mathbf{P}(A)}{\mathbf{P}(B)}. \tag{7.1}$$

In Bayesian methods, this equation is used to formulate relationships between conditional and total probabilities, often leading to iterative procedures. Letting A be a model and B the observed data in Eq. (7.1), $\mathbf{P}(A|B)$ is the posterior probability, $\mathbf{P}(A)$ the prior probability, the probability of A before any data are observed, and $\mathbf{P}(B|A)$ and $\mathbf{P}(B)$ the likelihood and marginal likelihood, respectively. The main principle is therefore to incorporate a presumed distribution – the *prior* – when determining the posterior.

7.1 Bayesian Average

The Bayesian average uses weight given by the properties of the sample. Calculating the Bayesian average uses the prior mean m and a weight C assigned a value proportional to the size of the observed data set. We have

$$\bar{x} = \frac{Cm + \sum_{i=1}^{n} x_i}{C + n}. \tag{7.2}$$

We can see this as using the experiment twice, where the second round is an exact copy of the first. The first time, we determine the average number of voters C, the average number m of data points per voter, and the average number of points Cm per category. In the second, we use the partial sum $\sum_i x_i$. The overall mean $\bar{x}$ is then given by the scaled mean (7.2).

5G Networks. https://doi.org/10.1016/B978-0-12-812707-0.00012-7

Table 7.1: Bayesian average of rating votes.

Category	Points
A	10
B	6
B	5
B	4
C	3
C	10

Example 7.1.1. *Suppose we are given ratings of three categories A, B, and C (which could represent movies, for example), given in Table 7.1.*

Using Eq. (7.2) with $C = (1 + 3 + 2)/3 = 2$ and $m = (10 + 6 + 5 + 4 + 3 + 10)/6 = 6.33$, we obtain

$$m_A = \frac{Cm + 10/1}{C + 1} = 7.56,$$

$$m_B = \frac{Cm + (6 + 5 + 4)/3}{C + 3} = 5.53,$$

$$m_C = \frac{Cm + (3 + 10)/2}{C + 2} = 6.42.$$

7.2 The Gibbs Sampler

Markov chain Monte Carlo simulation methods, such as the Gibbs sampler, have proven very useful for studying complex processes. These methods are discussed in, for example, [60]. The idea is that a random variable for which the probability distribution cannot be explicitly formulated can instead be simulated by specifying a Markov chain with conditional probabilities. The conditions under which the so-obtained solution exists and is unique is discussed below, where also some fundamental definitions are stated.

Markov Chains

Definition 7.2.1 (Discrete-time Markov chain). *Let P be a $(k \times k)$-matrix with elements $\{P_{i,j} : i, j = 1, \ldots, k\}$. A random process $(X_0, X_1, \ldots)$ with finite state space $S = \{s_1, \ldots, s_k\}$ is said to be a* (homogeneous) Markov chain *with transition matrix P, if, for all n, all $i, j \in \{1, \ldots, k\}$, and all $i_0, \ldots, i_{n-1} \in \{1, \ldots, k\}$, we have*

$$\mathbf{P}(X_{n+1} = s_j | X_0 = s_{i_0}, X_1 = s_{i_1}, \ldots, X_{n-1} = s_{i_{n-1}}, X_n = s_i) = \mathbf{P}(X_{n+1} = s_j | X_n = s_i) = P_{i,j}.$$

Every transition matrix satisfies

$$P_{i,j} \;\geq\; 0, \text{ for all } i, j \in \{1, \ldots, k\},$$

$$\sum_{j=1}^{k} P_{i,j} \;=\; 1, \text{ for all } i \in \{1, \ldots, k\}.$$

The *initial distribution* is represented as a row vector $\mu^{(0)}$ given by

$$\mu^{(0)} = (\mu_1^{(0)}, \mu_2^{(0)}, \ldots, \mu_k^{(0)}) = (\mathbf{P}(X_0 = s_1), \mathbf{P}(X_0 = s_2), \ldots, \mathbf{P}(X_0 = s_k))$$

and

$$\sum_{i=1}^{k} \mu_i^{(0)} = 1.$$

Theorem 7.2.1. *For a Markov chain* $(X_0, X_1, \ldots)$ *with state space* $\{s_1, \ldots, s_k\}$, *initial distribution* $\mu^{(0)}$, *and transition matrix P, the distribution* $\mu^{(n)}$ *at any time n satisfies*

$$\mu^{(n)} = \mu^{(0)} P^n.$$

The *initiation function* $\psi : [0, 1] \to S$ is a function from the unit interval to the state space S used to generate the starting value X_0, such that

(i) ψ is piecewise constant and
(ii) for each $s \in S$, the total length of the intervals on which $\psi(x) = s$ equals $\mu^{(0)}(s)$, that is,

$$\int_0^1 \mathbf{I}_{\{\psi(x)=s\}}(x)\mathrm{d}x = \mu^{(0)}(s)$$

for each $s \in S$. The function $\mathbf{I}_{\{\psi(x)=s\}}$ is the *indicator function* of $\{\psi(x) = s\}$,

$$\mathbf{I}_{\{\psi(x)=s\}}(x) = \begin{cases} 1, & \text{if } \psi(x) = s, \\ 0, & \text{otherwise.} \end{cases}$$

To compute from X_n to X_{n+1}, an *update function* $\phi : S \times [0, 1] \to S$ is used, which takes as input a state $s \in S$ and a random number U between 0 and 1 and produces another state $s' \in S$ as output. It is necessary that ϕ obeys the following properties:

(1) For fixed s_i, the function $\phi(s_i, x)$ is piecewise constant (when viewed as a function of x).
(2) For each fixed $s_i, s_j \in S$, the total length of the interval on which $\phi(s_i, x) = s_j$ equals
     $P_{i,j}$. This can be written as

$$\int_0^1 \mathbf{I}_{\{\phi(s_i,x)=s_j\}}(x)\mathrm{d}x = P_{i,j},$$

for all $s_i, s_j \in S$.

The simulation of the Markov chain is accomplished by letting

$$
\begin{aligned}
X_0 &= \psi(U_0), \\
X_1 &= \phi(X_0, U_1), \\
X_2 &= \phi(X_1, U_2), \\
&\vdots
\end{aligned}
$$

A state s_i is said to communicate with another state s_j, written $s_i \rightarrow s_j$, if the chain has a (strictly) positive probability of ever reaching s_j when starting from s_i, in other words, if there exists an n such that

$$\mathbf{P}(X_{m+n} = s_j | X_m = s_i) > 0.$$

If $s_i \rightarrow s_j$ and $s_j \rightarrow s_i$, then the states s_i and s_j are said to intercommunicate, which is written $s_i \leftrightarrow s_j$.

Definition 7.2.2 (Irreducible Markov chain). *A Markov chain* $(X_0, X_1, \ldots)$ *with state space* $S = \{s_1, \ldots, s_k\}$ *and transition matrix P is said to be* irreducible *if* $s_i \leftrightarrow s_j$ *for all* $s_i, s_j \in S$. *Otherwise the chain is said to be* reducible.

The *period* $\mathrm{d}(s_i)$ of a state $s_i \in S$ is defined as

$$\mathrm{d}(s_i) = \gcd\{n \geq 1 : (P^n)_{i,i} > 0\},$$

where "gcd" is the greatest common divisor. If $\mathrm{d}(s_i) = 1$, then the state s_i is called aperiodic.

Definition 7.2.3 (Aperiodic Markov chain). *A Markov chain is said to be* aperiodic *if all its states are aperiodic. Otherwise the chain is said to be* periodic.

For any Markov chain $(X_0, X_1, \ldots)$ it can be shown that the *distribution* of X_n settles down to a limit if the Markov chain is irreducible and aperiodic.

Definition 7.2.4 (Stationary distribution). *Let $(X_0, X_1, \ldots)$ be a Markov chain with state space $S = \{s_1, \ldots, s_k\}$ and transition matrix P. A row vector $\pi = (\pi_1, \ldots, \pi_k)$ is said to be a stationary distribution for the Markov chain, if it satisfies*

(i) $\pi_i \geq 0$ for $i = 1, \ldots, k$ and $\sum_{i=1}^{k} \pi_i = 1$ and
(ii) $\pi P = \pi$, meaning that $\sum_{i=1}^{k} \pi_i P_{i,j} = \pi_j$ for $j = 1, \ldots, k$.

Theorem 7.2.2 (Uniqueness of stationary distribution). *Any irreducible and aperiodic Markov chain has exactly one stationary distribution.*

Definition 7.2.5 (Total variation distance). *If $v^{(1)} = (v_1^{(1)}, \ldots, v_k^{(1)})$ and $v^{(2)} = (v_1^{(2)}, \ldots, v_k^{(2)})$ are probability distributions on $S = \{s_1, \ldots, s_k\}$, then the* total variation distance *between $v^{(1)}$ and $v^{(2)}$ is defined as*

$$d_{TV}(v^{(1)}, v^{(2)}) = \frac{1}{2} \sum_{i=1}^{k} |v_i^{(1)} - v_i^{(2)}|.$$

If $v^{(1)}, v^{(2)}, \ldots$ and v are probability distributions on S, $v^{(n)}$ is said to converge *to v in total variation as $n \to \infty$, written $v^{(n)} \overset{TV}{\to} v$, if*

$$\lim_{n \to \infty} d_{TV}(v^{(n)}, v) = 0.$$

Theorem 7.2.3 (Markov chain convergence). *Let $(X_0, X_1, \ldots)$ be an irreducible aperiodic Markov chain with state space $S = \{s_1, \ldots, s_k\}$, transition matrix P, and arbitrary initial distribution $\mu^{(0)}$. Then, for any distribution π which is stationary for the chain,*

$$\mu^{(n)} \overset{TV}{\to} \pi.$$

Theorems 7.2.2 and 7.2.3 state that if it can be shown that the simulated Markov chain is irreducible and aperiodic, then a stationary solution will be reached (after a sufficient number of simulation steps) and it is unique. A proof of the theorems can be found in, for example, [61].

Definition 7.2.6 (Reversible distribution). *Let $(X_0, X_1, \ldots)$ be a Markov chain with state space $S = \{s_1, \ldots, s_k\}$ and transition matrix P. A probability distribution π on S is said to be* reversible *for the chain if, for all $i, j \in \{1, \ldots, k\}$,*

$$\pi_i P_{i,j} = \pi_j P_{j,i}.$$

The Markov chain is called reversible *if there exists a reversible distribution for it.*

Theorem 7.2.4. *Let $(X_0, X_1, \ldots)$ be a Markov chain with state space $S = \{s_1, \ldots, s_k\}$ and transition matrix P. If π is a reversible distribution for the chain, then it is also a stationary distribution for the chain.*

If a random variable Z has a small state space S and a probability distribution π on S, it can be simulated by enumerating the states $s_1, \ldots, s_k$ and letting $Z = \psi(U)$, where U is a uniform $[0, 1]$ random variable and $\psi : [0, 1] \to S$ is given by

$$
\psi(z) = \begin{cases}
s_1 & \text{for } z \in [0, \pi(s_1)), \\
s_2 & \text{for } z \in [0, \pi(s_1), \pi(s_1) + \pi(s_2)), \\
\vdots & \vdots \\
s_i & \text{for } z \in [\sum_{j=1}^{i-1} \pi(s_j), \sum_{j=1}^{i} \pi(s_j)), \\
\vdots & \vdots \\
s_k & \text{for } z \in [\sum_{j=1}^{k-1} \pi(s_j), 1].
\end{cases}
$$

However, for large S, this method is infeasible. The idea with *Markov chain Monte Carlo methods* is to utilize the fact that an irreducible and aperiodic Markov chain has a unique stationary distribution π and that it converges to π for an arbitrary initial distribution as the simulation time $n \to \infty$. The transition probabilities of the Markov chain are given by the conditional probability distribution, given the state of the chain. For state spaces of the form S^V where S is the (finite) number of possible states of a node and V is the (finite) number of nodes, the Gibbs sampler is particularly useful. In each simulation step, the Gibbs sampler goes through the following cycle:

(1) Choose randomly a node $v \in V$ according to a uniform distribution.
(2) Determine $X_{n+1}(v)$ according to the conditional π-distribution of the value at v given that all other nodes take the values given by X_n.
(3) Let $X_{n+1}(w) = X_n(w)$ for all nodes $w \in V$ except v.

This Markov chain is aperiodic and has π as a reversible distribution. If the chain is irreducible as well (which depends on which elements have nonzero probabilities), then the Markov chain can be used to simulate the random variable Z to obtain an approximation of π.

Example 7.2.1. *Suppose that we have ratings for a product from star ratings by on-line users. Let the star ratings be based on a Likert scale of order 5, so the answers can take on values 1 to 5, and there are $N_q = 6$ questions. We want to find distribution parameter values for the overall rating of the product based on these sources using the Gibbs sampler.*

We assume that the rating is normally distributed with mean μ and variance σ^2. Let n be the number of user votes and denote by $X = (X_1, \ldots, X_q)$ the samples, where

$$
\hat{\mu} = \frac{1}{N_q} \sum_{i=1}^{N_q} X_i.
$$

That is, each X_i is the mean value of n ratings of question i.

The prior distribution is

$$\pi(\mu, \sigma^2) \sim \frac{1}{\sigma^2},$$

that is, it is uniformly distributed (noninformative). The likelihood function is

$$f(X|\mu, \sigma^2) \sim \left(\frac{1}{\sigma^2}\right)^{n/2} \exp\left(-\frac{1}{2\sigma^2} \sum_{i=1}^{N_q} (X_i - \mu)^2\right).$$

We define $\tau = 1/\sigma^2$, and so

$$\pi(\mu|\sigma^2, X) = \mathcal{N}(\bar{X}, \sigma^2/n),$$

$$\pi(\tau|\mu, X) = \Gamma\left(\frac{n}{2}, \frac{1}{2} \sum_{i=1}^{n} (X_i - \mu)^2\right).$$

The Gibbs sampler can now be written as

$$\mu_{t+1} \sim \mathcal{N}\left(\bar{X}, (n \cdot \tau_t)^{-1}\right),$$

$$\tau_{t+1} =\sim \Gamma\left(\frac{n}{2}, \frac{1}{2} \sum_{i=1}^{n} (X_i - \mu_{t+1})^2\right),$$

where $\sigma_{t+1}^2 = 1/\tau_{t+1}$ and t indexes the iterations.

7.3 The Expectation-Maximization Algorithm

The expectation-maximization (EM) method is an iterative technique to determine maximum likelihood parameter estimates of models when there are missing data formalized by Dempster et al. [62]. A thorough discussion can be found in, for example, [63]. It is often used in machine learning, e.g., data clustering and pattern recognition.

It consists of two steps: expectation, where the data are estimated given an assumed model, and maximization, aiming at maximizing the model probabilities using the likelihood function.

We find the idea of iterating between two computationally separate steps in many algorithms. A somewhat related algorithm is the k-means clustering algorithm. It consists of an assignment step and an update step. Initially, we choose k center points, preferably well separated from each other. In the assignment step, each element is connected to its nearest center, forming k clusters. In the update step, the k centers are recomputed to minimize the weighted distance within each cluster. Next, elements are reassigned to the updated center point, etc.

Maximum likelihood estimation (MLE) is a widely used method for estimating the parameters in a probabilistic model taking a parameter θ. For many analytic models, we can compute the parameter θ^{MLE} using the expression

$$\theta^{\text{MLE}} \arg\max_{\theta \in \Theta} \mathcal{L}(\mathbf{X}|\theta),$$

where $\mathcal{L}(\mathbf{X}|\theta)$ is the empirical likelihood, that is, observable data. MLE is normally performed by taking the derivative of the data likelihood $\mathcal{L}(\mathbf{X})$ with respect to the model parameter θ, setting the expression equal to zero and solving the equation for θ^{MLE}. However, there are cases where there are hidden (unobservable) variables in the model, so that the derivative cannot be written in closed form.

The EM algorithm generalizes the MLE to cases with incomplete data. It is an estimation technique to find the model parameters $\hat{\theta}$ that maximize the logarithmic probability $\log P(\mathbf{X}; \theta)$ of obtaining the actually observed data. In general, the maximization step in the EM-algorithm is more difficult than the MLE based on complete data. In the latter case, the objective function $\log P(\mathbf{X}, \mathbf{z}; \theta)$ has a single global optimum, which often can be written in closed form. However, when data are incomplete, we can have multiple local maxima in the expression $\log P(\mathbf{X}; \theta)$.

The algorithm divides the task of optimizing $\log P(\mathbf{X}; \theta)$ into an iterative sequence of simpler optimization problems, whose objective functions have unique global maxima that can often be written in closed form. These subproblems are chosen so that corresponding solutions $\hat{\theta}^{(1)}, \hat{\theta}^{(2)}, \ldots$ converge to a local optimum of $\log P(\mathbf{X}; \theta)$.

More specifically, the EM algorithm alternates between the steps expectation (the E-step) and maximization (the M-step). In the E-step, we choose a function g_t that is a lower bound to $\log P(\mathbf{x}; \theta)$ everywhere and for which $g_t(\hat{\theta}^{(t)}) = \log P(\mathbf{x}; \theta)$. In the M-step, the algorithm uses a new parameter set $\hat{\theta}^{(t+1)}$ that maximizes g_t. Since the value of the lower bound g_t matches the objective function at $\hat{\theta}^{(t)}$, it follows that $\log P(\mathbf{x}; \hat{\theta}^{(t)}) = g_t(\hat{\theta}^{(t)}) \le g_t(\hat{\theta}^{(t+1)}) = \log P(\mathbf{x}; \hat{\theta}^{(t+1)})$. The objective function therefore increases monotonically with each iteration of the algorithm. It is only guaranteed to converge to a local maximum. It is advisable to try running it with different start values, and these should be chosen to avoid symmetric and degenerate cases. We illustrate the EM-algorithm with an example, borrowed from [64].

Example 7.3.1. *Assume that we are given two coins A and B and we perform the experiment of flipping either coin in 5 sets of 10 flips. We wish to find the probability of the event of heads*

Table 7.2: Observed events in the coin example.

Events	Coin A	Coin B
HTTTHHTHTH		5H;5T
HHHHTHHHHH	9H;1T	
HTHHHHHTHH	8H;2T	
HTHTTTHHTT		4H;6T
THHHTHHHTH	7H;3T	
Sum:	24H;6T	9H;11T

showing for the two coins. Suppose the outcomes are as shown in Table 7.2, where "H" denotes heads and "T" tails.

We let the probability of coin A showing heads be θ_A (with $1 - \theta_A$ for tails), and we let θ_B be the probability of heads for coin B. The observations are recorded in a vector $\mathbf{x} = (x_1, \ldots, x_5)$, not knowing which coin, A or B, generated the sequence. The number of times x_i that heads show in ten flips is distributed as a binomial variable. The likelihood function for the binomial distribution is

$$\mathcal{L}(p|x, n) = \binom{n}{x} p^x (1 - p)^{n-x}$$

and the log-likelihood function is

$$l(p|x, n) = \log \binom{n}{x} + x \log p + (n - x) \log(1 - p). \tag{7.3}$$

In ordinary maximum likelihood, assuming that we know which of the coins A and B is flipped in each experiment, as shown in Table 7.2, we have the maximum likelihood estimators

$$\hat{\theta}_A = \frac{\text{number of heads using coin } A}{\text{total flips of coin } A},$$
$$\hat{\theta}_B = \frac{\text{number of heads using coin } B}{\text{total flips of coin } B},$$

or using the numerical values

$$\hat{\theta}_A = \frac{24}{24 + 6} = 0.80,$$
$$\hat{\theta}_B = \frac{9}{9 + 11} = 0.45.$$

Table 7.3: Outcomes and expected events for the coin example.

q_i	$1 - q_i$	x_i^H	x_i^T	y_A^H	y_A^T	y_B^H	y_B^T
0.45	0.55	5	5	2.25	2.25	2.75	2.75
0.80	0.20	9	1	7.20	0.80	1.80	0.20
0.73	0.27	8	2	5.84	1.46	2.16	0.54
0.35	0.65	4	6	1.40	2.10	2.60	3.90
0.65	0.35	7	3	4.55	1.95	2.45	1.05
				21.24	8.56	11.76	8.44

Now, however, suppose that we do not know which coin is flipped in each trial, but we suspect that one of the coins is biased and shows heads with $p_1 = 0.60$, whereas the other is most likely unbiased, with $p_2 = 0.50$. We use this as our initial values. In this case, we say that the data are incomplete when some data are missing. The variable containing the missing information of which coin is used is known as a hidden (or latent) variable. Consequently, we cannot use the MLE on the data directly, since that would give an average (implicitly assuming that the two coins have equal bias).

Next, we calculate the log-likelihood (7.3), knowing the events x_i, $i = 1, \ldots, 5$, $n = 10$, and using p_1 and p_2, respectively. We have (omitting the constant, since we only need the proportions)

$$5 \cdot \ln(0.6) + 5 \cdot \ln(1 - 0.6) = -7.14,$$
$$5 \cdot \ln(0.5) + 5 \cdot \ln(1 - 0.5) = -6.93.$$

Since these are the log likelihoods, we take the exponentials to get the probability weight

$$q_1 = \frac{\exp(-7.14)}{\exp(-7.14) + \exp(-6.93)} = 0.45 \tag{7.4}$$

and similarly

$$q_2 = \frac{\exp(-5.51)}{\exp(-5.51) + \exp(-6.93)} = 0.80,$$

$$q_3 = \frac{\exp(-5.92)}{\exp(-5.91) + \exp(-6.93)} = 0.73,$$

$$q_4 = \frac{\exp(-7.54)}{\exp(-7.54) + \exp(-6.93)} = 0.35,$$

$$q_5 = \frac{\exp(-6.32)}{\exp(-6.32) + \exp(-6.93)} = 0.65.$$

Next, we scale the recorded outcomes by $q_1, \ldots, q_5$ to obtain the expected events y_A^H, y_A^T, y_B^H, and y_B^T, shown in Table 7.3, which gives the new parameter estimates $p_1 = 0.71$ and

$p_2 = 0.58$. *Continuing the iteration, the probabilities settle around* $p_1 = p_A = 0.80$ *and* $p_2 = p_B = 0.52$.

Mixtures of Bernoulli Distributions

A mixture model is a probabilistic model for identification of subsets in a general population. A Bernoulli distribution is a model where a given sample has a probability q of attaining 1 (representing "success") and probability $1 - q$ of 0 (representing "failure"). This is a particularly suitable model in image recognition, where a pixel only can take a value of $\{0, 1\}$ [65].

In a mixture model, we assume that a sample may come from different subsets, where each subset k has a probability distribution $p(k)$. In other words, there is a certain probability of a sample **x** belonging to a subset k, and within the subset it has probability $p(k)$ to occur. We can write the mixture model as

$$p(\mathbf{x}) = \sum_{k=1}^{K} p(k) p(\mathbf{x}|k),$$

where K is the number of mixture components, and for each component k, $p(k)$ is its prior which can be interpreted as the relative frequency of subset k and $p(\mathbf{x}|k)$ is its conditional probability density function. First, we select subset k with probability $p(k)$, and within k the sample is generated with probability $p(\mathbf{x}|k)$. For a Bernoulli mixture model each subset k has a D-dimensional Bernoulli probability function determined by $\mathbf{p}_k(p_{k1}, \ldots, p_{kD}) \in [0, 1]^D$, so that

$$p(\mathbf{X}|k) = \prod_{d=1}^{D} p_{kd}^{x_d}(1 - p_{kd})^{1-x_d},$$

that is, a product of independent single Bernoulli probability functions. For any fixed k, the pixels are modeled as independent variables.

The probability of selecting the kth Bernoulli component out of K subsets is $p(k)$, which is also called the mixing proportion. Since we do not know which out of the K subsets each data point is drawn from, this variable is hidden in this mixture model. The log-likelihood function is

$$l(\theta|\mathbf{x}) = \sum_{n=1}^{N} \log \left(\sum_{k=1}^{K} p(k) p(x_n|k) \right).$$

We introduce a vector $\mathbf{Z}_n = (Z_{n1}, \ldots, Z_{nK})$ with 1 in the position of the subset generating $\mathbf{x}_n$ representing the missing data. Then the log-likelihood function can be written

$$\mathcal{L}(\theta|\mathbf{x}, \mathbf{x}) = \sum_{n=1}^{N} \sum_{k=1}^{K} z_{nk} (\log p(k) + \log p(\mathbf{x}_n|k))$$

$$= \sum_{n=1}^{N} \sum_{k=1}^{K} z_{nk} \left(\log p(k) + \sum_{d=1}^{D} x_{nd} \log p_{kd} + (1 - x_{nd}) \log(1 - p_{kd}) \right).$$

Having the log-likelihood function, we can formulate the EM algorithm for the mixture of Bernoulli distributions.

E-step: Compute z_{nk} using the current parameters, i.e.,

$$z_{nk} = p(z_{nk} = 1|\mathbf{x}_n, \theta) = \frac{p(k) \prod_{d=1}^{D} p_{kd}^{x_{nd}} (1 - p_{kd})^{(1-x_{nd})}}{\sum_{j=1}^{K} p(j) \prod_{d=1}^{D} p_{jd}^{x_{nd}} (1 - p_{jd})^{1-x_{nd}}}.$$

M-step: Update $p(k)$ and $\mathbf{p}_k$ as follows:

$$p(k) = \frac{\sum_{n=1}^{N} z_{nk}}{N}, \quad k = 1, \ldots, K,$$

$$\mathbf{p}_k = \frac{\sum_{n=1}^{N} z_{nk} \mathbf{x}_n}{\sum_{n=1}^{N} z_{nk}}.$$

Example 7.3.2. *We present some of the techniques used for clustering of handwritten digits. The data consists of 28×28 pixel images, which can be found on the MNIST web page [66]. We use a Bernoulli mixture model, and we attempt to cluster the data using the EM-algorithm.*

Here, we have $K = 10$, representing the digits 0–9. We initialize the cluster probabilities as $\mathbf{p} = (\frac{1}{K}, \ldots, \frac{1}{K})$ and $\mathbf{p} = (p_{kd}) = 0.25 + 0.5u$, where $u = \mathcal{U}(0, 1)$ is a uniformly distributed random variable, which is normalized so that $\sum_{d=1}^{D} p_{kd} = 1$.

Next, we run the EM-algorithm iterating 1000 times. The resulting clusters stored in $\mathbf{z}$ are shown in Fig. 7.1 using singular value decomposition for dimensionality reduction to two dimensions.

7.4 t-Distributed Stochastic Neighbor Embedding

The t-distributed stochastic neighbor embedding (t-SNE) is a technique for dimensionality reduction and visualization of data, first published by van der Maaten and Hinton [67]. Data are often characterized by a high number of dimensions. To visualize high-dimensional data, dimensionality reduction techniques can be used.

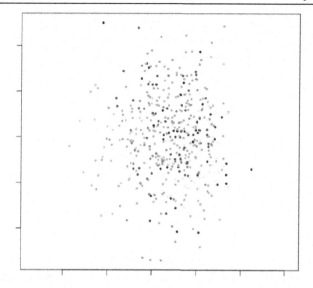

Figure 7.1: Singular value decomposition of MNIST digit data.

Dimensionality reduction methods convert a high-dimensional data set $\mathbf{x} = \{x_1, x_2, \ldots, x_n\}$ into two- or three-dimensional data $\mathbf{y} = \{y_1, y_2, \ldots, y_n\}$ that can be displayed in a scatterplot. We call the low-dimensional data representation $\mathbf{y}$ a *map*, and the individual data points y_i *map points*. We wish to preserve as much of the significant structure of the original data as possible in the low-dimensional map.

There are many widely used dimensionality reduction techniques, such as principal component analysis and various clustering techniques. The t-SNE method is a visualization technique for high-dimensional data, presenting it in clusters while largely preserving its local structure. It is capable of revealing clusters on various scales, making it suitable for fractal data sets.

Stochastic neighbor embedding (SNE) transforms high-dimensional Euclidean distances between data points into conditional probabilities as similarities. The similarity between two points x_j and x_i is the conditional probability, $p_{j|i}$, that x_i is a neighbor to x_j in proportion to a normal probability centered at x_i. For close points, $p_{j|i}$ is relatively high, whereas for distant points, $p_{j|i}$ is close to zero for a properly chosen variance of the normal distribution, σ_i. The conditional probability $p_{j|i}$ is given by

$$p_{j|i} = \frac{\exp(-||x_i - x_j||^2/2\sigma_i^2)}{\sum_{k \neq i} \exp(-||x_i - x_k||^2/2\sigma_i^2)}, \tag{7.5}$$

$$p_{i|i} = 0, \tag{7.6}$$

where σ_i is the variance of the Gaussian centered on data point x_i. For the low-dimensional counterparts y_i and y_j of the high-dimensional data points x_i and x_j, we define a similar conditional probability, denoted $q_{j|i}$, using the variance $1/\sqrt{2}$, so that

$$q_{j|i} = \frac{\exp(-||y_i - y_j||^2)}{\sum_{k \neq i} \exp(-||y_i - y_k||^2)},$$

$$q_{j|i} = 0.$$

When the map points y_i and y_j correctly model the similarity between the high-dimensional data points x_i and x_j, the conditional probabilities $p_{j|i}$ and $q_{j|i}$ should be equal. We therefore try to find a low-dimensional data representation that minimizes the discrepancy between $p_{j|i}$ and $q_{j|i}$.

The degree of agreement between $q_{j|i}$ and $p_{j|i}$ can be computed by the overall Kullback–Leibler divergence, giving the discrepancy (interpreted as cost) to be minimized as

$$C = \sum_i KL(P_i||Q_i) = \sum_i \sum_j p_{j|i} \log \frac{p_{j|i}}{q_{j|i}}, \tag{7.7}$$

where P_i is the conditional probability distribution over all points $j \neq i$ given point x_i and Q_i is the conditional probability distribution over all map points $j \neq i$ given map point y_i.

Now, the Kullback–Leibler divergence is asymmetric, so different discrepancies in the low-dimensional map are given unequal weights. It assigns a large cost when widely separated map points represent nearby data points, but only a small cost when nearby map points represent widely separated data points. This means that the cost function boosts the local data structure in the map.

We now turn to the selection of σ_i, the variance of the normal distribution centered on each original data point x_i. The variance varies with the data density, so that σ_1 is smaller in dense regions. Each chosen variance σ_i induces a probability distribution P_i over all data points $j \neq i$.

The entropy of the distribution increases with σ_i, that is, the distribution becomes more "random". First, we define a parameter related to the distribution P_i, called the *perplexity*, whose value is specified as input, i.e.,

$$\mathrm{Perp}(P_i) = 2^{H(P_i)},$$

where $H(P_i)$ is the Shannon entropy of P_i measured in bits, that is,

$$H(P_i)- = \sum_j p_{j|i} \log_2 p_{j|i}.$$

The perplexity can be interpreted as a continuous estimate of the effective number of neighbors. Given the amount of perplexity, we perform a binary search for the value of σ_i that generates a P_i with this value. Typical values have been reported to be a perplexity between 5 and 50. To minimize the cost function (7.7), we use the gradient descent method, which can be written

$$\frac{\partial C}{\partial y_i} = 2 \sum_j (p_{j|i} - q_{j|i} + p_{i|j} - q_{i|j})(y_i - y_j).$$

We must initiate the gradient descent, and the start solution is found by randomly sampling the map points' Gaussian distribution with small variance centered around the origin. In order to accelerate the optimization and to avoid local minima, a momentum term is added to the gradient. At each step, a gradient is added to an exponentially decaying sum of previous gradients. The momentum is given by

$$\mathbf{y}^{(t)} = \mathbf{y}^{(t-1)} + \eta \frac{\partial C}{\partial \mathbf{y}} + \alpha(t) \left(\mathbf{y}^{(t-1)} - (\mathbf{t} - \mathbf{2}) \right),$$

where $\mathbf{y}^{(t)}$ denotes the solution, η is the learning rate, and $\alpha(t)$ is the momentum at iteration t.

In the beginning of the optimization, Gaussian noise is added to the map points, and the variance of the noise term is gradually decreased. This technique, similar to simulated annealing, helps avoiding local minima in the cost function. The algorithm may have to be run with different parameters to obtain the best possible result.

With some modifications of the SNE, the method is called t-SNE. The cost function used in t-SNE is made symmetric, and simpler gradients are used in the optimization. In addition, the student's t-distribution is used rather than a normal distribution to determine the stochastic similarity between map points. The t-distribution has heavier tails than the Gaussian, which improves cluster separation and speeds up the optimization.

The cost function (7.7) is modified so that the sum of the Kullback–Leibler divergences between the conditional probabilities $p_{j|i}$ and $q_{j|i}$ are replaced by a single Kullback–Leibler divergence between joint probability distributions, in the high-dimensional space P and in the low-dimensional space Q, so that

$$C = KL(P||Q) = \sum_i \sum_j p_{ij} \log \frac{p_{ij}}{q_{ij}},$$

where p_{ii} and q_{ii} are zero. This function is symmetric, as opposed to (7.7), since $p_{ij} = p_{ji}$ and $q_{ij} = q_{ji}$ for all i, j. The pairwise similarities in the low-dimensional map q_{ij} are given by

$$q_{ij} = \frac{\exp(-||y_i - y_j||^2)}{\sum_{k \neq l} \exp(-||y_k - y_l||^2)}.$$

To avoid problems with outliers, we define the pairwise similarities in the high-dimensional space p_{ij} as

$$p_{ij} = \frac{p_{j|i} + p_{i|j}}{2n}, \tag{7.8}$$

so that $\sum_j p_{ij} > 1/2n$ for all data points x_i. The simpler gradient of the cost function is given by

$$\frac{\partial C}{\partial y_i} = 4 \sum_j (p_{ij} - q_{ij})(z_i - y_j).$$

In t-SNE, we employ a student t-distribution with one degree of freedom (known as the Cauchy distribution) as heavy-tailed distribution in the low-dimensional map. Using this distribution, the joint probabilities q_{ij} can be written as

$$q_{ij} = \frac{(1 + ||y_i - y_j||^2)^{-1}}{\sum_{k \neq l}(1 + ||y_k - y_l||^2)^{-1}}. \tag{7.9}$$

The gradient of the Kullback–Leibler divergence between P (7.8) and the student t-based joint probability distribution Q (7.9) can be shown to be

$$\frac{\partial C}{\partial y_i} = 4 \sum_j (p_{ij} - q_{ij})(y_i - y_j)(1 + ||y_i - y_j||^2)^{-1}. \tag{7.10}$$

Algorithm 7.4.1 (t-SNE (simplified)).
Given a data set $\mathbf{x} = \{x_1, x_2, \ldots, x_n\}$, parameters perplexity Perp, total number of iterations T, learning rate η, and momentum $\alpha(t)$.

STEP 0:
        Compute pairwise affinities $p_{j|i}$ (7.5)–(7.6) with perplexity Perp,
        Set $p_{ij} = (p_{j|i} + p_{i|j})/2n$,
        Sample initial solution $\mathbf{y}^{(0)} = \{y_1, y_2, \ldots, y_n\}$ from $N(0, 10^{-4}I)$.
STEP 1:T
    for $t = 1$ to T **do**
      Compute low-dimensional affinities q_{ij} (7.9),
      Compute gradient $\partial C/\partial \mathbf{y}$ (7.10),
      Set $\mathbf{y}^{(t)} = \mathbf{y}^{(t-1)} + \eta \frac{\partial C}{\partial \mathbf{y}} + \alpha(t)(\mathbf{y}^{(t-1)} - \mathbf{y}^{(t-2)})$.
    end

Output: low-dimensional data representation $\mathbf{y}^{(T)} = \{y_1, y_2, \ldots, y_n\}$. ●

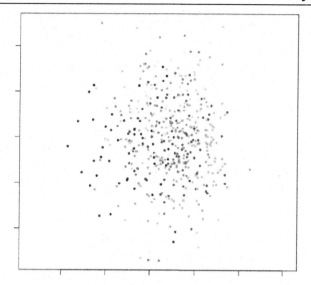

Figure 7.2: Result from the EM algorithm applied to MNIST digit data.

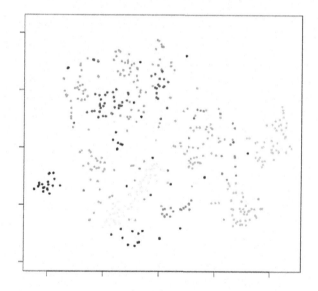

Figure 7.3: Result from the Rtsne algorithm applied to MNIST digit data.

7.5 Methods for Image Recognition

We use some of the techniques for clustering of handwritten digits. The data consists of 28 × 28 pixel images, which can be found on the MNIST web page [66]. The model we use is a

Bernoulli mixture model, and we attempt to cluster the data using the EM-algorithm and by using t-SNE.

We implement the EM algorithm for Bernoulli mixture models (BMMs) as described in Section 7.3. Fig. 7.2 shows the clusters as defined by the EM-algorithm iterated 10 times. The predicted cluster identities are found from the raw data and the Bernoulli probability distribution μ.

There are at least two implementations of t-SNE available for R. The package `Rtsne` is according to the CRAN source a wrapper for the C++ implementation by van der Maaten [68]. The second implementation, however, called `tsne`, is implemented purely in R and seems to be considerably slower than `Rtsne`.

Fig. 7.3 shows the result from `Rtsne` for 500 examples from the MNIST database. The result clearly shows clusters, well separated and with high precision.

CHAPTER 8

Data Centers and Clouds

One of the first steps in network planning is to determine locations for service facilities, data centers, and large support deployments, chosen to be as efficient as possible. The efficiency is often expressed as some type of cost. Mathematically, this is known as a facility location problem.

A simplified description of this problem is as follows. Given a number of customers (or stores) distributed throughout an area, we wish to find locations for a number of facilities (or warehouses) so that the total cost of building the facilities and the distribution is minimized while satisfying all the customers' demands.

The facility location problem is clearly closely related to clustering. Indeed, the k-means and k-median algorithms can be seen as restricted versions of facility location. It is also very similar to network design, discussed in Chapters 9 and 10.

We focus on approximation algorithms to facility location problems, loosely following the discussion by Shmoys et al. [69]. We are given a number of possible locations i to set up a facility (warehouse, service, or data center) and a number n client locations where the clients $j = 1, 2, \ldots, n$ need to be served by any of the facilities satisfying a demand d_j. A facility can be established at location i at a cost f_i, and a client j is served by facility i at a cost c_{ij}. A facility may or may not have a capacity limit, say u clients. The former case is known as the capacitated and the latter as the uncapacitated facility location problem. In addition, we may or may not allow clients to be served by more than one facility.

There are many possible approaches to such problems, some of which are variants of the clustering algorithms discussed in Chapter 6. Here, we describe α-approximation algorithms, polynomial-time algorithms that always produce a solution with a cost within a factor of α of the optimal.

8.1 Uncapacitated Facility Location

The uncapacitated facility location can be formulated as an integer program. Let N be a set of given locations $N = \{1, \ldots, n\}$, with distances c_{ij}, $i, j = 1, \ldots, n$, between them. Let a subset $F \subseteq N$ of locations be possible sites to set up a facility and another subset $D \subseteq N$ of locations represent clients that need to be assigned to some facility, having a demand $d_j \in \mathbb{Z}^+$, $j \in D$, that must be transported from facility i to the client's location j. For each location $i \in F$ there is a (nonnegative) cost $f_i \in \mathbb{R}^+$ of opening a facility at i. The cost of assigning a client at i to a facility at j is $c_{ij} \in \mathbb{R}^+$ per unit client demand. In addition, we assume that costs are symmetric, i.e., $c_{ij} = c_{ji}$, and satisfy the triangle inequality $c_{ij} + c_{jk} \geq c_{ik}$ for all $i, j, k \in N$.

We want to find a feasible solution, assigning each client i to a facility j satisfying all demands d_i, at minimum cost. The problem can conveniently be expressed as an integer program, where the variable $y_i, i \in F$, assuming either 0 or 1, indicates whether a facility is located at i or not, and the variable $x_{ij}, i \in F, j \in D$, takes value 1 if and only if the client at location j is assigned to a facility at location i. We have

$$\min \quad \sum_{i \in F} f_i y_i + \sum_{i \in F} \sum_{j \in D} d_j c_{ij} x_{ij} \tag{8.1}$$

$$\sum_{i \in F} x_{ij} = 1, \quad \text{for each } j \in D, \tag{8.2}$$

$$x_{ij} \leq y_i, \quad \text{for each } i \in F, j \in D, \tag{8.3}$$

$$x_{ij} \in \{0, 1\}, \quad \text{for each } i \in F, j \in D, \tag{8.4}$$

$$y_i \in \{0, 1\}, \quad \text{for each } i \in F. \tag{8.5}$$

The constraints ensure that each client $j \in D$ is assigned to some facility at location $i \in F$ and that, whenever a client j is assigned to a facility location i, a facility must have been set up at i. We let $x_{ij} = 0$ whenever $i \notin F$ or $j \notin D$, and $y_i = 0$ for each $i \notin F$. These cases are impossible from the definition of the problem.

Even if the problem with integer variables y_i and x_{ij} is $\mathcal{NP}$-hard, when relaxing these restrictions, the resulting linear program can be solved efficiently. In theory, an integer approximate solution can then be found from the solution to the linear program using branch-and-bound, but as the number of variables increases, so does the number of cases that need to be examined.

Jain and Vazirani [70] proposed a 3-approximation for the uncapacitated metric facility location problem. It has two phases: in the first phase all clients are assigned to at least one facility, and in the second the solution is trimmed so that any client is assigned to one single facility only.

We can interpret the problem in terms of the dual parameters, which we call prices. Let a price p_j be associated with each client. It then has the following properties:

(1) Every client $j \in D$ is assigned to some facility $i \in F$, and $p_j \geq 0$, $\forall j \in D$.
(2) If a client j is assigned to facility i, then $p_j \geq c_{ij}$.

(3) Let S_i be the subset of clients assigned to facility i, so the following holds:

$$\sum_{j \in S_i} (p_j - c_{ij}) = f_i.$$

This suggests that the cost of setting up a facility at location i is paid for by the clients assigned to it.

(4) For any facility $i \in F$ and any subset of clients $S \subseteq D$,

$$\sum_{j \in S} (p_j - c_{ij}) \leq f_i.$$

The inequality can be interpreted so that no subset of clients can set up a new facility at a lower cost than contributing to the facility at i. This can be seen as a sort of "economy of scale".

Assignment

The first phase of the algorithm is run until all clients are assigned to at least one facility. Initially, all clients $j \in D$ are unassigned and the prices they pay increase linearly. When the price client j pays exceeds its connection cost c_{ij} to facility i, it starts contributing to the set-up cost f_i. Whenever f_i is fully paid, that is, $\sum_{j \in S} (p_j - c_{ij}) = f_i$, the facility is opened at location i, and all clients contributing to its set-up cost f_i are assigned to it, after which the prices p_j, $j \in S$, remain unchanged.

If an unassigned client reaches an existing facility as a result of increasing its prices, it is assigned to it after which its price remains unchanged.

Algorithm 8.1.1 (Phase I – assignment).
Set $p_j = 0$, for all $j \in D$.

STEP 1:$|\mathcal{D}|$
  while there are unassigned clients:
   All unassigned clients j linearly increase their prices p_j until they reach
    a facility i at $p_j \geq c_{ij}$,
    Then they start contributing to the facility's opening cost f_i by $p_j - c_{ij}$
    When the facility's opening cost f_i is covered, the prices remain unchanged
   if an unassigned client j reaches an open facility i **then**:
    assigned j to i without contributing to its opening cost f_i
   end
  end

Output the set of coefficients C.

Pruning

After phase I, all clients are assigned to at least one facility. In the pruning (or cleanup) phase, we trim the solution of excess cost. The total cost can be shown to be $\hat{C} \leq 3 \sum j \in D p_j$.

We say that two facilities i and i' are in conflict if there exists a client j such that $p_j - c_{ij} > 0$ and $p_j - c_{i'j} > 0$. Consider the facilities in the reverse order they were opened and close any conflicting facility i that was added after i'.

After the conflict resolution step, for any client $j \in D$ there can be at most one open facility such that $p_j >= c_{ij}$, and we assign j to i. If there are any unassigned clients, assign them to facilities in the cheapest way. The cost is bound by $c_{ij} \leq 3 p_j$.

To see this, we note that if client j is unassigned, then facility i that it was connected to was closed due to conflict. But this means that there exists some other client j' that had i and some other facility i' in conflict. Since facility i was closed to resolve this conflict, i' must have been set up first, which implies that $p_{j'} \leq p_j$. For all connected clients, we have $c_{ij} \leq p_j$, and by the triangle inequality $c_{ji} \leq c_{ji} + c_{ij'} + c_{j'i'} \leq p_j + 2 p_{j'} \leq 3 p_j$. Using this argument for all clients, we have the following theorem.

Theorem 8.1.2. *The cost of the approximation is $\hat{C} \leq 3 \sum_{j \in D} p_j$.*

Proof. Let S_F be the set of clients assigned by the end of the first phase. These clients pay for the set-up cost of all facilities and their connection costs. For any client j assigned in the second phase, its connection cost can be at most $3 p_j$. Therefore, we have

$$\sum_{j \in S_F} p_j = \sum_{i \in F} f_i + \sum_{j \in S_F} c_{ij},$$

$$3 \sum_{j \notin S_F} p_j \geq \sum_{j \notin S_F} c_{ij},$$

which implies that the total solution cost $\sum_{i \in F} f_i + \sum_{j \in D} c_{ij} \leq 3 \sum_{j \in D} p_j$. $\square$

Algorithm 8.1.3 (Phase II – pruning).
Resolve conflicts among facilities set up in phase I.

STEP 1: $\mathcal{D}$
> **for** $i = 1$ **to** $|\mathcal{D}|$
>> **if** there exists an open facility i such that $p_j > c_{ij}$:
>> assign j to i
>> **else if** there exists an open facility i such that $p_j = c_{ij}$:
>> assign j to i

Figure 8.1: Uncapacitated facility location problem with its initial centroid.

> **else**
>> assign j to facility i that was in conflict with the facility
>> previously assigned to j at the end of phase I
> **end**
> **end**

Output the set of coefficients C. ●

Example 8.1.1. *Consider a clustering problem with the following parameters. Let $n = 21$ and let the points be distributed as shown in Fig. 8.1. The costs c_{ij} are the Euclidean distances and the demand is uniform. The establishment cost f_i at each location is calculated as*

$$f_i = \sum_{l=1}^{\alpha} \tilde{c}_{il},$$

where $\alpha \in \{1, \ldots, n\}$ and $\tilde{c}_{i\cdot}$ are the sorted distances from each i. The variable α determines how many of the clients' costs to include in the establishment cost. The costs f_i therefore increases with the area the facility covers. This strategy allows for a systematic approach to adjust the establishment costs. When the cost is too low, each client gets a facility, and when too expensive, only one facility is set up for all clients. The result, shown in Fig. 8.2, shows the resulting three clusters.

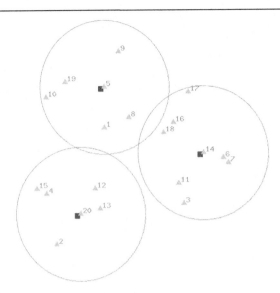

Figure 8.2: Near-optimal solution to an uncapacitated facility location problem with Euclidean distances and uniform demand.

8.2 A Primal-Dual Algorithm

Suppose we have a graph $G = (V, E)$, a set of potential facilities $F \subseteq V$, and a set of clients $D \subseteq V$. Now, we also have a set of resource types R and each client d_i requires a resource of type $r_{d_i} \in R$. We wish to set up a number of the facilities and connect the clients to these facilities via Steiner trees so that the total cost is minimized. Each potential facility i is associated with a fixed cost f_i to set it up, and edge e has a linear cost of $n_e c_e$ if it ships n_e different resource types a distance c_e, so that c_e is the length of the edge e.

We can express this problem as an integer program. Let the variables y_i represent whether the facility i is set up or not and let $x_{e,r} \in \{0, 1\}$ be a variable, taken as one if the edge e ships the resource type r and zero otherwise. Let $\delta(S)$ be the set of edges with one end in the set $S \subset V$ and the other in $V - S$. We have

$$\min \quad \sum_i f_i y_i + \sum_{e,r} c_e x_{e,r} \tag{8.6}$$

$$\sum_{e \in \delta(S)} x_{e,r} + \sum_{i \in S} y_i \geq 1, \quad \text{for all } r, S, \tag{8.7}$$

$$r_{d_j} = r, d_j \in D, \tag{8.8}$$

$$y_i, x_{e,r} \in \{0, 1\}. \tag{8.9}$$

The constraint (8.7) states that for all sets S containing a client with a demand of resource type r, either there is an established facility in S or there exists an edge shipping resource type r which crosses S. The linear relaxation to the integer program is obtained by replacing the last constraint (8.9) by

$$y_i, x_{e,r} \geq 0. \tag{8.10}$$

To formulate the dual of the linear program (8.6), (8.7), (8.8), and (8.10), we introduce a variable $\gamma_{S,r}$ for each set S and resource type r such that again there exists a client S with a demand for r. Then the dual is

$$\max \ \sum_{S,r} \gamma_{S,r}, \tag{8.11}$$

$$\sum_{S:e \in \delta(S)} \gamma_{S,r} \leq c_e, \quad \text{for all } e, r, \tag{8.12}$$

$$\sum_{S,r:i \in S} \gamma_{S,r} \leq f_i, \quad \text{for all } i, \tag{8.13}$$

$$\gamma_{S,r} \geq 0. \tag{8.14}$$

The following algorithm is based on the primal-dual formulation and consists of an assignment phase, a pruning phase, and a conflict resolution phase.

Assignment Phase

We call an edge e *tight for resource type r* if the dual constraint (8.12) is satisfied with equality for the pair (e, r). We say that a set S is *active for resource type r* if it contains a spanning tree of tight edges supplying r.

For each resource type r, we uniformly increase the dual variables for all active sets. Initially these sets contain the single clients with demand for r. Whenever the constraint (8.12) becomes tight for an edge e and a resource type r, we say that the edge e was *bought for shipping r*.

Denote by S_i the set of clients that contributed to make inequality (8.12) tight. We continue to increase the dual variables $\gamma_{S,r}$ at the same uniform rate, where $S = S_i$. When the constraint (8.13) becomes tight for facility i, it is temporarily set up and freezes the dual variables that contributed to make the inequality tight. We also associate facility i with the clients in S having a demand for r corresponding to frozen variables $\gamma_{S,r}$ (that is, clients contributing to setting up facility i). We continue the process for all S and r until all dual variables have been frozen.

Pruning Phase

Given the temporarily set up facilities from the assignment phase, suppose that an edge is shipping resource type r. Should removing r from the edge still satisfy all the client demands, we can prune r from the set of resources shipped by e.

Conflict Resolution Phase

We say that two facilities are at *conflict at time t* if they are within the same set S at time t, whose dual variable $\gamma_{S,r}$ is not frozen and the length of the path in S does not exceed t.

Two facilities are said to be at conflict if there exists a time t when they are in conflict. From the facilities in conflict, we try to open only one. To resolve a conflict, we create a graph G_F with the potential facilities F as a node and an edge between two facilities i and j whenever they are in conflict.

We greedily find a largest independent set F' in G_F, where facilities are considered in the order they were temporarily set up, and the facilities in F'. For a facility j in F that was not set up, we inspect the sets (S_j, r_j) that facility j was assigned to. It follows from the conflict resolution strategy that there exists a set up facility i, previously in conflict with j, that can be reached by a path π. We can therefore increase some γ_{S_j,r_j} to reconnect (or reroute) the sets (S_j, r_j) to facility i.

We can solve the problem by rounding of the relaxed integer program. The technique of creating an approximate integer solution by rounding of a linear program is described in Chapter 5.

This leads to a different approximation algorithm, which can be extended in several interesting ways.

We base the algorithm on the filtering and rounding technique of Lin and Vitter [41], where we solve the linear relaxation of the integer program, and by filtering we obtain a new fractional solution. In this solution, a client can be connected by different fractions to several facilities, and facilities can be partially set up.

The solution to the filtered problem has the property that whenever a location j is fractionally assigned to a (partially) set up facility i, the cost c_{ij} associated with that assignment is not too big, and it is referred to as the *closeness property*. Finally, a fractional solution with this closeness property is rounded to a near-optimal integer solution. Some proofs are omitted; they can be found in [69].

Consider the linear relaxation to the integer program (8.1)–(8.5), with the integer constraints (8.4) and (8.5) replaced by

$$x_{ij} \geq 0, \quad \text{for each } i \in F, j \in D,$$
$$y_i \geq 0, \quad \text{for each } i \in F.$$

Given a number g_j for each $j \in D$, we say that a feasible solution (x, y) to the linear relaxation is g-close if it satisfies

$$x_{ij} > 0 \Rightarrow c_{ij} \le g_j.$$

This means that a fractionally assigned client i only can be assigned to "reasonably close" facilities j.

Given a feasible fractional solution (x, y), we define an α-point, $c_j(\alpha)$, for each client location $j \in D$. To do this, fix a client location $j \in D$ and let π be a permutation such that $c_{\pi(1)j} \le c_{\pi(2)j} \le \cdots \le c_{\pi(n)j}$. We use the initial assumption that if $i \notin F$, then $x_{ij} = 0$. We then set $c_j(\alpha) = c_{\pi(i^*)j}$, where $i^* = \min\{i' : \sum_{i=1}^{i'} x_{\pi(i)j} \ge \alpha\}$.

Lemma 8.2.1. *Let α be a fixed value in the interval $(0, 1)$. Given a feasible fractional solution (x, y), we can find a g-close feasible fractional solution $(\bar{x}, \bar{y})$ in polynomial time, such that*

(1) $g_j \le c_j(\alpha)$, *for all $j \in D$,*
(2) $\sum_{i \in F} f_i \bar{y}_i \le (1/\alpha) \sum_{i \in F} f_i y_i$.

If we let $S = \{i : c_{ij} \ge c_j(\alpha)\}$, then the definition of $c_j(\alpha)$ gives $\sum_{i \in S} x_{ij} \ge 1 - \alpha$. We have

$$\sum_{i \in F} c_{ij} x_{ij} \ge \sum_{i \in S} c_{ij} x_{ij} \ge (1 - \alpha) c_j(\alpha),$$

or

$$c_j(\alpha) \le \frac{1}{1 - \alpha} \sum_{i \in F} c_{ij} x_{ij}.$$

Lemma 8.2.2. *Given a feasible fractional g-close solution $(\bar{x}, \bar{y})$, we can find a feasible integer $3g$-close solution $(\hat{x}, \hat{y})$ such that*

$$\sum_{i \in F} f_i \hat{y}_i \le \sum_{i \in F} f_i \bar{y}_i.$$

The lemma follows from the rounding algorithm. Given g_j, $j \in D$, and a fractional solution $(\bar{x}, \bar{y})$ which is g-close. The algorithm iteratively converts this initial solution into a $3g$-close integer solution $(\hat{x}, \hat{y})$ without increasing the total cost.

Let $(\hat{x}, \hat{y})$ be a feasible fractional solution, initially set $(\hat{x}, \hat{y}) = (\bar{x}, \bar{y})$. Let $\hat{F}$ be the set of partially set up facilities in the current solution, that is, $\hat{F} = \{i \in F : 0 < \hat{y}_i < 1\}$, and $\hat{D}$ the set of clients j that are assigned to facilities only in $\hat{F}$. It follows that if $\hat{x}_{ij} > 0$, then $i \in \hat{F}$.

In each iteration, we find the client location $j \in \hat{D}$ for which g_j is minimum; call this client j', so

$$S = \{i \in \hat{F} : \hat{x}_{ij'} > 0\}.$$

The client j' is assigned to the facility $i \in S$ with the smallest f_i; call this facility i'. Now, we round the values of $\{\hat{y}_i\}_{i \in S}$ by setting $\hat{y}_{i'} = 1$ and $\hat{y}_i = 0$ for all $i \in S - \{i'\}$.

Let T denote the set of locations that are partially assigned by $\hat{x}$ to locations in S, that is,

$$T = \{j \in D : \text{there exists } i \in S \text{ such that } \hat{x}_{ij} > 0\}.$$

We assign each client location $j \in T$ to the facility set up at i', that is, $\hat{x}_{i'j} = 1$ and $\hat{x}_{ij} = 0$ for $i \neq i'$. When $\hat{D}$ becomes empty, for each client location $j \in D$ there exists an i' such that $\hat{x}_{i'j} > 0$ and $\hat{y}_{i'} = 1$, so j can be assigned to i'.

The solution has the following properties:

(1) $(\hat{x}, \hat{y})$ is a feasible fractional solution,
(2) $\sum_{i \in F} f_i \hat{y}_i \leq \sum_{i \in F} f_i \bar{y}_i$,
(3) $\hat{x}_{ij} > 0$ and $i \in \hat{F} \Rightarrow c_{ij} \leq g_j$,
(4) $\hat{x}_{ij} > 0$ and $i \notin \hat{F} \Rightarrow c_{ij} \leq 3g_j$.

Starting with a feasible fractional solution (x, y), first using Lemma 8.2.1 to obtain $(\bar{x}, \bar{y})$ and then Lemma 8.2.2 to round $(\bar{x}, \bar{y})$ to obtain a feasible integer solution $(\hat{x}, \hat{y})$ with facility cost at most

$$\sum_{i \in F} f_i \hat{y}_i \leq \sum_{i \in F} f_i \bar{y}_i \leq (1/\alpha) \sum_{i \in F} f)i y_i.$$

For each location client $j \in D$, its assignment cost as determined by $\hat{x}$ is at most $3g_j \leq 3c_j(\alpha) \leq \frac{3}{1-\alpha} \sum_{i \in F} c_{ij} x_{ij}$. Combining the two bounds, we obtain the total cost of $(\hat{x}, \hat{y})$ as

$$\hat{c} = \sum_{i \in F} f_i \hat{y}_i + \sum_{i \in F} \sum_{j \in D} d_j c_{ij} \hat{x}_{ij}$$

$$\leq \frac{1}{\alpha} \sum_{i \in F} f_i y_i + 3 \sum_{i \in D} d_j c_j(\alpha)$$

$$\leq \max\left\{\frac{1}{\alpha}, \frac{3}{1-\alpha}\right\} \left(\sum_{i \in F} f_i y_i + \sum_{i \in F} \sum_{j \in D} d_j c_{ij} x_{ij}\right).$$

With $\alpha = 1/4$, both of the two factors in the maximum expressions above are 4, so the total cost of the approximation $(\hat{x}, \hat{y})$ is guaranteed to be within a factor of 4 of the cost of (x, y). We therefore have the following theorem.

Theorem 8.2.3. *For the metric uncapacitated facility location problem, filtering and rounding yields a 4-approximation algorithm.*

8.3 Capacitated Facility Location

In the capacitated facility location, we consider the case where each facility can be assigned to serve a total demand that is at most u, where u is a positive integer. The approximation algorithm based on filtering and rounding for the uncapacitated facility location can be adapted to the capacitated facility location.

In the uncapacitated case, if the optimal values of y are given, then it is a trivial task to find the corresponding x. We simply assign each client location $j \in D$ to the facility location i for which c_{ij} is the smallest among all possibilities where $y_i = 1$.

In the capacitated facility location, the assignment of clients to facilities is more involved. There are two variants of the problem, depending on whether each client must be assigned to one single facility, or the client's demand may be split between several facilities.

First, we study the variant with splittable demands. The integer program for the capacitated facility location is the same as for the uncapacitated case, with the additional constraint

$$\sum_{j \in D} d_j x_{ij} \leq u y_i, \quad \text{for each } i \in F.$$

We assume that each facility $i \in F$ is either open or closed and modeled by a 0-1 integer variable. In the relaxation of the integer program, we let

$$0 \leq y_i \leq 1, \quad \text{for each } i \in F.$$

We consider the situation of the capacitated facility location in which we must decide on an integer number y_i of facilities to build at any location $i \in F$.

When we allow the client demands to be served by more than one facility, we only need to find a solution where each value $y_i, i \in F$, is integer. We note that Lemma 8.2.1 is still valid and state a modified version of Lemma 8.2.2.

Lemma 8.3.1. *Given a fractional g-close solution $(\bar{x}, \bar{y})$, we can find an integer 3g-close solution $(\hat{x}, \hat{y})$ in polynomial time, such that*

$$\sum_{i \in F} f_i \hat{y}_i \leq 4 \sum_{i \in F} f_i \bar{y}_i.$$

Just like in the uncapacitated case, a solution $(\hat{x}, \hat{y})$ is created iteratively by rounding each $0 < \hat{y}_i < 1$ to an integer. Initially, we set $\hat{x} = \bar{x}$, we set $\hat{y}_i = \lceil \bar{y}_i \rceil$ for each i such that $\bar{y}_i \geq 1/2$, and we set $\hat{y}_i = \bar{y}_i$ otherwise. We also maintain a set $\hat{F} \subseteq F$ of facilities i for which $0 < \hat{y}_i < 1/2$.

In the capacitated case, we need to keep track of the assignments at each step. For each client location $j \in D$, we compute the fraction of the demand for client location j that is satisfied by locations in $\hat{F}$. This is stored in $\beta_j = \sum_{i \in \hat{F}} \hat{x}_{ij}$ for each $j \in D$. We also let $\hat{D} \subseteq D$ be the set of client locations j for which $\beta_j > 1/2$.

In each iteration, we find the client location $j \in \hat{D}$ for which g_j is minimum; call this client j', so

$$S = \{i \in \hat{F} : \hat{x}_{ij'} > 0\}$$

and

$$T = \{j \in D : \text{ there exists } i \in S \text{ such that } \hat{x}_{ij} > 0\}.$$

We do not open just one facility in S, but we open the cheapest $\lceil \sum_{i \in S} \hat{y}_i \rceil$ facilities in S instead; let O denote this set of facilities. For each $i \in O$, we update $\hat{y}_i = 1$, and for each $i \in S - O$, we update $\hat{y}_i = 0$. For each location $j \in T$, there is a total demand $\hat{d}_j$ currently assigned to locations in S, where

$$\hat{d}_j = d_j \sum_{i \in S} \hat{x}_{ij}.$$

This demand will be rerouted to go only to those facilities in O. For each $i \in O, j \in T$, let z_{ij} be the amount of j's demand that is assigned to i by an optimal solution. We update our solution by resetting $\hat{x}_{ij} = z_{ij}/\hat{d}_j$ for each $i \in O, j \in T$, and $\hat{x}_{ij} = 0$ for each $i \in S - O$, $j \in D$. All other components of $\hat{x}$ remain unchanged.

When $\hat{D}$ becomes empty, we have satisfied at least half of the demand for each client location $j \in D$, by assigning it to locations for which the component of $\hat{y}$ is integral. To compute the solution claimed by the lemma, we will simply ignore the β_j fraction of j's demand that is still assigned to the remaining facilities in $\hat{F}$ and rescale the part of $\hat{x}$ specifying the assignment to facilities not in $\hat{F}$. That is, for each $i \notin \hat{F}$, we reset $\hat{y}_i$ to be $2\hat{y}_i$, and we reset $\hat{x}_{ij}$ to be $\hat{x}_{ij}/(1 - \beta_j)$ for each $j \in D$. For each $i \in \hat{F}$, we set $\hat{y}_i = 0$, and we set $\hat{x}_{ij} = 0$ for all $j \in D$.

Let (x, y) be a feasible fractional solution to the linear relaxation of the capacitated facility location problem. We apply Lemma 8.2.1 to obtain $(\bar{x}, \bar{y})$ and Lemma 8.3.1 to obtain a feasible integer solution $(\hat{x}, \hat{y})$ with facility cost at most

$$\sum_{i \in F} f_i \hat{y}_i \leq 4 \sum_{i \in F} f_i \bar{y}_i \leq (4/\alpha) \sum_{i \in F} f)i y_i.$$

For each location client $j \in D$, its assignment cost is at most

$$\sum_{j \in D} c_{ij} d_j \hat{x}_{ij} \le 3 \sum_{j \in D} d_j g_j$$

$$\le 3 \sum_{j \in D} d_j c_j(\alpha)$$

$$\le \frac{3}{1-\alpha} \sum_{j \in D} d_j \sum_{i \in F} c_{ij} x_{ij}.$$

Combining the cost, we have

$$\frac{4}{\alpha} \sum_{i \in F} f_i y_i + \frac{3}{1-\alpha} \sum_{j \in D} d_j \sum_{i \in F} c_{ij} x_{ij}.$$

With $\alpha = 4/7$, the total cost of the approximation $(\hat{x}, \hat{y})$ is guaranteed to be within a factor of 7 of the cost of (x, y). We therefore have the following theorem.

Theorem 8.3.2. *For the metric capacitated facility location problem with splittable demands, filtering and rounding yields a 7-approximation algorithm.*

The more conventional definition of the capacitated facility location problem constrains each $y_i \in \{0, 1\}$. Suppose the optimal fractional solution (x, y) that we round also satisfies the constraint $y_i \le 1$, for each $i \in F$ (for example, because they were added to the linear programming relaxation).

The algorithm of Lemma 8.3.1 (with $\alpha = 4/7$) returns a solution in which $0 \le \bar{y}_i \le 7/4$, for each $i \in F$. For those i for which $1 \le \bar{y}_i \le 7/4$, we obtain the value $\hat{y}_i = 4$. For those i for which $0 \le \bar{y}_i < 1$, we get $\hat{y}_i \le 2$. Thus, if we apply the algorithm of Theorem 8.3.2 to an optimal solution to LP relaxation of the 0-1 capacitated facility location problem, we find an integer solution of cost with an approximation factor of 7, but which requires a small number of facilities at each opened site (that is, at most 4).

Next we turn our attention to the model in which the entire demand of each location must be assigned to the same facility. We shall call this problem the metric capacitated location problem with unsplittable demands. We will show that the solution found by the algorithm of Theorem 8.3.2 can be adjusted to satisfy this more stringent condition, while only slightly increasing the performance guarantees.

The extension to the model with unsplittable demands is based on a rounding theorem of Shmoys and Tardos [75] for the generalized assignment problem. This theorem can be explained as follows. Suppose that there is a collection of jobs J, each of which is to be assigned to exactly one machine among the set M; if job $j \in J$ is assigned to machine $i \in M$, then it requires p_{ij} units of processing, and it incurs a cost r_{ij}. Each machine $i \in M$ can be assigned jobs that require a total of at most P_i units of processing on it, and the total cost of the assignment must be at most R, where R and P_i, for each $i \in M$, are given as part of the input. The aim is to decide if there is a feasible assignment. If there is such an assignment, then there must also be a feasible solution to the following linear program, where x_{ij} is the relaxation of a 0-1 variable that indicates whether job j is assigned to machine i:

$$\sum_{i \in M} x_{ij} = 1, \quad \text{for all } j \in J, \tag{8.15}$$

$$\sum_{j \in J} p_{ij} x_{ij} \leq P_i, \quad \text{for all } i \in M, \tag{8.16}$$

$$\sum_{i \in M} \sum_{j \in J} r_{ij} x_{ij} \leq R, \quad \text{for all } i \in M, j \in J, \tag{8.17}$$

$$x_{ij} \geq 0, \quad \text{for all } i \in M, j \in J. \tag{8.18}$$

Shmoys and Tardos [75] show that any feasible solution x can be rounded, in polynomial time, to an integer solution that is feasible if the right-hand side of (8.16) is relaxed to $P_i + \max_{j \in J} p_{ij}$.

We show next how to apply this rounding theorem to produce a solution for the capacitated version with unsplittable demands. Consider the algorithm of Theorem 8.3.2 without specifying the choice of α. Suppose that we apply the algorithm starting with an optimal solution (x, y) to the linear relaxation of the capacitated facility location problem (that is, the linear program given by (8.6), (8.7), (8.8), and (8.10)). The algorithm delivers an integer solution $(\hat{x}, \hat{y})$, where the facility cost and the assignment cost are, respectively, within a factor of $4/\alpha$ and $3/(1 - \alpha)$ of the analogous costs for (x, y). Let O denote the set of facilities opened by the solution $(\hat{x}, \hat{y})$, that is,

$$O = \{i \in F : \hat{y}_i \geq 1\}.$$

We can view each facility $i \in O$ as a machine of processing capacity $\hat{y}_i u$ and each location $j \in D$ as a job that requires a total of d_j units of processing (independent of the machine to which it is assigned) and incurs a cost $d_j c_{ij}$ when assigned to machine i. Therefore, if we set $M = O$, $J = D$, $P_i = \hat{y}_i u$ for each $i \in M$,

$$R = \sum_{i \in F} \sum_{j \in D} d_j c_{ij} \hat{x}_{ij},$$

as well as $p_{ij} = d_j$ and $r_{ij} = d_j c_{ij}$ for each $i \in M$, $j \in J$, then $\hat{x}$ is a feasible solution to the linear program (8.15)–(8.18). The rounding theorem for the generalized assignment problem implies that we can round $\hat{x}$ to an integer solution $\tilde{x}$ such that each facility $i \in O$ is assigned a total demand at most $P_i + \max_{j \in D} d_j$ and the assignment cost of this solution is

$$\sum_{i \in O} \sum_{j \in D} d_j c_{ij} \tilde{x}_{ij} \leq \sum_{i \in F} \sum_{j \in D} d_j c_{ij} \hat{x}_{ij} \leq \frac{3}{1 - \alpha} \sum_{i \in F} \sum_{j \in D} d_j c_{ij} x_{ij}.$$

Note that, in order for a feasible solution with unsplittable demands to exist, the demand d_j must be at most u, for each $j \in D$; hence, we assume that our instance has this property. We can conclude that the rounded solution $\tilde{x}$ assigns a total demand to each facility $i \in O$ that is at most

$$\max_{i \in D} d_j + \hat{y}_i u \leq (1 + \hat{y}_i)u.$$

Hence, if we consider the solution $(\tilde{x}, \tilde{y})$ where $\tilde{y}_i = \hat{y}_i + 1$, for each $i \in O$ and $\tilde{y}_i = \hat{y}_i$ otherwise, then we see that it is a feasible integer solution to the unsplittable demand problem. Finally, since $\hat{y}_i \geq 2$ for each $i \in O$ (due to the final doubling when $\hat{D}$ becomes empty), we see that $\tilde{y}_i \leq (3/2)\hat{y}_i$, for each $i \in D$. This implies that the facility cost of $(\tilde{x}, \tilde{y})$ is

$$\sum_{i \in F} f_i \tilde{y}_i \leq (3/2) \sum_{i \in F} f_i \hat{y}_i \leq \frac{6}{\alpha} \sum_{i \in F} f_i y_i.$$

Thus, if we compare the solution $(\tilde{x}, \tilde{y})$ to the optimal fractional solution (x, y) from which we started, we have shown that the facility cost increases by at most a factor of $6/\alpha$, and the assignment cost increases by at most a factor of $3/(1 - \alpha)$. If we set $\alpha = 2/3$, then both of these bounds are equal to 9, and so we obtain the following theorem.

Theorem 8.3.3. *For the metric capacitated facility location problem with unsplittable demands, filtering, and rounding yields a 9-approximation algorithm.*

8.4 Resilient Facility Location

In many facility location problems, the supply should be fault-tolerant, that is, resistant to link and node failures. We described the situation where a client j is specified to be assigned to r_j facilities. The cost of assignment of this location is a weighted combination of these r_j assignments.

We define the fault-tolerant facility location problem as follows. We are given a graph $G = (V, E)$ with a distance function c defined on the edges, a set of possible facility locations $F \subseteq V$, and a set of client locations $D \subseteq V$ with certain demands. The cost of opening a facility at location i is f_i. Every client j must be connected to r_j open facilities, and let the weights corresponding to these assignments be $w_j^{(1)} \geq w_j^{(2)} \geq \cdots \geq w_j^{(r_j)}$.

Naturally this would ensure that the open facilities to which j is connected to would be ordered according to the (increasing) distance from j. The goal is to optimize the sum of the cost of open facilities and the weighted sum of the routing costs of each demand to the closest open facilities. We assume unit demands. This is no loss of generality, since general demands can be incorporated in the weights $w^{(r)}$.

This problem can be formulated as an integer program. Here, y_i is the variable signifying whether facility i is open or not, $x_{ij}^{(r)}$ contains one if client j is assigned to facility i, and facility i is the rth closest open facility to j and zero otherwise. The distance between i and j is c_{ij}. We have

$$\min \ \sum_i \sum_j \sum_r c_{ij} w_j^{(r)} x_{ij}^{(r)} + \sum_i f_i y_i,$$

$$\sum_i x_{ij}^{(r)} \geq 1, \forall j, r,$$

$$\sum_r x_{ij}^{(r)} \leq y_i, \forall i, j,$$

$$y_i \leq 1, \forall i,$$

$$x_{ij}^{(r)}, y_i \in \{0, 1\}, \forall i, j, r.$$

The relaxation will involve relaxing the last constraints to $0 \leq x_{ij}^{(r)}, y_i \leq 1$. The upper bound is only relevant for y_i and ensures that more than one facility is not built at a location. Define C^* to be the optimal fractional assignment cost and F^* to be the optimal fractional facility cost. That is,

$$\sum_i \sum_j \sum_r c_{ij} w_j^{(r)} x_{ij}^{(r)} = C^*$$

and

$$\sum_i f_i y_i = F^*,$$

where (x, y) denotes the optimal fractional solution of the above linear programming relaxation. The linear relaxation of the abovementioned integer program gives us a fractional solution. We will convert the solution (x, y) to a solution $(\bar{x}, y)$ such that the cost of the new solution does not increase, and the new solution has certain useful properties.

We will treat a client j as having r_j copies under the constraint that no two copies of any client are assigned to the same facility. In the fractional setting this reduces to the condition $\sum_r x_{ij}^{(r)} \leq yi \leq 1$. The converted solution will ensure that the set of facilities to which a copy

$j^{(r_1)}$ is fractionally assigned are closer to j than any facility to which the copy $j^{(r_2)}$ is assigned fractionally for $r_1 < r_2$. For every client j, we reassign it to facilities, fractionally, as follows. Order the facilities in nondecreasing distance from j, breaking ties arbitrarily. The ordering for a specific client j is fixed throughout the rest of the algorithm. The first client copy $j^{(1)}$ is assigned to the initial set of facilities whose fractional contributions sum up to 1. The last facility i in this set can be incompletely assigned, that is, $\bar{x}^{(1)} < y_i$. For the second copy, we start from this facility i, setting $\bar{x}_{ij}^{(2)} = y_i - \bar{x}_{ij}^{(1)}$. After that, we have $\sum_i \bar{x}_{ij}^{(2)} = 1$. We repeat this process for all the copies of the client.

Definition 8.4.1. Define $C_j^{(r)} = \sum_i x_{ij}^{(r)} c_{ij}$ and $C_j^{(r)}(\beta)$ to be the distance at which the rth copy of the demand point j assigns at least a fraction β of a facility. Then we have $\int_0^1 C_j^{(r)}(\beta) d\beta = C_j^{(r)}$.

The following is true by construction.

Proposition 8.4.1. *The cost of the solution does not increase, that is,*

$$\sum_{j,r} w_j^{(r)} C_j^{(r)} = C^*.$$

Proposition 8.4.2. *For any facility i and demand j, there exist at most two values of r such that $\bar{x}_{ij}^{(r)} > 0$. Further, if two such values exist they must be consecutive.*

Once the (fractional) facilities are fixed, it is simple to see that the above reassignment is (one of) the best possible. Intuitively, the copies of the demand j with larger weight w_j^r (and thus smaller r) are assigned to the closer open facilities.

The algorithm rounds the fractional solution in two phases. The algorithm uses the filtering technique of Lin and Vitter combined with reassignment of the fractional demands, such that each copy of the demand is assigned to a different facility. We treat the different copies of a demand as separate, and we denote the rth copy of client j by $j^{(r)}$. Fix the parameter $\alpha \in (0, 1)$.

In this section we will modify the fractional solution $(\bar{x}, y)$ to create a new solution $(\hat{x}, \hat{y})$, which we will round in the next phase. Let us fix a demand point j. We will perform the following operations for the copies $j^{(r)}$ in increasing order of $r = 1, 2, \ldots$. For every demand $j^{(r)}$, we consider the facilities to which it is fractionally assigned in increasing order of distance (the same ordering used in the previous Section 8.3). Let i be the first facility in the ordering of $j^{(r)}$ (therefore $x_{ij}^{(r)} > 0$) such that

$$\sum_{i':c_{i'j}<c_{ji},x_{i'j}^{(r)}>0} \hat{x}_{i'j}^{(r)} \geq 1 - \alpha.$$

For all i' appearing before i in our ordering, we set $\hat{x}_{i'j}^{(r)} = \bar{x}_{i'j}^{(r)}$. We set $\hat{x}_{ij}^{(r)}$ so that the total assignment of $j^{(r)}$ is exactly $1 - \alpha$. For all i' appearing after i in the ordering, we set $\hat{x}_{i'j}^{(r)} = 0$. We scale the $\hat{x}_{ij}^{(r)}$ by $1/(1-\alpha)$ so that $\sum_i \hat{x}_{ij}^{(r)} = 1$ for all $j^{(r)}$. Subsequently for all i we set $\hat{y}_i = \min\left\{\frac{y_i}{1-\alpha}, 1\right\}$.

Lemma 8.4.3. *If $\hat{x}_{ij}^{(r)} > 0$, then $c_{ij} \leq \frac{1}{\alpha}C_j^{(r)}$.*

We first show that $(\hat{x}, \hat{y})$ is feasible. For this, it is enough to show the following lemma.

Lemma 8.4.4. *For all i, j, we have $\sum_r \hat{x}_{ij}^{(r)} \leq \hat{y}_i$.*

Proof. Before filtering, by Proposition 8.4.2, we knew that at most two copies of a demand went to any one facility. We consider facility i and demand j. If exactly one copy, say r, is assigned to i, the inequality trivially holds, as $\hat{x}_{ij}^{(r)} \leq \hat{y}_i$.

We therefore assume that two copies of j are assigned to i. Let $j^{(r)}$ and $j^{(r+1)}$ be assigned to i. Note that by construction, i is the furthest assigned facility to $j^{(r)}$ and the closest to $j^{(r+1)}$.

The interesting case is when $y_i \geq 1 - \alpha$; otherwise $\sum_r \hat{x}_{ij}^{(r)} \leq y_i$ was true before scaling, and the lemma follows as we scale both the left- and the right-hand side by the same amount.

Let us look at the $\hat{x}_{ij}^{(r)}$ value before scaling (but after filtering). Therefore we need to show $\sum_r \hat{x}_{ij}^{(r)} \leq 1 - \alpha$. Then scaling cannot increase this value beyond 1. When we consider $j^{(r)}$ for filtering, we must set $\hat{x}_{ij}^{(r)} = \max\{0, \bar{x}_{ij}^{(r)} - \alpha\}$, as i is the furthest assigned facility to $j^{(r)}$. We now consider two cases.

Case 1: $\hat{x}_{ij}^{(r)} = 0$. Then, $\hat{x}_{ij}^{(r+1)} \leq 1 - \alpha$ because of filtering on $j^{(r+1)}$.
Case 2: $\hat{x}_{ij}^{(r)} = \bar{x}_{ij}^{(r)} - \alpha$. This implies $\hat{x}_{ij}^{(r)} + \hat{x}_{ij}^{(r+1)} = \bar{x}_{ij}^{(r)} + \bar{x}_{ij}^{(r+1)} - \alpha \leq 1 - \alpha$, as $\bar{x}_{ij}^{(r)} + \bar{x}_{ij}^{(r+1)} \leq y_i \leq 1$.

This completes the proof. □

Lemma 8.4.5. *Let $r_1 < r_2$. For any demand j, the furthest (from j) facility to which $j^{(r_1)}$ is assigned to (fractionally) is at a distance no greater than the closest (from j) facility to which $j^{(r_2)}$ is assigned to (fractionally) in the filtered and scaled solution.*

The next phase is the rounding of the fractional solution $(\hat{x}, \hat{y})$ from the previous phase. We describe this in steps.

Step 1: *Ordering the demands.* Arrange all copies of all clients in increasing order of the distance to the farthest fractional facility serving it. We will process the copies in this order and repeatedly apply steps 2–5. Note that copies of j will be picked in increasing order.

Step 2: *Choosing a facility.* Assume we consider $j^{(r)}$, the rth copy of the demand point j. Let the set of facilities serving it be $P_j^{(r)}$. We will build a facility at the cheapest facility i in $P_j^{(r)}$.

Step 3: *Merging facilities.* We now specify a set $\hat{P}$ of (fractional) facilities which will be closed down in exchange for the facility to be opened at i. In other words, we can view this set as a set of fractional facilities to be merged into i. The set will have the property that $\sum_{i' \in \hat{P}} \hat{y}_{i'} = 1$.

(a) We select facilities i' with $\hat{x}_{i'j}^{(r)} > 0$ starting with i (order does not matter) until the total fraction by which the selected facilities are open is at least 1. Let $Y = \sum_{i'} \hat{y}_{i'} \geq 1$ be the total fraction by which these facilities are open.

(b) If $Y > 1$ we will have to use the last selected facility, say i'', partially. Make two copies, i_1 and i_2, of facility i''. Set $\hat{y}_{i_2} = Y - 1$ and $\hat{y}_{i_1} = \hat{y}_{i''} - \hat{y}_{i_2}$. For any other demand $j'^{(r')}$ the assignment $\hat{x}_{i''j'}^{(r')}$ is distributed arbitrarily between the two facility copies i_1 and i_2, maintaining $\sum_r \hat{x}_{i'j}^{(r)} \leq \hat{y}_{i'}$ for both $i' = i_1$ and $i' = i_2$. The facility (copy) i_1 is selected and i_2 is not. Denote the set of picked facilities by $\hat{P}^1$.

We open a facility completely at i and close the rest of the facilities in the set $\hat{P}$.

Step 4: *Assignment of clients.* For any demand j' (inclusive of j), consider its copies $r_1, r_2, \ldots r_k$ served at least fractionally by $\hat{P}$. If $\hat{P}$ serves any copy of j' fractionally, we assign the smallest numbered copy (r_1) of j' to be completely served by i. Note that the assignment distance for $j'^{(r_1)}$ has at most tripled as compared to $C_{j'}^{(r)}(1 - \alpha)$.

Step 5: *Uncrossing neighborhoods.* We now reassign the remaining copies of j' (i.e., $j'^{(r_2)}, \ldots, j'^{(r_k)}$) completely to facilities outside the set $\hat{P}$ by performing an uncrossing step. For j', we compute $X_{j'}^{(1)} = \sum_{i'} \hat{P} \hat{x}_{i'j'}^{(r_1)}$ and $X_{j'}^{(2)}, \ldots, X_{j'}^{(k)}$ likewise. These quantities denote the fractions to which the copies of j' are assigned to the facilities in $\hat{P}$. Define $Y_{j'}^{(1)} = \sum_{i' \notin \hat{P}} \hat{x}_{i'j'}^{(r_1)} = 1 - X_{j'}^{(1)}$ and similarly $Y_{j'}^{(2)}, \ldots, Y_{j'}^{(k)}$. These quantities denote the fractions by which the copies of j' are assigned to facilities outside the set $\hat{P}$, respectively. For any j' which is fractionally assigned to the facilities in set $\hat{P}$, we have

$$X_{j'}^{(t)} + Y_{j'}^{(t)} = 1 \text{ for all } 1 \leq t \leq k,$$

$$\sum_t X_{j'}^{(t)} \leq \sum_{i' \in \hat{P}} \hat{y}_{i'} = 1.$$

We have assigned the copy $j'^{(r_1)}$ to i. But in this process it may be that $X_{j'}^{(r')} > 0$, that is, $\hat{P}$ serves some other copy $j'^{(r')}$ of j'. If we use the fractional facility of $\hat{P}$ (which amounts to 1), then we need to ensure that the copy $j'^{(r')}$ gets assigned (fractionally) to facilities outside $\hat{P}$; and the fraction is $X_{j'}^{(r')}$. It follows that

$$X_{j'}^{(r')} + X_{j'}^{(1)} \leq 1 = X_{j'}^{(1)} + Y_{j'}^{(1)}.$$

We consider the fraction $Y_{j'}^{(1)}$ by which the copy $j'^{(r_1)}$ was assigned to facilities not in $\hat{P}$, and we reassign this to the other copies of j' that were originally assigned to the set $\hat{P}$ as follows. Consider the fraction by which $j'^{(r_1)}$ was assigned to the facility closest to j but not in $\hat{P}$. We assign this fraction to $j_0^{(r_2)}$ until either $j_0^{(r_2)}$ is completely satisfied or we have assigned the fraction completely. In the former case, we move to $j'^{(r_3)}$; in the latter case, we consider the next closest facility not in $\hat{P}$ that was previously connected to $j'^{(r_1)}$, and repeat. During uncrossing, we maintain the invariants $\sum_{i'} \hat{x}_{i'j'}^{(r_t)} = 1$ and $\sum_t \hat{x}_{i'j'}^{(r_t)} \leq \hat{y}_{i'}$ for all $1 \leq t \leq k$.

At the end of one iteration of steps 2–5, we have opened facility i completely. For every demand fractionally assigned to the set $\hat{P}$, the smallest assigned copy is completely assigned to i. Every other copy is fractionally reassigned completely outside set $\hat{P}$. We drop the set $\hat{P}$ and the copies $j'^{(r_1)}$ from further consideration. Setting $\alpha = \frac{3}{4}$ we have the following theorem.

Theorem 8.4.6. *A fault-tolerant facility location has a 4-approximation in polynomial time.*

8.5 One-Dimensional Binpacking

A common problem is job assignment or scheduling, where we are given a number of jobs and a number of hosts. Any host can be assigned for a job, incurring some cost and/or profit. The hosts also have some cost budget, or alternatively, size limitation, that must be obeyed. We are interested to find a feasible assignment minimizing the cost (or maximizing the benefits).

The binpacking problem is one of the best-known and widely studied $\mathcal{NP}$-hard problems. In its basic form, we have a list $L = (a_1, a_2, \ldots, a_n)$ of real numbers in $(0, 1]$, and we want to place the elements of L into a minimum number L^* of "bins" so that no bin contains numbers whose sum exceeds 1. Motivated by the need to make fast scheduling decisions of jobs online, we will look at some simple heuristic algorithms and their performance; see for example Johnson et al. [76].

Algorithm 8.5.1 (First-fit (FF)). Let the bins be indexed as $B_1, B_2, \ldots$, with each initially filled to level zero. The numbers $a_1, a_2, \ldots, a_n$ will be placed in that order. To place a_i, find the lowest j such that B_j is filled to level $\beta \le 1 - a_i$ and place a_i in B_j. B_j is now filled to level $\beta + a_i$. •

Algorithm 8.5.2 (Best-fit (BF)). Let the bins be indexed as $B_1, B_2, \ldots$, with each initially filled to level zero. The numbers $a_1, a_2, \ldots, a_n$ will be placed in that order. To place a_i, find the lowest j such that B_j is filled to level $\beta \le 1 - a_i$ and β is as large as possible and place a_i in B_j. B_j is now filled to level $\beta + a_i$. •

Algorithm 8.5.3 (First-fit decreasing (FFD)). Arrange $L = (a_l, a_2, \ldots, a_n)$ in nonincreasing order and apply first-fit to the list. •

Algorithm 8.5.4 (Best-fit decreasing (BFD)). Arrange $L = (a_l, a_2, \ldots, a_n)$ in nonincreasing order and apply best-fit to the list. •

Denote by $FF(L)$, $BF(L)$, $FFD(L)$, and $BFD(L)$ the number of bins used by each of the four algorithms to the list L. The performance measure we are interested in is the ratio of the number of bins used by a particular algorithm on L to the optimal number of bins L^*. We use $R_{FF}(k)$ to denote the maximum value achieved by the ratio $FF(L)/L^*$ over all lists with $L^* = k$, with similarly defined ratios $R_{BF}(k)$, $R_{FFD}(k)$, and $R_{BFD}(k)$ for the other algorithms.

The algorithms have the following performance ratios:

(1)

$$FF(L) \le \frac{17}{10}k + 2, \tag{8.19}$$

$$\lim_{k \to \infty} R_{FF}(k) = \frac{17}{10}, \tag{8.20}$$

(2)

$$BF(L) \le \frac{17}{10}k + 2, \tag{8.21}$$

$$\lim_{k \to \infty} R_{BF}(k) = \frac{17}{10}, \tag{8.22}$$

(3)

$$FFD(L) \le \frac{11}{9}k + 4,$$

$$\lim_{k \to \infty} R_{FFD}(k) = \frac{11}{9},$$

(4)

$$\mathrm{BFD}(L) \leq \frac{11}{9}k + 4,$$

$$\lim_{k \to \infty} R_{\mathrm{BFD}}(k) = \frac{11}{9}.$$

All these ratios are achieved for small values of k, so that these asymptotic results actually reflect the performance for essentially all values of k. In addition, similar results are obtained for certain restricted lists L.

Both FFD and BFD can be implemented in $O(n \log n)$ time. It has been reported that, on average, BFD performs slightly better than FFD [77]. It should be pointed out that the performance bounds are valid only for one-dimensional knapsack problems, and for the decreasing versions, when all items are known in advance.

8.6 Multi-Dimensional Resource Allocation

In data centers, we may be interested in how to assign virtual machines (or jobs) to hosts so that the equipment utilization is as high as possible. This can be interpreted as a knapsack problem, for which we can use some approximation algorithms and their bounds. However, with multiple resource types, the problem is much harder.

By resource allocation in a cloud we mean the allocation of virtual machines to physical nodes, or hosts, where the hosts are characterized by the number of processors and the amount of processing capacity and RAM (and possibly other parameters). The scheduling of tasks is not considered part of the project, but it is desirable that the framework should be possible to extend to this situation, if desired.

An important step in defining an algorithm for the resource allocation problem is defining an objective of the optimization. We can identify two different objectives of the customers and the cloud operator. In a simplified setup, we may here assume that for the customer, the cost is fixed, so we try to maximize the utility of resources. For the operator, the total amount of resources is fixed, and the potential for adding more customer requests to the cloud depends on the assignment of the existing requests.

We consider an IT cloud consisting of physical resources. The cloud is managed by the cloud operator, who fully controls the resources and how requests are assigned. Requests for resources are specifically virtual machines (VMs), and the requesting entity (not necessarily human) is called a customer. The customer may have preferences where its request should be allocated. The final resource allocation, however, is determined by the cloud operator. We wish to optimize the assignment by means of the ant colony algorithm, as described in Chapter 5. In the context of the algorithm, the customer is also referred to as an *ant*.

The physical resources are residing in several levels. The lowest unit is referred to as a host. Hosts are aggregated into clusters – typically residing in the same location – and several clusters form a cloud. We do not consider geographically distributed clusters, so a cluster and a cloud are essentially the same.

Resource allocation in clouds can be divided into different categories or steps. Mills et al. [78] evaluate different allocation heuristics. They categorize the optimization type based on *initial placement*, where new requests are allocated subject to available resources, *consolidation*, where a new request may modify existing allocation to achieve lower cost, and *trade off* between service level agreements (SLA) and cost. The cost/demand structure is also categorized into *reservations*, where the customer pays a fixed price for a service running for a specified amount of time, *on-demand access*, where customers place requests and the cost is dependent on utilization, and *spot market*, where the price of a service also depends on demand.

Cloud Resources and Descriptors

Both the cloud resources and the requests are assumed to be described in sufficient detail so that an assignment can be made, and this information is available to the algorithm at all times. The resources are quantifications of available physical properties, such as number of processors in the CPU core, CPU speed, amount of RAM, amount of disk storage space, and network bandwidth.

The idea of using a constantly updated database with a resource vector for each node [74] has been adopted as a general and technically appealing solutions to manage cloud state data. The network is clearly dynamic, so rather than allocating according to the physical resources of a node, it should be done with respect the instantaneously available *free* resources of a node. The result of the optimization is an assignment of VM–node pairs.

Some of the resources are static, such as the CPU core, and can be included as side constraints (or rather, an infeasible assignment results in a zero probability). Other properties are dynamic and change with each assignment. Also here, we can distinguish between resources that set a definite limitation on the service capability, such as the amount of memory available. Should the VM requirements exceed a node's capabilities, the node is considered to not be able to host that VM. The second type of resource – for example processing power or network bandwidth – scales gradually with the number of VMs. Service quality can then be seen as the expected average processing time (or throughput).

We may assume that a virtual machine can be specified in the same terms as a physical node. For simplicity, we let the resource vector be the triple (core, cpu, memory), as described in [74]. Thus, a VM requirement can be directly compared to a node's available resources and allocated to any physical node having sufficient resources.

We assume that a host can have a number of VMs, and this number is limited by the aggregate requirements on the node resources. We also assume that a VM occupies the resources it specifies [72]. Should a node not have a sufficient amount of free resources, the assignment is infeasible and will be disregarded.

Optimization Criteria

In the scenario under consideration – initial placement of VM requests – customers would only benefit from trying to gain a service quality as good as possible, since the price is fixed for a specified time. The cloud operator, on the other hand, may save cost by assigning resources optimally.

The optimization goal is to find in some respect the best trade-off between cost and quality. A multioptimal approach seems infeasible due to the hardness of the problem. Solutions optimal with respect to different criteria will tend to be vastly different, and there is no way to find a trade-off by interpolation due to the discrete nature of resource assignment. A second question is how to define service quality from a system perspective. With a central function for resource allocation, requests are assigned one by one, and simple heuristics would give no guarantee of fairness in service quality. In this chapter, the minimum level of quality is therefore determined by the SLA of each request.

In ant colony optimization (ACO), the ants' movements are governed by target probabilities that are a product of two parts. The first is an assignment probability that is proportional to the *attractiveness* of a match from the customer point of view (called *visibility* in [45]), and the second is a memory of the best past assignments represented by the fictitious pheromone trail. As long as the SLA of a request can be fulfilled, the attractiveness is nonzero, otherwise it is zero.

The probability of transition to another (including itself) node is

$$p_{ij} = \frac{(\tau_{ij}(t))^{\alpha} (\eta_{ij}(t))^{\beta}}{\sum (\tau_{ij}(t))^{\alpha} ((\eta_{ij}(t))^{\beta}}, \tag{8.23}$$

evaluated for feasible assignments; otherwise $p_{ij} = 0$. The pheromone $\tau_{ij}(t)$ and the attractiveness $\eta_{ij}(t)$ are time-dependent, which is indicated by the argument in t. The first property changes with each cycle, and the second changes with each move within a cycle.

From a global optimization perspective, the system lets the customers find an assignment according to their preferences and the given constraints and then select the best assignment out of a number N of trials.

Attractiveness

From the customers' point of view, it is reasonable that they will try to maximize their own benefit at each step. Assuming that the service comes at a fixed price per VM configuration, the customer would try to maximize the service quality accordingly. This is likely represented by response time, that is, the sum of CPU processing time and network transmission delay. The transmission delay is not considered here, as it depends mainly on the infrastructure outside the cloud.

The attractiveness refers to the property of the algorithm on which the probability of choosing a server is based. The free capacity measure is used to describe this property based on the principle that the more free the capacity in a server, the more "attractive" it is for the customer to be allocated to.

The CPU capacity is typically measured in millions of instructions per second (MIPS). The effective processing power available to applications is dependent on system configuration and simultaneously running processes, etc. In [74], the authors propose measuring the available CPU and RAM capacity by performing a matrix inversion operation and measuring the execution time. This method would give an accurate instantaneous measurement on which the attractiveness can be based. For this discussion, however, it is sufficient to assume that this information is available.

Thus, we assume that a customer selects the server based on the available processing power of the CPU (the total processing power adjusted for system processes and other VMs using it). The customer then sees the available processing power as a resource potentially available to itself.

Cost

Cost can be defined in terms of idle capacity, that is, unoccupied capacity that cannot be assigned to another VM due to limitations in some other resource type. The cost will depend on the applications, or in other words, the distribution of demands of arriving requests.

The cost for a cloud operator can be expressed in the degree of infrastructure utilization, or equivalently, return on investments. The operator wishes to allocate requests to resources in a "best-fit" manner, so that not more resources than necessary are occupied by an allocated request.

The greedy principle from the cloud operator's perspective is that the more VMs can be allocated, the higher the utilization and the return on investment. As a metric for system efficiency, the *energy* of the relative free resources is used here. The optimization then follows the principle of minimum energy. The system energy is defined as

$$E = \sum_{i=1}^{n} (C_i - \sum_{j=1}^{v_i} r_{ij})^2, \tag{8.24}$$

where C_i is the capacity of the server and r_{ij} is the VM capacity requirement of VM j on host i. The total requirement sums over the v_i VMs allocated to host i. Rescaling Eq. (8.24) gives

$$E = \sum_{i=1}^{n} (1 - \sum_{j=1}^{v_i} r_{ij}/C_i)^2, \tag{8.25}$$

which is the objective function that we will minimize. Under this metric, it is more efficient to allocate available resources occupying the best fit between request and resource. The best fit is when the resource matches the VM specification exactly. Then, the energy of the match is zero.

Algorithm for Resource Optimization

The algorithm is an adaptation of the ACO algorithm for solving the traveling salesman problem (TSP), described in [45]. The assignment problem is modeled as a complete graph on the set n of nodes. Initially, the ants are distributed between the nodes in a round-robin fashion. They could also originate from a source node (a "nest"), but this is not necessary, as the algorithm only performs a single iteration in each cycle. The ants could also be distributed randomly, which would affect the order of the assignment. In the example below, however, this has little or no effect.

The ants move according to a matrix of transition probabilities, where self-loops are allowed, so that an ant may request its job to be assigned to the node it originally occupies. The probabilities are proportional to the attractiveness of the target and the pheromone level of the edge between the origin and the target. The algorithm therefore has to keep track of the resource requirements of each ant and the instantaneous amount of free resources at each node.

The system state changes with the assignment of a new job (the property of the ant or customer). Therefore, after each move, the transition probabilities change and must be recalculated. The attractiveness of a server to a given customer decreases when resources are assigned to another customer. As a measure of attractiveness, the (possibly scaled) available CPU processing power of the host is used.

The objective here is to find an assignment. Since the constraints ensure that all allowed assignments are feasible, the algorithm only runs for one iteration, where each ant is moved (including possibly back to its origin) exactly once. The number of cycles will have to be fairly large though, in order to find an optimum. After an iteration, a candidate assignment has been generated. The algorithm has to keep track of the so-far best assignment throughout the cycles.

We introduce constraints forbidding an ant to attain infeasible solutions by a *taboo list*. In the present case, the taboo list simply consists of the nodes that with remaining resources cannot accommodate the ant, that is, the VM. This includes the host it is initially starting from.

Next, the system cost c_N is calculated according to Eq. (8.24). This energy can be a composite measure including other resources, such as RAM, as suggested in [74]. For the purpose of describing the algorithm, however, the energy is only based on the CPU processing power. The deposited amount of pheromone, $\Delta\tau$, on each edge is dependent on system cost. This quantity is given by

$$\Delta\tau = Q/c_k,$$

where Q is a scaling constant and c_k the cost in cycle $k \in \{1, 2, \ldots, N\}$. Since c_k can be zero, a maximum limit on $\Delta\tau$ is set to one. This limit is rather arbitrary, and it is an additional system parameter that may affect the convergence properties of the algorithm.

Since the attractiveness changes dynamically throughout the algorithm, the transition probabilities are given by two matrices: the attractivity matrix A, which changes throughout an iteration but is reset for each cycle, and the pheromone level matrix P, which remains constant throughout a cycle, but is updated after each cycle.

The cost is used to update matrix P, first by multiplying all previous pheromone levels p_{ij} by the evaporation constant $(1 - \rho)$ and then by adding $\Delta\tau$ onto edges describing assignments made in the iteration.

The minimum cost $c_{\min}$ and the corresponding assignment obtained so far is recorded after each cycle, and the matrix A and the vectors of free-node resources and assignments are restored to their initial values, corresponding to not yet assigned VMs.

Algorithm 8.6.1 (Resource allocation).

Given matrices of server capabilities S and VM requirements V.

STEP 0: (initialize)
Let J be a list of initial node assignments and set algorithm parameters α, β, τ_0, ρ, and K, the number of cycles. Set the matrix of free resources to $A = S$ and the matrix of pheromone concentrations $P = (\tau_0)$, the matrix where all entries equal τ_0. Set $c_{\min} = \infty$

STEP $k = \{1, 2, \ldots, N\}$:

    while $k < K$ (the number of cycles) **do**:

        Randomly select a node i and a customer request,

        Divide a customer request into tasks (ants). For each ant j:

            Calculate the probabilities p_{ij} (Eq. (8.23)) based on A and P,

            By simulation, select a move of ant j, assign to the selected

            target node if resources are available; assign resources, update A.

        When all ants have been moved once:

            Calculate the cost c_k (defined by the energy in Eq. (8.24)) of the assignment

            and update matrix P by $\Delta \tau$

            If $c_{\min} < c_k$, let $c_{\min} = c_k$ and $J_{\min} = J_k$. Reset J, $A = S$,

            and let $P = (1 - \rho)P$

        end

        Delete customer request

        end

    end

Output $c_{\min}$ and $J_{\min}$, the optimal assignment.　　　　　　　　　　　　　•

It should be noted that the algorithm assigns VMs all at once, so for the CloudSim experiment described below, the assignment rule has to be formulated so that VMs can be assigned sequentially. Also, the result of the algorithm depends on the random number generator, and so to find an optimum the algorithm may have to be run several times, possibly with different seeds. The memory induced by the pheromone trails may therefore take many runs to change from a suboptimal assignment to an optimal one.

8.7 An Example

To test the algorithm, a small cluster with hosts similar to the one described in [74] (Section 6, Experiments, Tables 5 and 6) is used. The simplicity of this scenario with five servers having different characteristics and a single type of VM makes a manual comparison with other assignment schemes straightforward. To evaluate different assignment strategies, it is a good idea to have hosts with different characteristics but identical VMs. That shows clearly how they are assigned by different strategies. The algorithm as such could easily be extended to larger and more general cases. Like in [45], the number of ants (VMs) is set equal to the number of nodes. The properties of the host servers in the cluster are listed in Table 8.1, and those of the VMs in Table 8.2.

To compare the algorithm with other assignment schemes, we compare the result with the round-robin and with a customer greedy heuristic scheme. In the round-robin scheme, the

Table 8.1: Cluster specification: MIPS and RAM capacities.

Host ID	Core	MIPS	RAM
0	1	1000	2048
1	2	500	2048
2	2	300	2048
3	1	2000	2048
4	2	300	2048

Table 8.2: Virtual machine specification: requirements on MIPS and RAM.

VM ID	Core	MIPS	RAM
0-4	1	300	512

Table 8.3: Efficiency of round-robin assignment.

Host ID	No. VMs	Free cap. w/o PEs	Free cap. with PEs
0	1	0.7	0.7
1	1	0.4	0.7
2	1	0.0	0.5
3	1	0.85	0.85
4	1	0.0	0.5
Energy		1.37	2.20

Table 8.4: Efficiency of greedy assignment.

Host ID	No. VMs	Free cap. w/o PEs	Free cap. with PEs
0	1	0.7	0.7
1	0	1.0	1.0
2	0	1.0	1.0
3	4	0.4	0.4
4	0	1.0	1.0
Energy		3.65	3.65

VMs are simply distributed with one at each node, and the relative free capacity is shown in Table 8.3. In Tables 8.3, 8.4, and 8.5 the entries for each host are the percentage of free capacity, calculated as $1-$(occupied capacity)/(total host capacity). The cost as defined in Eq. (8.24), that is, the sum of the squared entries, is 1.3725. Taking the number of processors into account, the energy is 2.2025.

The round-robin and the greedy algorithms are deterministic, whereas the ACO algorithm is random. The algorithm may therefore give a different result at each run. This depends on the seed used in the random number generator.

By letting each customer choose the server according to the largest amount of available processing capacity, the assignment is as shown in Table 8.4. The cost in this case is 3.65. The same value is achieved when taking the number of processors into account, since there is one

Table 8.5: Efficiency of AOC assignment.

Host ID	No. VMs	Free cap. w/o PEs	Free cap. with PEs
0	2	0.4	0.4
1	1	0.4	0.7
2	1	0.0	0.5
3	0	1.0	1.0
4	1	0.0	0.5
Energy		1.32	2.15

VM per host. The greedy scheme is essentially what would be expected from a single iteration of the algorithm.

The algorithm applied to the same problem gave the assignment shown in Table 8.5. It should be noted that lower-capacity hosts (1, 2, and 4) are assigned VMs, but not node 3. The minimum energy obtained is 1.32. After having reached the minimum energy, the algorithm was run for up to $N = 10,000$ without showing any further improvement. Taking the number of processors into account, the energy is 2.15 for this policy.

The parameter values used are $\alpha = 0.5$, $\beta = 0.5$, $\rho = 0.1$, and $\tau_0 = 0.1$. The cut-off limit for the inverse of the cost is set (rather arbitrarily) to unity. Elaborating on the system parameters affects the convergence of the algorithm noticably.

CloudSim Implementations

In the CloudSim experiment the goal was to keep things as simple as possible apart from the hosts and VMs. Only one user, one data center, and one broker were therefore inserted. The VMs represent the ants and the cloudlets jobs assigned to the VMs. The implementation uses 10 cloudlets, like in the code in [73]. These can of course be chosen differently, but the cloudlets are basically just a test to see whether the cloud works.

The CPU capacity is measured in MIPS in CloudSim – per processor (Pe). Thus, MIPS and RAM are properties used of the hosts in the data center (bandwidth and storage are not used). It does not really matter for the algorithm which unit is used, GHz or MIPS, so they are treated as exchangeable and give the same results. Host IDs are set manually, so these are defined according to Table 8.1.

The three assignment strategies were implemented in CloudSim. The cloud was defined as described in Tables 8.1 and 8.2. The round-robin assignment was implemented by [73] and has been used for comparison. The simulation uses ten cloudlets of equal small sizes to illustrate that the clouds with the given assignment policies work properly. The round-robin assignment is shown in Table 8.3.

For the greedy algorithm, the assignment policy is implemented so that each VM is assigned the host with the largest available million instructions per second (MIPS, a measure of CPU capacity), possibly after some VM already has been assigned to the host. The assignment is in agreement with Table 8.4.

The optimal assignment is implemented so that VMs can be assigned sequentially. This is necessary, because the algorithm described makes a repeated assignment of all VMs at once, whereas in CloudSim, VMs are assigned on a first-come first-served basis. Therefore, we need to reformulate the optimal result of the algorithm as a policy. In order to do so, we may use dynamic programming.

Consider the energy equation (8.25). This is the cost we wish to minimize by assignments of VMs to hosts. We also have the rather obvious restrictions on the problem

$$\sum_{j=1}^{v_i} r_{ij} \leq C_i,$$

$$r_{ij} \geq 0,$$

for all i and j. The dynamic program can be written

$$v_k(y) = \min\{v_{k-1}(y), v_{k-1}(y) + (1 - r_k/C_k)^2\},$$

where $y = \sum_{i=1}^{n}((1 - \sum_{j=1}^{v_i} r_{ij}/C_i)^2)$ is the relative free capacity at each instant and r_k is a new VM to be allocated to a host with capacity C_k in step k. The dynamic programming formulation of the problem is the base for the implementation of the policy in CloudSim, since it is sequential as compared to the algorithm which is parallel. Thus, the energy equation (8.25) is therefore implemented only implicitly in the simulation.

The algorithm described assigned VMs so that the largest decrease in energy occurs for each VM assignment to a host. For example, the decrease is much greater when the free capacity decreases from 1 to 0.5 than when the capacity decreases from 0.5 to 0. Thus, heuristically (and knowing the capacities of the hosts), if the proportion of available MIPS goes below the threshold (50%), the second-best match is used.

The policy aims at minimizing unutilized capacity in a host. It also aims at distributing VMs so that the load is less than unity. Since the energy in Eq. (8.25) is a sum of squares, this leads to a minimum energy assignment. The result of the simulation is in agreement with Table 8.5.

The assignment is in practice made as follows. For the first VM, the host with the lowest capacity able to accommodate the VM is found, and an assignment is made. In this case, the free capacity decreases from 1 to 0.5, or the energy contribution (which is the squared value) from 1 to 0.25. The second VM could be assigned to the same host. Then the energy contribution

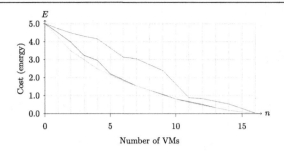

Figure 8.3: Comparison of the efficiency of the algorithms; the greedy (uses the most energy), round-robin (middle line), and AOC (the most efficient).

would decrease from 0.25 to 0. But assigning it to another host with the same capacity would give an energy decrease from 1 to 0.25 as well, which lowers the energy sum more than the decrease from 0.25 to 0, which is achieved when the VM is assigned to the first host.

Continuing in this way, we may find a situation where the decrease in energy is larger when assigning a VM to a host which already has a VM assigned, rather than assigning it to another host with larger free capacity. So for a host with ID 0, the decrease in energy is larger by assigning two VMs to this host, as compared to assigning one VM to host 0 and one VM to host 3.

The cost in the implementation in CloudSim therefore looks for the host for which the VM allocation gives the largest decrease in energy. This is done in the CloudSim policy as an optimization parameter to find the best match.

Fig. 8.3 shows the energy for each of the three assignment strategies for an increasing number of standard-size VMs in Table 8.2. The algorithm described here has lower energy than the other two, although the round-robin strategy is close to optimal.

8.8 Optimal Job Scheduling

We consider the problem of scheduling jobs with time constraints, as proposed by Chuzhoy et al. [71]. Given a set $J = \{1, \ldots, n\}$ of jobs, for each job $j \in J$, there is a set $I(j)$ of time intervals, called job intervals, during which they must be scheduled. Scheduling a job j means choosing one of its associated time intervals from $\mathcal{I}(j)$. The goal is to schedule all jobs so that the number of machines is minimized, and so that no two jobs assigned to the same machine overlap in time. This implies that the maximum number of chosen job intervals at any point of time must not exceed the number of machines.

There are two variations of the problem: discrete and continuous job times. In the discrete version, the job intervals $I(j)$ are given explicitly. In the continuous version, each job j has a release date r_j, a deadline d_j, and a processing time p_j. The time interval $[r_j, d_j]$ is called the *job window*. The set of job intervals $I(j)$ is generated by these parameters, being all time intervals of length p_j that are contained inside the window $[r_j, d_j]$.

Here we consider the continuous version of the machine minimization problem. The linear programming formulation can be strengthened by adding constraints that explicitly forbid certain configurations which cannot arise in any integral solution. We deploy a rounding scheme that allows us to transform the resulting fractional schedule into an integral solution by using a constant number of machines. This provides an approximation algorithm that achieves a solution of cost $O(\text{OPT})$.

Extending the idea of forbidden configurations to instances where an optimal schedule itself requires multiple machines is technically difficult, since the configurations that need to be forbidden have a complex nested structure and we need to discover them by recursively solving linear programs on smaller instances. Specifically, the strengthened linear programming solution for a given time interval is computed via a dynamic program that uses a linear programming subroutine to compose recursively computed solutions on smaller time intervals. We believe this novel idea for strengthening a linear programming relaxation is of independent interest.

Let L denote the set of all the left endpoints of the job intervals. For each job $j \in J$, for each job interval $I \in \mathcal{I}(j)$, we define a variable $x(I, j)$ indicating whether j is scheduled on interval I. Our constraints guarantee that every job is scheduled and that the number of jobs scheduled at each point of time does not exceed the number of available machines. The linear programming formulation is as follows:

$$\min \quad z \tag{8.26}$$

$$\sum_{I \in \mathcal{I}(j)} x(I, j) = 1, \quad \text{for all } j \in J, \tag{8.27}$$

$$\sum_{j \in J} \sum_{I \in \mathcal{I}(j): t \in I} x(I, j) \leq z, \quad \text{for all } t \in L, \tag{8.28}$$

$$x(I, j) \geq 0, \quad \text{for all } j \in J, I \in \mathcal{I}(j). \tag{8.29}$$

The number of machines we need for the fractional schedule is $k = \lceil z \rceil$. Consider the case where the input is continuous and OPT $= 1$. Let $j \in J$ be any job and let $I \in \mathcal{I}(j)$ be one of its intervals (see Fig. 8.4). Suppose there is some other job $j' \in J$, whose time window is completely contained in I, that is, $[r_{j'}, d_{j'}] \subseteq I$. Since the optimal solution can use only one machine and all the jobs are scheduled, job j cannot be scheduled on interval I in the optimal solution. We call such an interval I a forbidden interval of job j. All the other job intervals of job j are called allowed intervals.

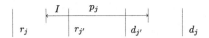

Figure 8.4: Time instants defining the job window.

In the linear program, we can either set a priori the values $x(I, j)$, where I is a forbidden interval of job j, to 0, or, alternatively, add the following set of valid inequalities:

$$x(I, j) + \sum_{I' \in \mathcal{I}(j'):I' \subseteq I} x(I', j') \leq 1, \quad \text{for all } j, j' \in J, I \in \mathcal{I}(j). \tag{8.30}$$

Note that if I is a forbidden interval of job j, then for some job j', $\sum_{I' \in \mathcal{I}(j'):I' \subseteq I} x(I', j') = 1$, and thus the value of $x(I, j)$ is going to be 0 in the linear program solution.

We now turn our attention to the scenario where the optimal solution uses multiple machines. Obviously, inequality (8.30) is not valid anymore. Indeed, suppose interval I is a forbidden interval of some job j, $I \in \mathcal{I}(j)$, and let j' be some job whose window is contained in I. Suppose now the optimal solution uses two machines. Then, job j can be scheduled on interval I on one machine and job j' can be scheduled inside its window on the other machine, thus (8.30) does not hold anymore. Now let T be any time interval containing the window of job $j' \in J$. We know that we need at least one machine to accommodate jobs whose windows are contained in T. Therefore, we can schedule at most one job on an interval that contains T. So for the case of two machines, we can add the corresponding inequality for every interval T that contains some job window. This idea extends to any arbitrary number of machines.

For each time interval T (not necessarily belonging to any job), we define a function $m(T)$ that is, intuitively, a lower bound on the number of machines needed to accommodate all the jobs whose windows are contained in T. We compute the value of $m(T)$ recursively by the means of dynamic programming, going from smallest to largest intervals.

For intervals T of length 0, set $m(T) = 0$. Given a time interval T, let $J(T)$ be the set of jobs whose time window is completely contained in T. The value of $m(T)$ is defined to be $\lceil z \rceil$, where z is the optimal solution to the linear program, which is defined as follows:

$$\min \quad z \tag{8.31}$$

$$\sum_{I \in \mathcal{I}(j)} x(I, j) = 1, \quad \text{for all } j \in J(T), \tag{8.32}$$

$$\sum_{j \in J(T), I \in \mathcal{I}(j):t \in I} x(I, j) \leq z, \quad \text{for all } t \in T, \tag{8.33}$$

$$\sum_{j \in J(T), I \in \mathcal{I}(j): T' \subseteq I} x(I, j) \leq z - m(T'), \quad \text{for all } T' \subset T, \tag{8.34}$$

$$x(I, j) \geq 0, \quad \text{for all } j \in J(T), I \in \mathcal{I}(j). \tag{8.35}$$

The first two sets of constraints are similar to those of (8.26)–(8.29), except they are applied to the time interval T and subset $J(T)$ of jobs. The third set of constraints models constraint (8.30) for the case of multiple machines. Suppose we are given some interval $T' \subset T$. As T' is smaller than T, we know the value of $m(T')$ from the dynamic programming table, and this value is a lower bound on the number of machines needed to accommodate jobs whose windows are contained in T'. Therefore, we have at most $z - m(T')$ machines available for scheduling jobs on intervals that contain T'. The third set of constraints ensures that the total number of jobs scheduled on intervals which contain T' does not exceed $z - m(T')$.

Note that constraints (8.34) can be omitted, since they are a special case of constraints (8.35), for intervals T' of length 0.

Rounding

In this section we show that, given a fractional solution to (8.32)–(8.29) that uses $k = m(T)$ machines, we can find an integral solution using at most $O(k^2)$ machines. The rounding will proceed iteratively: at each step we will identify a subset of jobs that can be scheduled on $O(k)$ machines, such that the remaining jobs will have a feasible fractional solution on at most $k - 1$ machines. Thus, all jobs will get scheduled on $O(k^2)$ machines.

Suppose we are given a solution to the linear program (8.32)–(8.29) for some time interval T and let $\mathcal{T}$ be some collection of disjoint subintervals of T, such that for each $T' \in \mathcal{T}$, $m(T') < m(T)$. We partition the set $J(T)$ of jobs into two subsets, J' and J'', where $j \in J''$ iff its window is completely contained in one of the intervals of $\mathcal{T}$ and $j \in J'$ otherwise. We say that $\mathcal{T}$ is good with respect to the LP-solution if for each job $j \in J'$ and each interval $I \in \mathcal{I}(j)$ where $x(I, j) > 0$, I overlaps with at most two intervals belonging to $\mathcal{T}$.

We will show that if the optimal solution cost of (8.32)–(8.29) is z and $\mathcal{T}$ is good with respect to the solution, then we can schedule the jobs J' on at most $O(k)$ machines where $k = \lceil z \rceil$. Before we formalize this argument, we define the *partition* algorithm, which receives as input an interval T, a set $J(T)$ of jobs, and a solution to (8.32)–(8.29), and it produces a collection $\mathcal{T}$ of subintervals of T which is good with respect to the linear program solution.

Algorithm 8.8.1 (Partition).

Input: Time interval T, set of jobs $J(T)$ whose windows are contained in T, and a solution to (8.32)–(8.29). Start with $\mathcal{T} = \emptyset$ and set t to be the left endpoint of T.

STEP 1: N

  while there are jobs $j \in J(T)$ such that the right endpoint of some interval
  $I \in \mathcal{I}(j)$ lies to the right of t and $x(I, j) > 0$ **do:**
   if no job j exists such that one of its intervals
   $I \in \mathcal{I}(j)$ contains t and $x(I, j) > 0$ **then:**
    move t to the right till the above condition holds.
   Among all the job intervals I that contain time point t such that
   for some $j \in J$, $I \in \mathcal{I}(j)$ and $x(I, j) > 0$,
   choose the interval with rightmost right endpoint and denote this endpoint by t'
   Add time interval $[t, t']$ to T and set $t \leftarrow t'$
   Let $J'' \subset J(T)$ denote the set of jobs whose windows are contained in one of
   the intervals of T, and let $J' = J(T) \setminus J''$
  end

Output the partition $\mathcal{T}$.

Theorem 8.8.2. *Suppose we are given a feasible solution to (8.32)–(8.29) on $k = m(T)$ machines, a collection $\mathcal{T}$ of disjoint subintervals of T, and a corresponding subset $J' \subset J(T)$ of jobs, and assume that $\mathcal{T}$ is good with respect to the linear program solution. Then we can schedule all the jobs in J' on αk machines in polynomial time, for some constant α.*

Corollary 8.8.3. *There is an $\mathcal{O}\left(\frac{\log n}{\log \log n}\right)$ approximation algorithm for machine scheduling.*

Definition 8.8.1. The job types are defined as follows.

Type 1: Denote by $\mathcal{I}^C$ the intervals which cross the boundaries of the intervals in $\mathcal{T}$. A job j is defined as a job of type 1 if

$$\sum_{I \in \mathcal{I}(j) \cup \mathcal{I}^C} x(I, j) \geq 0.2.$$

  The idea of scheduling jobs of this type on $\mathcal{O}(k)$ machines is to find a matching between the jobs and the boundaries of the intervals in $\mathcal{T}$. The LP-solution gives a fractional matching where for each job of type 1 at least a fraction 0.2 of the job is scheduled. Therefore, the integral matching gives a schedule of jobs of type 1, where on each boundary of an interval in $\mathcal{T}$, at most $5k$ jobs are scheduled. As the intervals of the jobs that have nonzero values in the LP-solution overlap with at most two intervals in $\mathcal{T}$, the maximum number of jobs running at any time t is at most $10k$.

Type 2: An interval I belonging to job j is called large, if it is completely contained in an interval $T' \in \mathcal{T}$ whose size is at most $2p_j$. Let $\mathcal{I}^L$ denote the set of the large intervals.

Jobs of type 2 are all the jobs j which are not of type 1 and for which the following property holds:

$$\sum_{I \in \mathcal{I}(j) \cup \mathcal{I}^L} \geq 0.2.$$

In order to schedule jobs of type 2, observe that in the LP-solution, for each interval $T \in \mathcal{T}$, the sum of $x(I, j)$ where I is a large subinterval of T and $j \in J'$ is at most $2k$. We perform a matching between jobs of type 2 and the intervals in T to determine the schedule of these jobs on $10k$ machines.

Type 3: For each job j, let $T^d(j)$ denote the interval in $\mathcal{T}$ that contains its deadline d_j. Job j is of type 3, if it does not belong to any of the previous types and

$$\sum_{I \in \mathcal{I}(j), I \subseteq T^d(j)} \geq 0.2.$$

Each job j of type 3 is going to be scheduled inside the interval $T^d(j)$. Consider some interval $T' \in \mathcal{T}$ and the subset of type-3 jobs j whose deadline belongs to T'. As the release dates of these jobs are outside T', this can be viewed as scheduling jobs with identical release dates. We solve this problem approximately and use the linear program solution to bound the number of machines we use.

Type 4: For each job j, let $T^r(j)$ denote the interval in $\mathcal{T}$ that contains its release time r_j. Job j is of type 4, if it does not belong to any of the previous types and

$$\sum_{I \in \mathcal{I}(j), I \subseteq T^r(j)} \geq 0.2.$$

The scheduling is performed similarly to the scheduling of the jobs belonging to type 3.

Type 5: This type contains all the other jobs. Note that for each job j of this type, the sum of fractions $x(I, j)$ for intervals I, such that I is not large, does not cross any boundary of intervals in $\mathcal{T}$, and the interval $T' \in \mathcal{T}$ which contains I does not contain the release date or the deadline of j is at least 0.2. The linear program solution ensures that all the jobs of this type can be (fractionally) scheduled inside intervals $T \in \mathcal{T}$ (i.e., without crossing their boundaries), even if we shrink the job windows so that their release dates and deadlines coincide with boundaries of intervals in $\mathcal{T}$, using $5k$ machines. This allows us to schedule all the jobs of type 5 on $\mathcal{O}(k)$ machines.

The Scheduler

In the final schedule, jobs of each type are scheduled separately. We now provide a formal description of the schedules of each type and prove that $\mathcal{O}(k)$ machines suffice for scheduling each type.

Type 1:   We claim that we can schedule all the jobs of type 1 on at most $10k$ machines. Construct a directed bipartite graph $G = (V, U, E)$, where V is the set of jobs of type 1 and U is the set of boundaries of the intervals of $\mathcal{T}$. There is an edge (j, b) of capacity 1 from $j \in V$ to $b \in U$ if and only if there is an interval $I \in \mathcal{I}(j)$ that crosses the boundary b and $x(I, j) > 0$. Add a source vertex s and an edge (s, j) of capacity 1 for each $j \in V$. Add a sink vertex t and an edge (b, t) of capacity $5k$ for each $b \in U$. The solution to the linear program defines a feasible flow in this graph of value at least $|V|$ as follows. For each $j \in V$ and $b \in U$, let $\mathcal{I}(j, b)$ be the subset of intervals $\mathcal{I}(j) \cup \mathcal{I}^C$ that cross a boundary b. Set the flow value on the edge (j, b) to be

$$\frac{\sum_{I \in \mathcal{I}(j,b)} x(I, j)}{\sum_{I \in \mathcal{I}(j) \cap \mathcal{I}^C} x(I, j)}.$$

Note that by the definition of type 1, the value of the flow on edge (j, b) is at most

$$5 \sum_{I \in I(j,b)} x(I, j),$$

and the total amount of flow leaving j is exactly 1. The total flow entering each $b \in U$ is at most $5k$. Therefore, there is an integral flow of value $|V|$. This flow defines a schedule of jobs of type 1 as follows. For each such job j, there is a unique boundary b, such that the flow on edge (j, b) is 1. By the construction of the network, there is an interval $I \in \mathcal{I}(j, b) \subseteq \mathcal{I}(j)$, such that $x(I, j) > 0$. We say that j is scheduled on boundary b.

The number of machines used in such a schedule is at most $10k$. Since each job interval $I \in \mathcal{I}(j)$ with $x(I, j) > 0$ overlaps with at most two intervals in $\mathcal{T}$, at any point t inside some interval $T' \in \mathcal{T}$, the jobs that run at time t are either scheduled on the left or on the right boundary of T'. Therefore, the number of jobs scheduled at any such point is at most $10k$.

Type 2:   In the fractional solution, for each interval $T' \in \mathcal{T}$, the sum of $x(I, j)$ where I is a large subinterval T' is at most $2k$. We show how to schedule all the jobs of type 2 on $10k$ machines. This is done in a similar fashion to type 1. We build a directed bipartite graph $G = (V, \mathcal{T}, E)$, where V is the set of all the jobs of type 2. There is an edge (j, T') of capacity 1 if and only if job j has a large interval in $T' \in \mathcal{T}$. Add a source vertex s and an edge (s, j) of capacity 1 for each $j \in V$. Add a sink vertex t and an edge (T', t) of capacity $10k$ for each $T' \in \mathcal{T}$. The value of the maximum flow in this network is exactly $|V|$. A fractional flow of this value is obtained as follows. For each $j \in V, T' \in \mathcal{T}$, set the flow on edge (j, T') to be

$$\frac{\sum_{I \in \mathcal{I}(j) \cap \mathcal{I}^L, I \subseteq T'} x(I, j)}{\sum_{I \in \mathcal{I}(j) \cap \mathcal{I}^L} x(I, j)}.$$

Note that by the definition of type 2, this value is at most

$$5 \sum_{I \in \mathcal{I}(j) \cap \mathcal{I}^L, I \subseteq T'} x(I, j),$$

and the total flow leaving j is 1. Since in the solution to the linear program, for each interval $T' \in \mathcal{T}$, the total fraction of all the large intervals inside T' is at most $2k$, the value of flow entering T' is at most $10k$.

Therefore, there is an integral flow of value $|V|$. For each job j of type 2, there is exactly one $T' \in \mathcal{T}$ such that there is a flow on edge (j, T'). We schedule j in this interval. In each interval T', we thus schedule at most $10k$ jobs (note that each such job has at least one interval in T'). We schedule these $10k$ jobs on $10k$ different machines. Observe that the intervals belonging to $T', T'' \in \mathcal{T}$, are disjoint, and thus we need $10k$ machines in all.

Type 3: Consider some $T' \in \mathcal{T}$ that starts at time $S_{T'}$ and finishes at time $F_{T'}$. Let $J^d(T')$ be the set of all jobs of type 3 whose deadlines are inside T'. We show how to schedule all these jobs on $10k$ machines. Once again, since the intervals $T' \in \mathcal{T}$ are disjoint, we can use the same $10k$ machines for all jobs of type 3. For each job j, define $t_j = d_j - p_j$. We build a new fractional schedule x' of jobs in $J^d(T')$ in T'. For each job $j \in J^d(T')$, for each $I \in \mathcal{I}(j), I \subseteq T'$, define

$$x'(I, j) = \frac{x(I, j)}{\sum_{I' \in \mathcal{I}(j), I' \subseteq T'}}.$$

Note that by the definition of type-3 jobs, $x'(I, j) \leq 5x(I, j)$. This new fractional solution has the following properties:

- The sum of fractions of intervals that cross any time point t is at most $5k$.
- Each $j \in J^d(T')$ is scheduled completely at subintervals of T' that start before or at t_j.

We now show a greedy schedule of job set $J^d(T')$ on $10k$ machines, in interval T'. We proceed from left to right on all the $10k$ machines simultaneously, using the following greedy rule: whenever any machine becomes idle, schedule any available job $j \in J^d(T')$ that has minimal t_j.

We claim that all the jobs in $J^d(T')$ are scheduled at the end of this procedure. Suppose this is not the case and let j be the job with minimal t_j that the procedure does not schedule. Let B be the set of all the jobs that are scheduled by the greedy procedure at intervals that start before or at time t_j. By the definition of greedy algorithm,

- all the $10k$ machines are busy during the time interval $[S_{T'}, t_j]$ and
- the set of jobs that are scheduled at intervals that start before or at time t_j is exactly B (because j is the first job that we were unable to schedule).

Let Z_{5k} denote the sum of lengths of $5k$ longest jobs from B. Then

$$\sum_{j' \in B} p_{j'} > Z_{5k} + 5k(t_j - S_{T'}). \tag{8.36}$$

We now bound the left-hand expression in Eq. (8.36) using the new fractional solution. As already mentioned, all the jobs in B must be completely scheduled at intervals starting before or at time t_j in the fractional solution. Given an interval $I \in \mathcal{I}(j')$, the volume of the interval is defined to be $x'(I, j')p_{j'}$. The sum of lengths of jobs in B equals exactly the total volume of their intervals, which is at most the total volume of intervals belonging to jobs from B that finish before time t_j plus the total volume of intervals crossing the time point t_j. The former is bounded by $5k(t_j - S_{T'})$, and the latter is at most Z_{5k}. This is a contradiction. Thus, all the jobs in $J^d(T')$ are scheduled by the greedy procedure inside T' on $10k$ machines.

Type 4: This type is defined exactly like type 3, only with respect to the release dates of jobs. As in type 3, all jobs of this type can be scheduled on $10k$ machines.

Type 5: These are all the other jobs. Let G be the set of all the intervals $I \in \mathcal{I}(j)$, $j \in J$, such that I does not cross boundaries of any interval in $\mathcal{T}$, and if $I \subseteq T' \in \mathcal{T}$, then T' does not contain the release date or the deadline of j and the length of T' is more than twice the length of I. Note that each job j of type 5, in the fractional solution, has

$$\sum_{I \in \mathcal{I}(j) \cap G} \geq 0.2.$$

Furthermore, if $I \in \mathcal{I}(j)$ and $I \in G$, then job j can be scheduled anywhere inside the interval $T' \in \mathcal{T}$ that contains I. We divide the jobs of this type into size classes. Class J_i contains all jobs j of type 5 such that $2^{i-1} < p_j \leq 2^i$. For each interval T' and for each i, let $X(T', i)$ be the total fraction of intervals of size $(2^{i-1}, 2^i]$ that belong to set G and that are contained in T'. We are going to schedule at most $\lceil 5X(T', i) \rceil$ jobs from J_i inside T'.

Proposition 8.8.4. *Each interval T' can accommodate, on $22k$ machines, $\lceil 5X(T', i) \rceil$ jobs of size 2^i, simultaneously for all i.*

All we have to decide now is, for each job size, which job is scheduled in which block. Consider a set of jobs J_i. We want to assign all the jobs in J_i to blocks, such that:

- Job j can only be assigned to a block that is completely contained in j's window.
- At most $\lceil 5X(T', i) \rceil$ jobs are assigned to each interval $T' \in \mathcal{T}$.

Note that the solution to the linear program implies a feasible fractional solution to this problem: for each j and T' such that T' is contained in j's window, the fraction of j assigned to T' is 5 times the total fraction of intervals belonging to j inside block T' in the fractional solution. It is easy to see that an integral solution to the problem above can be obtained by the earliest deadline greedy assignment of jobs to blocks.

Finally, each job that is assigned to an interval T' can be scheduled anywhere inside the interval. Therefore, we can use the schedule from Proposition 8.8.4 to accommodate all the jobs.

CHAPTER 9

Access Networks

The role of the access network is to concentrate traffic from a more or less densely distributed field of sources (often referred to as customers) and to deliver its demand to an interconnection point in the core (aggregation layer) network for processing and further routing. Typically, very little logic is inherent in the access network. Large uptake areas and low traffic volumes in distant parts of the access network make the cost effectiveness of the network very important. Redundancy aspects play a subordinate role to link utilization; indeed, access networks are often designed for a lower grade of service than core networks.

The components of an access network are terminal nodes (or "clients"), concentrator nodes (such as multiplexing devices), and one (or more) interconnection point(s). All components are interconnected by links, which form the network topology. Concentrator devices aggregate two or more traffic streams into a traffic stream with higher bandwidth, requiring higher capacity links. The concentrators are relatively inexpensive equipment that allow for an efficient network topology with a low number of links. Without concentrators, the terminals would have to be directly connected to the access point, each by a dedicated link in a star topology.

Thus, access networks connect the terminals and route traffic to the backbone, or core, transportation network, possibly by performing traffic aggregation in one or more steps. The possibility of aggregation in the access network is a main enabler for constructing cost-efficient topologies.

We usually assume that the connection cost is proportional to the distance between the connected nodes and possibly proportional to the maximum capacity of the link. Different technologies can be used in the access network, such as optical fiber, microwave links, or copper lines, and given the size of access networks, we are interested in finding cost-efficient topologies, often under the assumption of restricted link capacity. Links may also have physical limitations in terms of distance.

The natural access topology is a tree, whereas capacity (or distance) restrictions suggest clustering of terminals satisfying given capacity limits. The resulting model is known as the capacitated minimum spanning tree (CMST) problem.

9.1 Capacitated Minimum Spanning Trees

In the capacitated minimum spanning tree (CMST) problem we seek a minimum-cost tree spanning a given set of nodes such that some capacity constraints are observed.

5G Networks. https://doi.org/10.1016/B978-0-12-812707-0.00014-0

To define the problem formally, we consider a connected graph $G = (V, E)$ with node set V and edge set E. Each node $i \in V$ is associated with a nonnegative node weight b_i with $b_0 := 0$. The node weights may be interpreted as capacity requirements. The edges are associated with nonnegative weights c_{ij} representing the cost of using edge $(i, j) \in E$ to connect nodes i and j. We identify a special node, say node r, called the root of the tree, usually having unlimited capacity.

We see the similarity to the minimum spanning tree problem, where we wish to find a minimum-cost tree spanning node set V. In the MST, however, node weights are not taken into account. The capacity restriction renders the CMST problem $\mathcal{NP}$-complete.

We identify the edge (r, i) connecting node i to the root, which is sometimes called a central edge. The capacity requirement of a subtree is the sum of the node weights of the included nodes. We let the capacity constraint at each node be that the weights of the nodes connecting to it must not exceed a given capacity limit K.

For the CMST as a cost minimization problem any feasible solution provides an upper bound on the optimal objective function value whereas lower bounds are obtained by solving relaxations, that is, treating discrete connection variables as real and solving the linear program thus obtained form the initial integer program.

The position of the root node seems to have a large impact on the performance of CMST algorithms. When it is in the middle of an area containing terminal nodes, they usually perform better than when the root node is in a border area.

A simple relaxation of the CMST is ignoring capacities and using the MST. Whenever the MST solution is feasible, it must be optimal for the CMST.

Esau–Williams Algorithm

We regard a solution to the CMST as consisting of components $G_i = (V_i, E_i)$ with node set V_i and edge set E_i, where G_i is connected – usually a spanning tree. If two different node sets V_i and V_j have a common node, this is the center node. Each component G_i can only contain one center edge; otherwise it is split into two or more components.

A component is called feasible if it satisfies the capacity constraints and infeasible otherwise. The component is called central if it includes the center node and noncentral otherwise.

The solution $G = (V, E) = \cup_i G_i$ is called:

- feasible if every component G_i is feasible and central,
- incomplete if every component G_i is feasible but at least one component is noncentral,
- infeasible if all components G_i are central but at least one is infeasible.

Using this characterization, we find that the empty tree solution having $n + 1$ components $G_i = (V_i, E_i)$, $V_i = \{i\}$, $E_i = \emptyset, \ldots, n$, and all components except G_r noncentral, is incomplete with a zero total cost.

A solution with n components $G_i = (V_i, E_i = \{(0, i)\})$, $V_i = \{i, r\}$, $E_i = \{(i, r)\}i = 1, \ldots, n$, is called a star tree. The star tree is a feasible solution since every component is central and feasible. Without loss of generality we assume that $b_i \leq K$ for all $i = 1, \ldots, n$. It usually has a high total cost.

We distinguish two possible optimization strategies: either we start from an incomplete initial solution and build a complete solution maintaining feasibility, or we start from a minimum-cost solution and modify it to become feasible.

Initial feasible solutions

The Esau–Williams algorithm starts with a feasible solution, for example the star tree, and finds the local best feasible change, that is, the change that yields the largest saving in cost. The procedure is repeated until no further cost saving is obtained.

The cost saving s_{ij} of joining two components G_i and G_j is defined as

$$
s_{ij} = \begin{cases} \max\{\xi_i, \xi_j\} - c_{ij}^*, & \text{if joining of } G_i \text{ and } G_j \text{ is feasible,} \\ -\infty, & \text{otherwise,} \end{cases}
$$

where ξ_i is the minimal cost of connecting the root to the nodes of G_i and c_{ij}^* is the minimal cost of an edge connecting G_i and G_j. Then the cost of the newly formed component has to be recalculated and the procedure is repeated. The algorithm runs in $\mathcal{O}(n^2 \cdot \log_2 n)$ time, which can be reduced to $\mathcal{O}(n \cdot \log_2 n)$ time when costs increase monotonically with the Euclidean distance.

Further savings may be obtained by allowing changes (deletions) of noncentral edges as well as central edges, leading to a larger number of possible recombinations, but also longer running times, with a complexity of $\mathcal{O}(n^5)$.

Improvement strategies

It should be noted that the ordering of the nodes affects the result in general. Therefore, a first improvement and verification procedure is to rerun the optimization algorithm on a permutation of the node ordering.

Improvement strategies in CMST algorithms can be classified as either local exchange procedures or second-order procedures. Local exchange procedures start with any feasible solution and seek an improvement by applying a transformation (exchange), verifying the feasibility and recalculating the cost. In a feasible solution we can use the following transformation. For every node i, connect i to its nearest neighbor j not yet connected to i and remove the edge with the largest cost from the resulting cycle, while maintaining feasibility [46]. The transformation leading to the largest cost decrease is selected, and the improvement procedure is repeated as long as the costs decrease.

Second-order algorithms iteratively apply an optimization subroutine to different start solutions, with some edges forced to be included or excluded in a modified initial solution. This can be managed by modifying the edge cost matrices, promoting edges by low costs and barring others by very large costs. Then, a standard CMST procedure is applied to the initial solutions, and the best solution is selected for further iterations. We may, for example, investigate edges of decreasing classes of distance.

Although there are many published algorithms for CMST, the Esau–Williams algorithm has been shown to perform well overall [80]. There are other possibilities to approach access network design as well. To create initial approximate components, we may apply a clustering algorithm. In the case there are optional nodes that may or may not be included in the solution, the problem is to find a minimum-cost Steiner tree, which may be approached by disabling edges and nodes as described above.

Amberg et al. [80] propose the exchange of nodes between the subsets V_i of the components, such that feasibility is maintained. Referring to the transformations as moves, the authors consider two cases:

(1) Shift moves a chosen node from its current component to another component.
(2) Exchange moves two chosen nodes from different subtrees and exchanges them.

After such a transformation, the Esau–Williams CMST algorithm can be used. The authors suggest controlling the node transformation by a metaheuristic, such as simulated annealing. In simulated annealing, a feasible move is chosen randomly and computes its change in cost. If the change decreases the cost the move is carried out. Otherwise, the change is performed with a certain probability. This probability limit decreases exponentially with the number of iterations.

9.2 Mixture of Microwave and Fiber Access Network

We consider a set of LTE sites, which constitute our total node set. Some of these sites are connected to an existing fiber infrastructure. Other sites are to be connected to any of the fiber-connected sites to form an access network. For additional sites, we have a choice be-

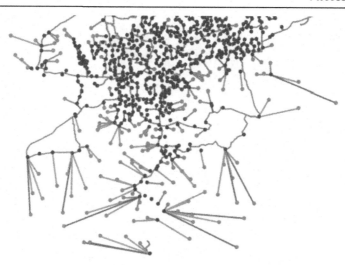

Figure 9.1: Initial CMST solution, i.e., star solution.

tween microwave links and fiber, where microwave links are associated with a unit cost, a maximal distance, and a capacity limit, whereas fiber can be deployed at a cost proportional to its distance and has no capacity limit (in practice). The challenge is to find a cost-efficient, "rational", topology. The "rationality" of the topology means that

(1) we connect component to the closest fiber-connected site, which will be the root for the component,

(2) we do not mix fiber segments with microwave links, so even if a fiber segment would be cheaper than a microwave link, we would rather deploy the latter if preceded or followed by microwave links in the component.

Following a pricing scheme of transmission resources, we initiate the Esau–Williams algorithm by separating all sites into two sets: those directly connected to fiber and those which are not. This separation is logical, because the former set has unlimited capacity and consists of all possible root nodes. The latter set consists of sites likely to be connected by microwave links and is therefore subject to capacity constraints.

We construct an access network topology following the Esau–Williams algorithm where we know the prices and constraints of different transmission technology alternatives. A tree topology can conveniently be represented by a list containing the parent node $P(u)$ of a given node u. For the initial solution we simply find the closest fiber-connected node as the parent of any terminal. This topology is shown in Fig. 9.1.

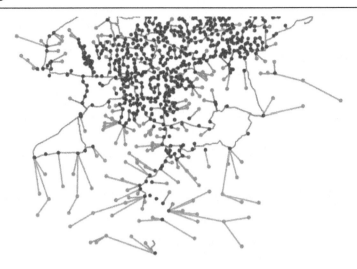

Figure 9.2: First optimization step.

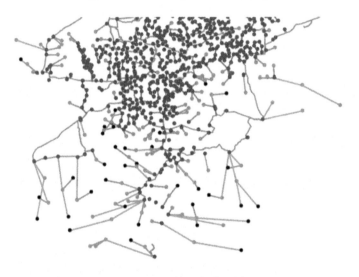

Figure 9.3: Second optimization step.

In the first optimization step, we search for the largest savings alternative, and this gives us the topology shown in Fig. 9.2. This step turns out to give approximately 15% savings in transmission compared to the initial solution.

Finally, a third step using an exchange principle gives some further improvement, with the topology shown in Fig. 9.3. This step yields a transmission solution some 20% cheaper compared to the initial solution.

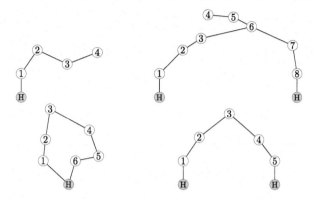

Figure 9.4: Typical connection scenarios in an access network.

9.3 Resilience in Access Networks

In legacy networks, base stations do not communicate directly with each other, but are logically connected to a controller. In such networks, base stations have little or no routing capability, and therefore there is no need for direct links between base stations apart from the purpose of traffic aggregation. Transmission resources are expensive, but usually more reliable than other equipment in the access network.

In 4G and 5G networks, however, direct communication between base stations is an important feature, and it actualizes the question of resilience in access networks. In particular, resilience is very important in C-RAN architectures, which are discussed later in this chapter.

We consider an access network where all sites are assumed to be connected by fiber, for simplicity. In this access network, we can identify the four elements (or scenarios) shown in Fig. 9.4. We may refer to them as "spur" and "loop" in the first column and as "half-bridge" and "bridge" in the second column. All sites are connected so that they eventually end up in a hub, marked by "H". We ignore the topology beyond the hub point and just take for granted that it is sufficient in terms of capacity and resilience.

When planning access networks, it is common to use elementary structures like loops (rings) and spurs, terminating in one or (for some loops) two hubs. The hubs provide interconnection to the aggregation layer or the backbone network. Loops are distinguished between those connected to a single hub (wrapped-around loops) and dual-homing hubs. From a resilience point of view, dual-homed loops provide the most and spur the least resilient structures of these scenarios. We may also have hybrids – spurs connected to a point on a loop other than the hub.

We let the cost of a connection be proportional to the Euclidean distance. We would also need a way to classify the improvement in resilience. To do this, let us assign each site one point

Table 9.1: Point scheme for degree of resilience in access scenarios.

Property	Spur	Loop	Partial bridge	Bridge
Number of nodes	10	10	5/5	10
Paths	1	2	1/2	2
Hubs	1	1	1/2	2
Points	10	20	30	40

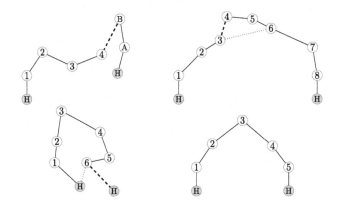

Figure 9.5: Improvement transformations for resilience scenarios.

for each transmission path $\{1, 2\}$ available and one point for each hub the scenario has $\{1, 2\}$. We then have the example cases in Table 9.1. Usually, the final solution also needs to satisfy some technical constraints, such as:

(1) An upper allowable limit of the number of sites per configuration type, N_1 (spur), N_{21} (wrapped-around loop), N_{22} (dual-homed loop), N_{h1} (wrapped-around hybrid), and N_{h2} (dual-homed hybrid).

(2) A maximum total distance of each configuration, L_1, L_{21}, L_{22}, L_{h1}, and L_{h2}.

(3) A maximum distance between sites, D_{sites}.

To quantify resilience, we adopt the following simple scheme: for any node, multiply the number of routes by the number of access points (hubs) to the aggregation layer. Thus for a spur with 10 nodes, we have $10 \cdot 1 \cdot 1 = 10$ points, or 1 point per node. For a wrapped-around loop having 10 nodes, the total score is $10 \cdot 2 \cdot 1 = 20$, or 2 points per node. For a dual-homed hybrid with 5 nodes in the loop and 5 nodes in the spur, we have $5 \cdot 1 \cdot 2 + 5 \cdot 2 \cdot 2 = 30$, or an average of 3 points per site.

Based on this, we can select a change based on the *relative improvement*, that is, the increase in resilience divided by the increase in cost. Next, we identify transformations that improve the resilience score, as shown in Fig. 9.5. We may combine two spurs by connecting their end points (labeled 4 and B). We can also wrap a spur around, either to its own hub to obtain a loop, or to a different hub to obtain a bridge. Similarly, we can rewire a loop to make it a bridge, or a partial bridge to make it a full bridge, by "inflating" it.

We assume that the base station and hub locations are given and no new nodes are to be deployed. The cost structure can be chosen as appropriate. Here, we let $c = a + bl$, where a is a fixed cost for each new link, and b is the cost per length unit l.

To solve this design problem, we can use a greedy approach. The design is restricted to predefined configurations, which limits the number of possibilities. For each configuration, we identify the transformations that lead to an improvement of resilience. We may

(1) combine two spurs terminating in different hubs to a dual-home loop,
(2) combine two spurs terminating in the same hub to a wrapped-around loop,
(3) wrap around a spur to a wrapped-around loop,
(4) change a wrapped-around hybrid to a wrapped-around loop,
(5) change a wrapped-around loop to a dual-homed loop.

Now, after these preparations, we can sketch a software structure. It seems from our problem definition that an exchange approach is appropriate here. The code must be able to

(1) identify scenarios and compute resilience points and costs (sum of fiber distances),
(2) identify edges to change out,
(3) search for strengthening edges.

There are a number of possible approaches to construct such a program, but a conceptually simple and efficient way is using depth-first search in identification of scenarios. We note that a spur and partial bridge both have a leaf, but the former a single hub and the latter two. Similarly, the difference between a loop and a bridge is also given by the identities of their start and end hubs.

We sketch an algorithm that has been used in access network design with highly satisfactory results. It makes sense to begin with spurs, followed by modification of hybrids and finally dual-homing, in order of the potential gains in resilience. With feasibility, we mean that the technical boundary conditions are satisfied.

Algorithm 9.3.1 (Greedy algorithm for resilience improvement).

Given an access network decomposable into configurations as shown in Fig. 9.4 with specified coordinates and a cost function $C(\cdot)$ for transmission links.

STEP 1:*N*

    for all cases **do**

      Find two close spurs configurations: we may estimate their lengths
      and the distance between the outmost base stations,
      or look for spurs originating from adjacent hubs,
      if feasible **then**:
        temporarily connect the outermost nodes of the spurs,
        compute the additional cost and total increase in resilience points
        Implement the change leading to the largest increase in resilience per cost.
      end
      for all cases **do**:
        Investigate hybrid configurations,
      end
      if feasible **then**:
        temporarily change to a loop configuration,
        compute the additional cost and total increase in resilience points,
        Implement the change leading to the largest increase in resilience per cost.
      end
      for all cases **do**
        Investigate loop and hybrid configurations,
      end
      if feasible **then**:
        temporarily change to a dual-homed configuration,
        compute the additional cost and total increase in resilience points,
        Implement the change leading to the largest increase in resilience per cost.
      end
    end

Output: cost-efficient resilience-optimized access network.   ●

When searching for strengthening edges, we note that for spurs, we are looking for edges from the leaf to either another leaf, another hub, or its own hub. In loops we look at replacing the outgoing edges from the hub by an edge from one of its terminal end points to another hub. For the partial bridge, we need to identify the branching node, which is the only nonhub node with degree 3 in the scenario. We seek to replace one of the edges from this node to a neighbor on the "bridge" part. These transformations should be fairly clear from Fig. 9.5.

We usually begin with transformations that give the largest improvement, that is, spurs. This greedy approach usually works well. We may also want to compare with the minimum distance solution, which is essentially given by a minimum spanning forest, that is, components

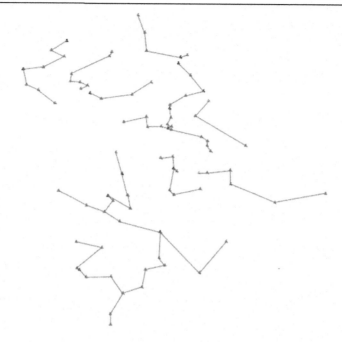

Figure 9.6: Minimum-cost (shortest-distance) resilient access network.

that are minimum spanning trees containing a root node. Such a solution is shown in Fig. 9.6. In this solution (which could be replaced by any existing topology we wish to improve), we identify scenarios that can be improved. Optimization using the principles outlined above result in a minimum-cost maximum-resilience solution, shown in Fig. 9.7. We have omitted some resilience improving links due to their prohibitively high cost.

In this optimization we can easily include technical constraints such as a maximum scenario distance. We have only allowed scenarios to connect to maximum two hubs, a restriction that limits the number of possibilities. We may also wish to set a limit of the number of sites in a scenario, particularly spurs.

9.4 Centralized Radio Access Networks

One of the most profound improvements to the radio access network is the Centralized or Cloud Radio Access Network (C-RAN), whereby signal processing baseband equipment is physically and logically separated from the actual radio and antenna systems and located in a resource pool referred to as a baseband hotel (BBH). The remote radio units (RRUs) are connected to the baseband hotels by optic fibers, or possibly by microwave links. The C-RAN

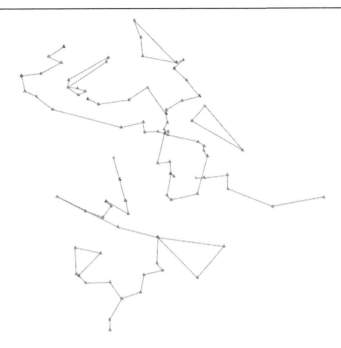

Figure 9.7: Maximum-resilience access network.

architecture offers a number of spectral efficiency and cost saving gains compared to traditional access networks.

By separating the energy-intensive baseband equipment in a relatively few well-chosen locations, energy consumption can be substantially reduced. The pool of resources makes geographical and temporal fluctuations in traffic spread out, leading to a higher degree of utilization of the equipment on average. Since the radio units are comparatively small and consume much less power than a traditional base station, the space requirements decrease dramatically – a radio unit can be mounted directly on a wall or pole, without the need for site infrastructure when hosting baseband equipment. At the same time, fiber or equipment capacity limitations may put an upper bound on the feasible size of a pool. In addition, by using sophisticated radio features controlling the baseband equipment, radio units belonging to the same pool can be made to cooperate to improve radio conditions, leading to higher coverage and throughput. The interdependence structure of C-RAN is shown in Fig. 9.8.

On the downside, C-RAN poses some rather strict requirements on the transmission infrastructure in terms of maximum transmission distance, bandwidth demand, and presence of redundant transmission paths. In the following discussion, we assume the following set-up (see Fig. 9.9). We have a set of base stations possibly connected by fronthaul fiber links and

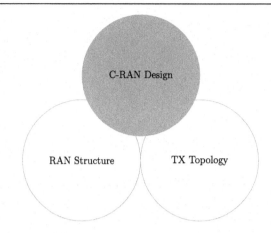

Figure 9.8: Centralized RAN interdependence structure.

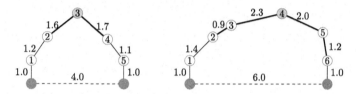

Figure 9.9: Chains showing hotel placement and additional fiber pairs in the case of Scenario A equipment.

backhaul links to a hub, or concentrator, responsible for traffic aggregation and its transportation to the backbone network.

In order to formulate and solve this complex optimization problem, first we must decide what we want to optimize – cost or performance or some combination of both. The total cost likely has many contributions; equipment investment, costs related to fiber jetting or digging ducts, site rental, and energy consumption. In the most likely scenario, many base stations and fiber links would already be in place, and a decision has to be made on whether to allow changes to existing infrastructure or not. Typically, we would only make changes when a substantial gain in overall cost or performance follows.

The performance, however, is a bit trickier to measure. We may assume that the gain from individual base stations varies, both due to strategic importance or carried traffic and radio cell overlap with neighboring base stations. Note that the overlap expressed as a percentage of the cell overage area, the degree of overlap between two sectors, is not symmetric. In the final solution, the cluster quality or geographic coverage may also be taken into account (see Fig. 9.10). To create a single optimization target function, we can choose a selection of variables, multiply these with some weight, and take the sum.

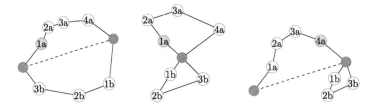

Figure 9.10: Combination of chains to form larger clusters are based on their *affinity*.

The C-RAN architecture is dependent on both the direct radio distance 'as-you-fly' and the transmission path carried by fiber between base stations. Since the quality of the solution is dependent on the relative proximity of neighboring base stations, C-RAN deployment is more advantageous in dense areas.

To find a strategy for C-RAN optimization, we notice that, for each solution, we have to be able to establish its feasibility and its cost. The feasibility check amounts to verifying that all technical restrictions are met. This may be time consuming when the distances of alternative paths need to be evaluated and compared. Computing the cost is normally less demanding, using cost functions or tables with some of the solution parameters as arguments (such as equipment type and quantity, number of fiber connections and their distances, and site rental costs).

To reduce the degrees of freedom, it is a good idea to assume a fixed number k of clusters. We can get an idea of this number by inspection of existing topology, or by using a heuristic. Making this assumption, we not only decrease the complexity of the optimization algorithm, but we have a simple way of initial verification and solution consistency check as well. Should they fail to produce a solution, or return a solution with anything by k clusters, we may safely conclude that either the problem is misspecified, there is no feasible solution for this k, or the algorithm does not work properly.

As clustering algorithm, we consider a heuristic due to its simplicity, spectral clustering for its high-quality cluster properties, and a genetic algorithm for its appropriateness for network design.

We may be tempted to use spectral clustering for C-RAN design. We here find that the base stations are separated by two distances – direct radio distance and admissible fiber transmission distance. These two distance measures lead to two different similarity matrices and clusterings. In the case the clusters overlap to a large degree, the cut between the sets could constitute a candidate solution.

Another efficient approach is a genetic algorithm. The advantage of a genetic algorithm is the relative ease of coding various constraints. We assume that the number of pools k is given.

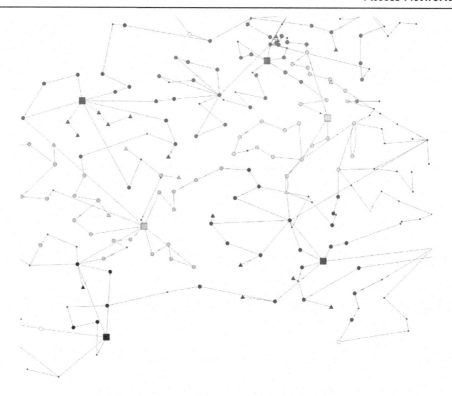

Figure 9.11: Design of C-RAN with six pools depicted.

Radio pools can be represented by vectors of length n, the number of nodes, which contain a 1 whenever a node is present in the pool, and 0 otherwise. Such a coding requires a consistency check so that a node is present in only one pool at any time.

We typically want to optimize with respect to both cost and performance. In particular, large sector overlap and traffic are factors which lead to substantial performance benefits with C-RAN. A solution found by a genetic algorithm is shown in Fig. 9.11.

9.5 Antenna Systems

Antenna arrays provide an efficient way to control the coverage and interference levels of a base station. It is considered one of the fundamental enabling techniques to achieve 5G bandwidth and coverage goals. Antenna arrays can be used to form directed beams by combining geometric and electric properties, leading to systems with high transmit power, low energy consumption, and efficient radio spectral utilization.

By dynamically exciting antenna elements, the signal strength in undesired directions can be minimized, at the same saving energy. The antenna elements can be arranged in different patterns, such as a circular or a hexagonal arrangement, which show to have specific properties.

Radiation Patterns

First, we need some definitions from electromagnetic wave theory.

An antenna radiation pattern (or antenna pattern) is defined as a mathematical function or a graphical representation of the radiation properties of the antenna as a function of space coordinates.

- Defined for the far-field.
- Expressed as a function of directional coordinates.
- There can be field patterns (magnitude of the electric or magnetic field) or power patterns (square of the magnitude of the electric or magnetic field).
- Often normalized with respect to their maximum value.
- The power pattern is usually expressed in decibels (dB).

Radiation patterns $E(\theta, \phi)$ are often represented in spherical coordinates. An infinitesimal area element is

$$dA = r^2 \sin\theta d\theta d\phi,$$

where ϕ is the azimuth and $\pi/2 - \theta$ is the elevation.

A radiation lobe is a portion of the radiation pattern intersected by regions of relatively weak radiation intensity. The pattern consists of main lobes, minor lobes, side lobes, and back lobes. Minor lobes usually represent radiation in undesired directions and should be minimized. Side lobes are normally the largest of the minor lobes. The level of minor lobes is usually expressed as a ratio of the power density, often termed the side lobe ratio or side lobe level (SLL).

The beamwidth of an antenna is a very important figure of merit and often is used as a trade off between it and the SLL; that is, as the beamwidth decreases, the side lobe increases and vice versa. The angular separation between the first nulls of the pattern is called the first null beamwidth (FNBW).

An isotropic radiator is a hypothetical lossless antenna having equal radiation in all directions.

Massive MIMO Antenna Arrays

Optimizing the excitation levels of antenna elements in isotropic antenna arrays is difficult due to strong correlations between the elements, and the resulting beam form depends on excitation levels, the element separation, and the arrangement pattern. The elements can for example be arranged in circular arrays (CAs) or hexagonal arrays (HAs), and the nominal excitation levels are 100%.

We show that the antenna performance can be optimized to obtain lower SLLs and narrower first null beamwidths (FNBWs) in a chosen direction. For the purpose of performance optimization, evolutionary methods have proven successful, such as particle swarm optimization [81][82], firefly optimization [83][84][85], or grey wolf optimization [86]. The optimization, or lowering of excitation levels, is referred to as thinning.

We study two cases, a 12-element hexagonal array with excitation levels {0, 1} and an 18-element hexagonal array with continuous excitation amplitudes, both for three different element spacings $\lambda \in \{0.50, 0.55, 0.60\}$. The performance is compared with fully excited antenna elements.

In an array, antenna elements are arranged in a geometric grid pattern – often circular or hexagonal – and fed the same properly phase-shifted signal by the base station. It is usually assumed that the only difference between the input to the antenna elements is a difference in phase. Thus, neglecting the effect caused by the phase shift, the elements are typically at the same excitation level.

It turns out, however, that if we allow the excitation levels to vary between antenna elements in an array, we may further improve the performance of the array. In effect, we let some of the elements be fed a copy of the signal with lower amplitude than in the standard set-up, a so-called thinning of the array. This constitutes an optimization problem, which we describe next.

In the optimization, one objective is to suppress the side lobes, which have low gains and point in various directions relative the main lobe. Lower SLLs improve the signal-to-noise ratio due to lower undesired interference.

A linear array has high directivity and can be used to form a narrow main lobe in a given direction, but it does not perform efficiently in all azimuth directions. Circular arrays can be rotated in the plane of the array without considerable change in beam shape due to the lack of edge elements, but it has no nulls in the azimuth plane. Nulls in the azimuth plane are important to reject unwanted signals.

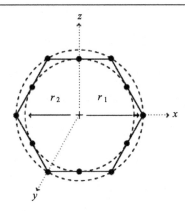

Figure 9.12: Hexagonal array pattern with 12 antenna elements.

Concentric arrays can be utilized to suppress the SLLs. Hexagonal antenna arrays have been shown to have good properties, and this is the constellation we wish to optimize.

Array thinning is the disabling of some radiating elements from an array to generate a pattern with low SLL. For the 12-element hexagonal array, we let the elements only have the two states active or disabled. All active elements are fed with the same signal amplitude, while the disabled elements are turned off.

The far-field pattern of a general array can be written

$$AF(\theta, \phi) = \sum_{n=1}^{N} A_n e^{j(\alpha_n + k R_n \cdot a_r)},$$

where N is the number of isotropic elements, A_n the excitation amplitude of the nth element, a_n the relative phase of the nth element, R_n the position vector of the nth element, which also depends on the geometry of the array, a_n the unit vector of an observation point in spherical coordinates, and k the wave number.

The hexagonal array can be described as consisting of two concentric N-element circular arrays of different radii r_1 and r_2, defining the pattern with an element on either circle.

Fig. 9.12 shows the geometry of a regular hexagonal array with $2N$ elements ($N = 6$), of which N elements are located at the vertices of the hexagon and the other N elements are located at the midpoints of the sides of the hexagon.

Considering the geometry, we have

$$AF(\theta, \phi) = \sum_{n=1}^{N} \left[A_n e^{jkr_1 \sin\theta(\cos\phi_{1n}\cos\phi + \sin\phi_{1n}\sin\phi)} + B_n e^{jkr_2 \sin\theta(\cos\phi_{2n}\cos\phi + \sin\phi_{2n}\sin\phi)} \right],$$

where

$$r_1 = d_e / \sin(\pi.N),$$
$$r_2 = r_1 \cos(\pi/N),$$

where d_e is the interelement spacing along any side of the hexagon, $\phi_{1n} = 2\pi(n-1)/N$ is the angle in the x-y plane between the x-axis and the nth element at the vertices of the hexagon, $\phi_{2n} = \phi_{1n} + \pi/N$ is the angle in the x-y plane between the x-axis at the midpoint of each edge of the hexagon, and A_n and B_n are the relative amplitudes of the nth element placed at the vertices and the midpoints of the hexagon, respectively.

There are two target functions reflecting the design objectives, the first T_1 related to the normalized SLL, which we wish to minimize, and the second T_2 related to the beam width expressed as the first null bandwidth (FNBW). We have

$$J_1 = \min_{\theta \in \{-90°, -|FN|°\} \wedge \{|FN|°, 90°\}} \{\max\{20\log|AF(\theta)|\}\}, \qquad (9.1)$$

$$J_2 = W_1[(\text{SLL}_i - \text{SLL}_0)] + W_2[(\text{FNBW}_i - \text{FNBW}_0)], \qquad (9.2)$$

where FN is the first null in degrees, which depends on the array pattern, SLL_i and SLL_0 are the desired and computed value of the SLLs, and FNBW_i and FNBW_0 are the desired and computed values of the FNBW, respectively. The desired values are the targets for the thinned array; W_1 and W_2 are importance weights used to control the relative importance of each term in (9.2). Since our primary goal is to minimize SLL, we choose $W_1 > W_2$.

We use the firefly algorithm to minimize (9.1)–(9.2), so that SLLs and FNBW are minimized. In the present experiment, a firefly algorithm is used to find an optimal antenna array excitation in 12-element and 18-element hexagonal arrays with uniformly distributed antenna elements, such that SLLs and FNBW are minimized compared to an antenna configuration with uniform excitation. The optimization problem is formulated to minimize the objective function (9.2) with $W_1 = 0.7$ and $W_2 = 0.3$. The optimization therefore prioritizes a minimization of the SLL over FNBW. The radius is set to unity and the cut-off signal strength is -20 dB. The element spacing d is set to $d = 0.50$, $d = 0.55$, and $d = 0.60$.

The firefly optimization is implemented with $n = 40$ fireflies and a maximum number of 100 generations. The numerical results are summarized in Tables 9.2 and 9.3. The signal strengths for a uniformly excited and an optimized 12-element array are shown in Fig. 9.13.

Table 9.2: Optimization of a hexagonal array with 12 elements.

Spacing d	SLL_0	$FNBW_0$	SLL_{opt}	$FNBW_{opt}$	Thinning
0.50	−8.72	22	−18.58	22	41.7%
0.55	−8.70	20	−18.56	20	50.0%
0.60	−8.70	20	−17.35	18	50.0%

Table 9.3: Optimization of a hexagonal array with 18 elements.

Spacing d	SLL_0	$FNBW_0$	SLL_{opt}	$FNBW_{opt}$
0.50	−18.30	16	−19.52	18
0.55	−18.24	14	−19.60	16
0.60	−18.30	14	−18.64	14

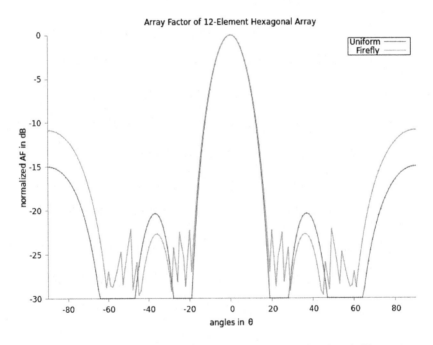

Figure 9.13: Hexagonal 12 pattern with spacing $\lambda = 0.50$.

CHAPTER 10

Robust Backbone Design

The backbone network is, as the name suggests, the main long-haul transport infrastructure. Due to its importance in keeping different parts of the network together, it is particularly important it is designed to be resilient and robust.

With network resilience we usually mean fault tolerance in one way or another. It is a difficult concept for two reasons. Firstly, there is no single universal definition of resilience that can be directly applied in network design. This follows from the many different possible failure events in a network and their dependence on internal or external factors. In addition, failures are rare events and difficult to measure. Lastly, not only is resilient network design $\mathcal{NP}$-hard in itself, but many definitions of resilience, such as path diversity, are $\mathcal{NP}$-hard to calculate too.

In 5G, network resilience can be regarded as an extended quality-of-service parameter, akin blocking in narrow-band transmission networks. We would here be interested in determining the *end-to-end resilience* between any two pairs of nodes in the network.

From a graph theoretical point of view, common measures of resilience include (degree) k-connectivity, link cuts, and node cuts. More elaborate concepts include the reliability polynomial and network strength. Operational measures include mean time between failure (MTBF) and mean time to repair (MTTR).

10.1 Network Resilience

To begin with, we need to quantify resilience, define it, and find some way of measuring it conveniently. This turns out to be far from trivial. Indeed, a network is normally assumed to be in a functioning state, and there are many potential reasons for the network to malfunction. This also implies that resilience cannot be reflected by a single network property. Even if we model failures in the network by independent probabilities – which of course is an approximation in most cases – it is difficult to find an efficient measure. Furthermore, empirical investigation is hardly a viable approach either. It is very hard to estimate rare events in a short time span. The events are bound to be rare in a successfully engineered network, and to be able to monitor the effect of architectural changes, the measurement period must be relatively short.

5G Networks. https://doi.org/10.1016/B978-0-12-812707-0.00015-2

We begin by trying to disentangle some terminology related to resilience, and we continue by sketching some resilience measures. We loosely follow definitions suggested in [87]. Resilience refers to a system's ability to maintain an acceptable level of service when exposed to internal and external disturbances.

We will call a flaw in a system a fault, which may cause an error and which may or may not be immediately observable. We let the term failure denote a system malfunction, that is, a state where the system does not operate as intended or expected, which is clearly observable.

We can broadly categorize resilience into the following disciplines. These are closely related topics but treated somewhat differently:

- Component reliability.
- System (network) design.
- Service quality under high load.
- Protection against external attacks.
- System recovery.

Fault tolerance can be described as providing resilience in terms of back-up or spare components in the case of faults. Such principles are usually based on the assumption of independent faults. A common rule of thumb for systems consisting of parallel components with low failure probability is to add M spare components to each system, known as the $N + M$ rule.

Survivability is a broader term referring to fault tolerance on node and link level, where the system is designed to cope with correlated faults. We therefore refer to the design of such networks as survivability network design. At this level, we distinguish between topological design, which is the hard-wired architecture aiming at minimizing the failure probability, and rerouting techniques with the purpose of avoiding or recovering from failure. In this chapter we will look in some detail at resilience diagnostics, topological design of survivable networks, and failure analysis of rerouting strategies.

Disruption tolerance is a network's ability to operate under variable and challenging conditions, which do not necessarily need to cause a failure. Networks are subject to changing environmental conditions, including varying traffic load, affecting its performance. Such conditions include channel quality, possibly affected by weather conditions, traffic distribution and delay, and variability in energy supply. Critical traffic load levels or delays can be reached as a consequence of self-similar traffic sources and so the network needs to be able to handle variable load conditions. Functionality for overload protection includes admission control, routing, and bandwidth allocation.

For users of network services, we can characterize its perceived quality by:

(1) dependability (availability and reliability),

(2) security, and

(3) performance.

Dependability can be viewed from two different aspects: availability and reliability. The measures defining dependability are mean time to failure (MTTF), which is the expectation of the failure probability density function, and MTTR, the expectation of the repair probability density function. The MTBF is the sum of MTTF and MTTR, so

$$\text{MTBF} = \text{MTTF} + \text{MTTR}.$$

Suppose a failure occurs. Then this failure is repaired on average in MTTR time, and from there to return to the operational state, the next failure occurs after MTTF time. The MTTF is also

$$\text{MTBF} = \frac{1}{\lambda},$$

where λ is the failure rate.

The availability of the system is the relative probability that the system is operable,

$$A = \text{MTTF}/\text{MTBF}.$$

Since the availability is expressed as MTTF and MTBF, we conclude that it depends on both how often a component fails and how fast it can be repaired or replaced. Expressed in probabilities, this can be written, as a function of time t,

$$A(t) = \mathbf{P}\{\text{operational at time } t | \text{no failure time } 0\}. \tag{10.1}$$

In contrast, reliability $R(t)$ is the probability that a system remains operable for a prescribed time period $[0, t]$. Given the failure cumulative distribution function $F(t)$, the reliability is defined as

$$R(t) = \mathbf{P}\{\text{no failure in } [0.t]\} = 1 - F(t). \tag{10.2}$$

We can immediately see from (10.1) and (10.2) that $A(t) \geq R(t)$.

Availability is thus the relative proportion of the time a connection is up, whereas reliability is the probability that a connection will be functioning continuously for a given time span. The importance of availability and reliability measures therefore depends on the service. For connectionless services such as web browsing the average uptime is the more important measure, and even frequent disruptions can be tolerated if they are of short duration. For connection-oriented services such as teleconferencing, a continuously operational connection is required.

Security is the system's ability to enforce, monitor, and report breaches of security policies. These may include procedures for authentication and authorization and protection of confidentiality and integrity. There are potential threats from both inside and outside an organization, and usually these are treated differently.

System performance is its operational conformance with a specified set of criteria. Network services need to fulfill requirements on throughput, packet loss, latency, and latency variation (jitter). The performance requirements are often specified so that services have to comply with critical levels for a minimum proportion of time. Performance is dependent on both load and traffic engineering and is supervised by congestion control.

Robustness is a system's ability to maintain an intended operational regime when the input and environmental parameters change. In terms of resilience, robustness can be seen as the collective improvement in avoiding, mitigating, or recovering from serious failures. These measures include topological design aspects, routing mechanisms, and congestion control.

10.2 Connectivity and Cuts

Firstly, we need to define what exactly is meant by "reliability", as well as some way to measure it (see Section 2.6). Measures that are studied in network reliability are $\{s, t\}$- or *two-terminal reliability*, *k-terminal reliability*, and *all-terminal reliability*. Using the undirected graph $G = (V, E)$ as a topological model, these measures correspond to the restriction of the connectivity analysis to subsets of the vertex set V of cardinality 2, $2 < k < |V|$, and $|V|$, respectively.

In other words, we select the appropriate measure depending on whether we want to study the reliability between two specific vertices, the mutual reliability in a k-subset of the vertices, or the mutual reliability between all vertices.

All-terminal reliability is appropriate when designing backbone networks having a minimum degree of resilience in some sense. In a wider context k-terminal reliability or two-terminal reliability are also of interest, in particular if we view the reliability of a link as an extended quality-of-service parameter.

The most basic reliability concept is to determine whether two vertices in a graph are at all connected or not. The connectivity of two vertices in a graph is defined in Definition 2.3.1. Generally speaking, we say that two vertices i and j in a graph are connected if there is a path from i to j.

Minimum Cuts

Suppose we are given a network $G(V, E)$ and a fixed small link failure probability p. From an all-terminal reliability point of view we are interested in finding the edge-cuts that disconnect any two nodes in the network. Since p is assumed to be small, cuts with a small number of edges have much larger probability for failure to occur than cuts having many edges. For this reason, we are interested in finding the minimum cut. We describe an implementation of a procedure by Karger [7].

We rephrase the minimum cut problem slightly. For a graph $G(V, E)$, let n be the number of nodes and m the number of links. We define a cut (A, B) as a partition of the vertices of G into two nonempty sets A and B. A link (v, w) crosses the cut (A, B) if one of the nodes v and w is in A and the other in B. If the links carry weights – assumed to be nonnegative – the value of the cut is the sum of the weights of links crossing the cut; otherwise, it is the number of crossing links. The minimum cut is the cut (A, B) in G having minimum value.

We assume that G is connected, because otherwise the problem is trivial. When the value of the cut is the number of links crossing it, this value is sometimes called the graph's connectivity, since it represents the smallest number of links that need to be removed to disconnect the graph.

Now, if we let f_k denote the number of link sets of size k, the minimum number of links whose removal disconnects the graph, then its disconnection probability is

$$P = \sum_{k=1}^{K} f_k p^k (1 - p)^{m-k}.$$

When p is very small, the sum P is approximated well for small K, motivating our focus on minimum cuts and near-minimum cuts. There are two principally different approaches to finding minimum cuts. These can be found either by solving a sequence of maximum flow problems, or by successive contraction of the graph topology. Karger presents an algorithm that potentially finds all approximately minimum cuts.

Since the number of links in a cut is likely to be a small fraction of the total number of links, a randomly chosen edge and contraction (merging of the nodes) would simplify the graph while leaving the cut intact.

First, we consider unweighted graphs, so the value of the cut is simply the number of links crossing it. Formally, we randomly select a link by choosing two neighboring nodes u_1 and u_2 (nodes having a link (u_1, u_2) between them). The graph is contracted by replacing u_1 and u_2 by a node u. All links incident on u_1 and u_2 are rewired to be links incident on u. Thus, the link (u_1, u_2) is deleted and each link (u_1, v) or (u_2, v) is replaced by a link (u, v). The rest of the graph remains unchanged. For a link $e = (u, v)$ in G, we denote by G/e the graph with link e contracted. If F is a set of links, G/F denotes the graph resulting from contracting all links $e \in F$.

The contraction is repeated until only two "metanodes" remain, which define the cut, representing the sets A and B.

We present the immediate results, the proofs of which can be found in [7].

Lemma 10.2.1. *A cut (A, B) is produced by the contraction algorithm if and only if no link crossing (A, B) is contracted by the algorithm.*

Theorem 10.2.2. *A particular minimum cut in G is returned by the contraction algorithm with probability at least*

$$\binom{n}{2}^{-1} = \Omega(n^{-2}).$$

In weighted graphs, the algorithm chooses a link (u, v) with probability proportional to the weight w_{uv} of (u, v).

The implementation of the contraction algorithm uses a weighted $n \times n$ adjacency A. We also keep a vector $\mathbf{d}$ of the weighted degrees. For a node u, the weighted degree $d_u = \sum_v A(u, v)$.

To randomly select links based on their weights, we can sample as follows. We first choose an endpoint u with probability proportional to d_u, fix u, and choose the second endpoint with probability proportional to $A(u, v)$.

The contraction is implemented by using the following procedures:

$$d_u \leftarrow d_u + d_v - 2A(u, v),$$
$$d_v \leftarrow 0,$$
$$A(u, v) \leftarrow 0,$$
$$A(v, u) \leftarrow 0$$

and for all nodes z except u and v,

$$A(u, z) \leftarrow A(u, z) + A(v, z),$$
$$A(z, u) \leftarrow A(z, u) + A(z, v),$$
$$A(v, z) \leftarrow 0,$$
$$A(z, v) \leftarrow 0.$$

We find that the two steps of the algorithm, the random link selection and the update step, are done in $\mathcal{O}(n)$ time, so the total time complexity is given as follows.

Corollary 10.2.3. *The contraction algorithm can be implemented to run in $\mathcal{O}(n^2)$ time.*

Figure 10.1: The complete four-node graph.

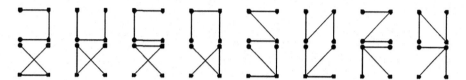

Figure 10.2: The 16 spanning trees of the complete four-node graph.

10.3 Spanning Trees

An alternative measure of graph G connectivity is the number $\tau(G)$ of spanning trees it has. Graphs with a high number of spanning trees are the most reliable for networks where the failure probability p of the links is high. This statement will be made more precise below. We recall the definition of spanning tree.

Definition 10.3.1 (Spanning tree). A spanning tree of a connected graph G is a connected, acyclic subgraph of G that spans every vertex. ●

Example 10.3.1. *Consider the very small network consisting of four nodes in Fig. 10.1. It has the 16 spanning trees depicted in Fig. 10.2.* □

The Kirchhoff Matrix Tree Theorem

The number of spanning trees can be computed efficiently for any given graph G by *Kirchhoff's theorem*, which relates the number of spanning trees of G to the eigenvalues of the Laplacian matrix of G, where the Laplacian matrix equals the difference between the degree matrix of G and its adjacency matrix [9].

Definition 10.3.2. Let $G(V, E)$ be a graph with $|V|$ labeled vertices and D its *degree matrix*, that is, the $|V| \times |V|$ matrix such that

$$D_{ii} = \deg(v_i) \quad \forall v_i \in V,$$

and A its adjacency matrix, that is,

$$A_{ij} = \begin{cases} 0, & \text{if } i = j, \\ 1, & \text{if } i \neq j \text{ and } v_i \text{ is adjacent to } v_j, \\ 0, & \text{otherwise.} \end{cases}$$

Then the *Laplacian matrix* of G is $L(G) = D - A$. Note that multiple edges are allowed in the adjacency matrix. ●

Example 10.3.2. *Let G be the small network as depicted in Fig. 10.1. The Laplacian matrix $L(G)$ is then the difference between the degree matrix and the adjacency matrix, that is,*

$$L(G) = \begin{pmatrix} 3 & 0 & 0 & 0 \\ 0 & 3 & 0 & 0 \\ 0 & 0 & 3 & 0 \\ 0 & 0 & 0 & 3 \end{pmatrix} - \begin{pmatrix} 0 & 1 & 1 & 1 \\ 1 & 0 & 1 & 1 \\ 1 & 1 & 0 & 1 \\ 1 & 1 & 1 & 0 \end{pmatrix} =$$

$$= \begin{pmatrix} 3 & -1 & -1 & -1 \\ -1 & 3 & -1 & -1 \\ -1 & -1 & 3 & -1 \\ -1 & -1 & -1 & 3 \end{pmatrix}.$$

☐

Note that each row or column of $L(G)$ sums to zero, so it is certainly true that $\det(L(G)) = 0$. An amazing result is therefore the following statement. For a proof, see for example [10].

Theorem 10.3.1 (The Kirchhoff matrix tree theorem). *Let $G(V, E)$ be a connected graph with $n = |V|$ labeled vertices and let $\lambda_1, \lambda_2, \ldots, \lambda_{n-1}$ be the nonzero eigenvalues of the Laplacian matrix Q of G. Then the number of spanning trees $\tau(G)$ of G is*

$$\tau(G) = \frac{1}{n}(\lambda_1 \lambda_2 \ldots \lambda_{n-1}).$$

Equivalently, the number of spanning trees is equal to the absolute value of any cofactor (signed minor) of the Laplacian matrix of G. This is obtained by choosing an entry a_{ij} in the matrix, crossing out the entries that lie in row i and column j, and taking the determinant of the reduced matrix. Thus, in practice, it is easy to construct the Laplacian matrix for the graph, cross out for example the first row and the first column, and take the determinant of the resulting matrix to obtain the number of spanning trees $\tau(G)$.

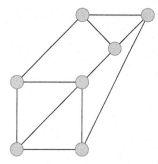

Figure 10.3: A network with seven nodes and 972 spanning trees.

Example 10.3.3. *Once again we let G be the graph in Fig. 10.1. We have already determined the Laplacian matrix in Example 10.3.2. By canceling the first row and the first column the reduced matrix is obtained as*

$$M_{11} = \begin{pmatrix} 3 & -1 & -1 \\ -1 & 3 & -1 \\ -1 & -1 & 3 \end{pmatrix},$$

and finally, taking the determinant of M_{11}, $\det(M_{11}) = 16$, gives the number of spanning trees for the graph. □

Example 10.3.4. *Let G be the seven-node network in Fig. 10.3. Its Laplacian matrix $L(G)$ is*

$$L(G) = \begin{pmatrix} 3 & -1 & -1 & -1 & 0 & 0 & 0 \\ -1 & 4 & -1 & 0 & -1 & 0 & -1 \\ -1 & -1 & 3 & 0 & 0 & 0 & -1 \\ -1 & 0 & 0 & 3 & -1 & -1 & 0 \\ 0 & -1 & 0 & -1 & 3 & -1 & 0 \\ 0 & 0 & 0 & -1 & -1 & 3 & -1 \\ 0 & -1 & -1 & 0 & 0 & -1 & 3 \end{pmatrix}$$

and the number of spanning trees is 972 (by the Kirchhoff matrix tree theorem). □

Graph Strength

The strength of a graph relates the concepts of spanning trees and edge connectivity of a graph. It can also be used as a reliability criterion. It is particularly useful for finding the connectivity "bottlenecks" in a graph. The *strength* $\sigma(G)$ of an undirected simple graph $G = (V, E)$ can be seen as a number corresponding to partitioning of the graph into subgraphs by disconnecting it where it is "weakest". The strength is the minimum ratio of deleted edges to the number of subgraphs created.

Definition 10.3.3. Let Π be the set of all partitions of V and $\partial\pi$ be the set of edges crossing over the sets of the partition $\pi \in \Pi$. Then

$$\sigma(G) = \min_{\pi \in \Pi} \frac{|\partial\pi|}{|\pi| - 1}.$$

●

There are several known algorithms for computing the strength of a graph. One of them, an approximate, but intuitively illuminating, algorithm [11] is presented here. It is based on an alternative definition, formulated in terms of spanning trees.

Definition 10.3.4. Let $\mathcal{T}$ be the set of spanning trees of G, Π the set of all partitions of V, and $\partial\pi$ the set of edges crossing over the sets of the partition $\pi \in \Pi$. Then

$$\sigma(G) = \max \left\{ \sum_{T \in \mathcal{T}} \lambda_T : \lambda_T \geq 0, \sum_{e : e \in T} \lambda_T \leq w(e) \right\},$$

where λ_T are real numbers and $w(e)$ is some edge weight. ●

The dual problem to the one formulated in Definition 10.3.4 is

$$\sigma(G) = \min \left\{ \sum_{e \in E} w(e) y_e : y_e \geq 0, \sum_{e \in T} y_e \geq 1 \right\}. \tag{10.3}$$

Theorem 10.3.2. *Given a connected graph G and a positive real $\epsilon \leq 1/2$, there exists an algorithm of computational time $O\left(m \log(n)^2 \log\left(\frac{m}{n}\right) / \epsilon^2\right)$ that returns a set of trees $T_1, \ldots .T_p$ of G, associated to real positive numbers $\lambda_1, \ldots, \lambda_p$, with*

$$\forall e \in E, \qquad \sum_{i \in \{1, \ldots, p\} : T_i \ni e} \lambda_i \leq 1$$

and

$$\sum_{i \in \{1, \ldots, p\}} \lambda_i \geq \frac{1}{1 + \epsilon} \sigma(G).$$

The expression in Eq. (10.3) can be used directly to formulate an algorithm for computing the strength.

Algorithm 10.3.3 (Strength).

Given a graph $G = (V, E)$ and a small number ϵ.

STEP 0:

    Initiate by assigning to each edge $e \in E$ a small weight $w(e) = \delta = O(n^{-3/\epsilon})$;

STEP 1

        while $w(T) < 1$ **do**

            For each $e \in T$, multiply $w(e)$ by $(1 + \epsilon)$

            Compute a minimum spanning tree T with respect to w

            Calculate $w(T) = \sum_{e \in T} w(e)$

        end

Output $w(G) = \sum_{e \in E} w(e)$, the strength of G.   ●

In step 2. we can alternatively update $w(e)$ by a small additive constant ε. This variation may be useful to obtain a different relation between speed of convergence and approximation error.

The algorithm determines, in each iteration, the minimum spanning tree of the graph, and whenever an edge e belongs to the minimum spanning tree, its weight is increased by a small amount. The edges in the "weakest cut" in the graph will consequently receive the largest weights, because they are used more often than other edges.

An important property of the strength is that each subgraph (which is not a singleton) generated by partitioning the graph according to Definition 10.3.3 has better strength than the original graph. Let $P = \{S_1, \ldots, S_p\}$ be a partition of G that achieves the strength of G, that is,

$$\sigma(G) = \frac{w(\delta\{S_1, \ldots, S_p\})}{p - 1},$$

then, for all $i \in \{1, \ldots, p\}$, we denote by $G(S_i)$ the restriction of G to S_i, and we have

$$\sigma(G(S_i)) \geq \sigma(G(S)).$$

If c is the minimum cut of G, the strength lies between $c/2$ and c. The upper bound follows immediately from the definition of the minimum cut: we have two disjoint sets if the edges in the minimum cut are removed, and Definition 10.3.3 requires $\sigma(G)$ to be smaller than or equal to this number.

The interpretation of the strength is readily provided by the *Tutte–Nash-Williams theorem*.

Theorem 10.3.4 (Tutte–Nash-Williams). *A graph G contains k edge-disjoint spanning trees if and only if the strength of G is larger than or equal to k, that is, $\sigma(G) \geq k$.*

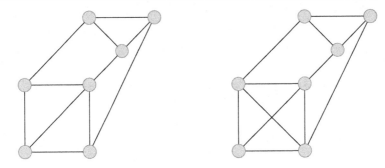

Figure 10.4: The optimal 3-connected solution of Example 10.4.1 (left). A network with strength 2 (right).

The theorem suggests that the strength of a graph may be an appropriate reliability measure in certain situations. This may be the case when failures are dependent along a route, or when a fast switchover to an independent circuit is desired for some other reason. It may also be desired to have a number of disjoint trees if one wishes to separate different types of traffic onto edges dedicated to particular traffic types (for example payload and signaling).

Example 10.3.5. *Let G_1 be the first graph in Fig. 10.4 and G_2 the second graph. Exact calculation of the strength of G_1 using Definition 10.3.3 gives the fractions*

$$\frac{3}{1}, \frac{5}{2}, \frac{6}{3}, \frac{8}{4}, \frac{10}{5}, \frac{11}{6},$$

so the strength $\sigma(G_1) = \frac{11}{6} = 1.83$. We can therefore conclude, referring to Theorem 10.3.4, that G_1 does not have two edge-disjoint spanning trees.

Similarly, for G_2 we have

$$\frac{3}{1}, \frac{6}{2}, \frac{8}{3}, \frac{9}{4}, \frac{11}{5}, \frac{12}{6},$$

so the strength $\sigma(G_1) = \frac{12}{6} = 2$, and so two edge-disjoint spanning trees exist in G_2. Note that how to find the two disjoint spanning trees is, however, not obvious. One possible pair of disjoint spanning trees is depicted in Fig. 10.5.

Computing the graph using Algorithm 10.3.3 yields the approximate values $\hat{\sigma}(G_1) = 1.77640$ and $\hat{\sigma}(G_2) = 1.93614$. We therefore conclude that $\hat{\sigma}(G_1) < \hat{\sigma}(G_2)$.

The Reliability Polynomial

The *reliability polynomial* is a probabilistic measure of network reliability. It is defined for a connected graph $G(V, E)$ in which each edge is associated with the probability p of operating

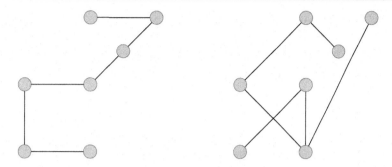

Figure 10.5: An edge-disjoint spanning tree (left). The complementary edge-disjoint spanning tree (right).

(failing with probability $q = 1 - p$) and all edges operate or fail independently of each other. This is a simple model which has been used extensively in network design and analysis.

The reliability polynomial ties together topological aspects (the coefficients) with the operational probability p. Often, the probability p is not known, but the polynomial can nevertheless be used to compare two network topologies and give insight into their reliability properties.

It should be kept in mind that the model is an idealization: faults are rarely independent or fault rates uniform across a network.

Consider a graph $G(V, E)$ in which each edge $e \in E$ is operational with the same probability p, independently of other edges. We introduce a *structure function*

$$\phi : 2^E \mapsto \{0, 1\},$$

which is a mapping from the *state space* – the set of all possible states (which is of size $2^{|E|}$) – to the binary set $\{0, 1\}$ representing "failed" and "operational" states of the network, respectively.

Let the set of operational edges $S \subseteq E$ be the *state* of the network. That is, the network is in state S when all edges of S are operational and all edges $E \ S$ (not in S) fail. The *state* S is then operational when $\phi(S) = 1$ and fails when $\phi(S) = 0$.

Consider a state S in which $|S| = i$ edges are operational out of a total $|E| = m$ edges. Then the probability of the network being in a state S is $\mathbf{P}(S) = p^i(1 - p)^{m-i}$. The reliability polynomial can formally be defined as

$$R(G; p) = \sum_{S \subseteq E} \mathbf{P}(S)\phi(S),$$

since the 2^m states are disjoint events covering all possibilities. Then $R(G; p)$ is a polynomial of degree at most m in one variable, the edge operation probability p. Thus, if every edge operates independently, the reliability polynomial $R(G; p)$ evaluated at p gives the probability that the graph is in an operating state.

There exist several forms of the polynomial. The two most useful in this context are the forms where the coefficients are expressed in the number of operational states and the number of cut-sets, respectively. Let N_i be the number of operational states with i edges in operation. Let $F_i = N_{m-i}$. Then the *all-terminal reliability polynomial* is

$$R(G; p) = \sum_{i=0}^{m} F_i (1 - p)^i p^{m-i},$$

which is called the *F-form* of the reliability polynomial. Similarly, if C_i is the number of sets consisting of i edges whose removal renders the network to fail, then

$$R(G; p) = 1 - \sum_{i=0}^{m} C_i (1 - p)^i p^{m-i}.$$

It follows that $R(G; 0) = 0$ and $R_\phi(G; 1) = 1$, provided that $\phi(\emptyset) = 0$ and $\phi(E) = 1$. Note that

$$F_i + C_i = \binom{m}{i}.$$

Bounds

Suppose that we were able to find lower and upper bounds, $N_i^{(L)} \leq N_i \leq N_i^{(U)}$, on each coefficient. Then

$$\sum_{i=0}^{m} N_i^{(L)} p^i (1 - p)^{m-i} \leq R(G; p) \leq \sum_{i=0}^{m} N_i^{(U)} p^i (1 - p)^{m-i}.$$

The exact calculation of all of the coefficients is #P-hard ("sharp P-hard"). However, some of them can easily be calculated.

If fewer than $n - 1$ edges are in operation, the graph is disconnected. Therefore, $F_i = 0$ for $i > m - n + 1$. If the smallest cut-set has size c (the edge connectivity is c), there is no way to remove fewer than c edges and disconnect the graph. Thus $F_i = \binom{m}{i}$ for $i < c$. The coefficient F_{m-n+1} is exactly the number of spanning trees of the graph, which can readily be calculated using Kirchhoff's theorem.

According to a lemma by Sperner ([9]), the following relation holds:

$$(m - i)N_i \le (i + 1)N_{i+1}.$$

Therefore, given F_i, a lower bound on F_{i-1} and an upper bound on F_{i+1} can be derived. Together with exact coefficients, we can compute lower and upper bounds on every coefficient in the F-form.

A Randomized Algorithm

Another approach for computation of the coefficients of the reliability polynomial approximately has been proposed by [12]. Given a graph G, we construct a tree where each node is a connected subgraph in G. The root consists of G itself and all children of a node in a given level are the possible connected subgraphs in G with one edge removed as compared with the previous level.

Each level i is composed of exactly $i!$ copies of each of the connected subgraphs having exactly $|E| - i$ edges. Each level of the tree has exactly $i! C_{|E|-i}$ nodes.

The number of nodes in the tree can be estimated by a randomized algorithm proposed by [12]. The counting of the number of subtrees is done by a procedure proposed by [13].

Algorithm 10.3.5 (Reliability polynomial coefficients).

Given a graph $G = (V, E)$.

STEP 0: Set $a_0 = 1$ and let C be an empty vector of coefficients
STEP 1: **for** $k = 1$ **to** $|E| - |V| + 1$ Let D_k be the set of all edges which if removed do not disconnect G Set $a_k = |D_k|$ Uniformly select and edge e from the set D_k Let the new graph G be $G - \{e\}$, that is, G with the edge e removed
STEP 2: **for** $k = 0$ **to** $|E| - |V| + 1$ Set $C_{|E|-k} = \prod_{0 \le i \le k} a_i / k!$
STEP 3: **for** $k = 0$ **to** $|V| - 2$ Set $C_k = 0$

Output the set of coefficients C.   ●

Algorithm 10.3.5 can be run N times to improve the coefficient estimates. Then, in step 1, we let a_k be an average of the magnitudes of $|D_k|$. Step 3 follows from the definition of the reliability polynomial; these coefficients are always zero.

Example 10.3.6. *Consider the graph G in Fig. 10.6, studied in [14]. The graph has the reliability polynomial coefficients $(1, 11, 55, 163, 310, 370, 224)$. Using Algorithm 10.3.5 with 400 iterations gives the coefficients $(1, 11, 55, 163, 309, 370, 224)$, which is a very good approximation.*   □

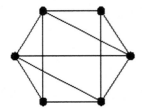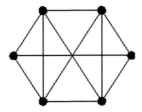

Figure 10.6: Two graphs with different reliability polynomials.

10.4 Minimum-Cost Survivable Network

In this section, a survivable network design problem is discussed where the design criterion is expressed as a minimum requirement on the vertex connectivity $\kappa(G)$ in the graph G. Associated with the problem is a cost matrix defining the cost of all potential edges in the graph. The optimal solution is the graph which fulfills the reliability requirement and has the lowest cost. A formal definition of the *minimum-cost survivable network (MCSN)* problem is given in [46].

A requirement connectivity matrix $R = (r_{ij})$ is a matrix defining the two-terminal connectivity requirement for each pair i, j of vertices. Similarly, the "actual" connectivity matrix $S = (s_{ij})$ is defined for each candidate solution.

The MCSN problem is $\mathcal{NP}$-complete, so the best we can hope for is to find an approximately optimal solution. A simple and yet often fruitful approach to find a good candidate is to search for a solution in a neighborhood of an initial "good guess". Such a *local search* procedure can be constructed given that:

(a)   A reasonably good initial solution can be found.
(b)   A transformation for generating similar – but slightly different – solutions by making changes to the initial solution can be defined. This transformation defines the *search neighborhood*.

In order to solve the problem, we need a way to generate candidate solutions and a way to test these candidates for *feasibility*, that is, verifying that the design criterion (10.5) is met. Clearly, the feasible solution with the lowest cost (10.4) is the solution to the problem. We conclude that a necessary condition for feasibility is that the minimum degree is at least k, where k is the prescribed (uniform) connectivity. This condition does not, however, guarantee the vertex connectivity of the graph. We also note that by the *handshaking lemma* it follows that if we have an odd number of vertices and k is odd, at least one vertex will have a connectivity larger than k.

Figure 10.7: Convert a node v in an undirected graph (left) to two nodes v_1 and v_2 in a directed graph (right).

Testing Feasibility

For each solution, its feasibility needs to be verified, that is, we need to make sure that (10.5) holds. For simplicity, we will let the connectivity requirement be $r_{ij} = k$ for all i, j. This uniform connectivity requirement leads to a more efficient algorithm than for the general case, although the principles of the method remain the same. The algorithm for verifying feasibility is based on the *max-flow min-cut* algorithm.

If the edges in a network are assigned unit capacities, the max-flow algorithm will give zero or unit augmenting capacities and thus provides a way to verify feasibility of the candidate solution. First, we introduce some results on the complexity of the max-flow min-cut algorithm. The following theorems (for proofs, see [46]) are valid for directed graphs $G = (V, E)$ with unit edge capacities.

Theorem 10.4.1. *For digraphs $G = (V, E)$ with unit edge capacities, the max-flow algorithm requires at most $O(|V|^{2/3} \cdot |E|)$ time.*

If, in a directed graph $G = (V, E)$ with unit edge capacity, each vertex has either indegree 1 or 0, or outdegree 1 or 0, the graph is called *simple*. An upper bound on the complexity of calculating the two-terminal connectivity is stated in the following theorem (see, for example, Papadimitriou and Steiglitz [46]).

Theorem 10.4.2. *For simple networks, the max-flow algorithm requires at most $O(|V|^{1/2} \cdot |E|)$ time, where $|E|$ is the number of edges.*

Now we arrive at the following theorem. The proof is very instructive and can be used to construct an algorithm for verification of solution feasibility.

Theorem 10.4.3. *Let $G = (V, E)$ be an undirected graph and i, j two distinct vertices in V. Then it is possible to find the vertex connectivity $s_{ij} \geq r_{ij}$ in $O(|V|^{2.5})$ time.*

Proof. Create a directed graph $G' = (V', E')$ from $G(V, E)$ by replacing every vertex $v \in V$ by two vertices $v_1, v_2 \in V'$ as shown in Fig. 10.7. Consider the flow network obtained by assigning a unit capacity to every arc of G'. Then the maximum flow between vertices i and j in this network is the vertex connectivity s_{ij}, because the unit capacity arc from v_1 to v_2 means that the vertex v in the original graph G can lie on at most one path between i and j. The flow network corresponding to G' is simple, so by Theorem 10.4.2 we can calculate the maximum flow in $O(|V|^{2.5})$ time.   □

Now we can formulate the following procedure for verification of the connectivity between two vertices.

Proposition 10.4.4. *Let $G = (V, E)$ be a network where the capacity of every edge is 1. Then, for each pair of vertices $i, j \in V$, the value of a maximum flow in G equals the number of edge-disjoint directed paths in G.*

Generating an Initial Solution

An initial solution can be constructed using an algorithm based on a greedy heuristic described in [46]. As noted earlier, the degree of vertex i must be at least $\max_j r_{ij}$. Define the *deficiency* of a vertex i in an undirected graph G as

$$\text{deficiency}(i) = \max_j r_{ij} - \deg(i).$$

Next, add edges to the graph G until all the deficiencies are nonpositive. We then need to test the resulting graph for feasibility using the max-flow algorithm with unit edge capacities.

Algorithm 10.4.5.

(0) Initiation: Order the vertices (randomly) and create an array of size $|V|$ containing the deficiency of each vertex.
(1) Starting from the left, add an edge between a vertex with the largest deficiency and one with the second-largest deficiency. Of all the vertices with second-largest deficiency, we choose the one that results in the smallest increase in cost; all other ties are resolved by choosing the earliest vertex in the array. Multiple edges are not allowed.
(2) If all deficiencies are equal to or less than zero, stop: an initial solution has been found; else go to 1.

The procedure is used and illustrated in detail in Examples 10.4.1 and 10.4.2 below.

The Search Neighborhood

We can create a neighborhood of a given initial feasible solution by using the *X-change* transformation. We will assume that the costs are proportional to distances, as no capacity is considered in this problem. Without loss of generality, we let $c_{ij} = d_{ij}$ for the edges (i, j).

Figure 10.8: Connections before the X-change (left) and connections after the X-change (right).

Definition 10.4.1 (The X-change neighborhood). Let the set of graphs feasible in an instance of the MCSN problem be denoted by F. That is, F consists of all graphs with a given number of vertices and vertex connectivity satisfying

$$r_{ij} \geq s_{ij} \quad \forall i, j, i \neq j.$$

Consider a graph $G = (V, E) \in F$ in which the edges (i, l) and (j, k) are present and the edges (i, k) and (j, l) are absent. Define a new graph $G' = (V, E')$ by removing edges (i, l) and (j, k) and adding edges (i, k) and (j, l). That is,

$$E' = E \cup \{(i, k)(j, l)\} - \{(i, l)(j, k)\}.$$

Then if $G' \in F$, we say it belongs to an X-*change neighborhood* of G, and the set of all X-changes of G defines the X-change neighborhood. If the new cost is less than the old solution, that is,

$$d_{ik} + d_{jl} < d_{il} + d_{jk},$$

then the X-change is called *favorable*. The transform is illustrated in Fig. 10.8. ●

The reason for selecting the X-change neighborhood here from many possible search neighborhoods is that the transformation preserves the degrees of the vertices. This property makes the feasibility test of the new graph more efficient. In fact, if we had to check the entire graph for feasibility after each favorable X-change candidate was discovered, the local search algorithm would be very slow, but it turns out that a complete check is not necessary. In the case that $r_{ij} = k$, it is only necessary to check that two vertex connectivities are preserved to establish the feasibility of the new graph.

Proposition 10.4.6. *If an X-change on a feasible network destroys feasibility by reducing s_{ab} below r_{ab}, where $a, b \in V$, then either*

$$s_{ik} < r_{ab} \ or \ s_{jl} < r_{ab},$$

where the X-change removes edges $\{i, k\}$ and $\{j, l\}$.

This follows immediately from the fact that adding new edges cannot decrease the connectivity, and thus we only need to check that the connectivity between the nodes i and k and between j and l has been preserved. Since the connectivity between any of i, j, k, and l and any of the rest of the nodes is satisfied before the X-change, the only way the connectivity between any other pair of nodes (a and b, say) can be affected is by changing paths, for example from $a \rightarrow i \rightarrow l \rightarrow b$ to $a \rightarrow i \rightarrow k \rightarrow b$.

Since the X-change has the property that it preserves the number of edges and the degree of every vertex, it is desirable to have an assortment of starting solutions in order to widen the search space, possibly with a different number of edges and different vertex degrees. This may sometimes be obtained by randomly reordering the vertices before applying the greedy heuristic to create an initial solution. These starting solutions tend to have low cost and a small number of edges [46].

Algorithm Summary

We summarize the discussion above as an algorithm, assuming that the cost matrix C and the uniform connectivity requirement $r_{ij} = k$ are given.

(1) Generate an initial solution using Algorithm 10.4.5.
(2) Test the initial solution for feasibility using the max-flow min-cut algorithm and Proposition 10.4.4. If the solution is infeasible, permute the nodes and go to step 1.
(3) Start local search; Compute the cost c_{init} of the initial solution. Find vertices for an X-change and transform the graph.
(4) Test the new graph for feasibility, using Proposition 10.4.6. If the new solution is infeasible, X-change back to the original solution and go to step 3.
(5) Compute the cost of the new solution c_{new}. If $c_{new} < c_{init}$, accept the new solution, let $c_{init} < c_{new}$, and go to step 3.

Note that we need to specify some sort of terminating criterion for the algorithm, for example the maximum number of iterations.

Example 10.4.1. *Consider the MCSN problem for the seven nodes depicted to the left in Fig. 10.9, an example borrowed from Steiglitz et al. [88][46].*

Let the edge costs be the integer part of the Euclidean distances between the corresponding vertices. Let, for example, the distance between vertex i and vertex j be 20 units and suppose the vertices are enumerated as in Fig. 10.9. The cost matrix is then

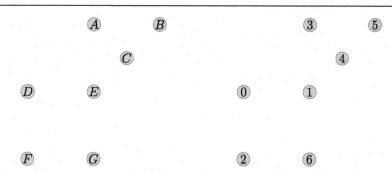

Figure 10.9: Seven given vertices that are to be connected so that the resulting network is 3-vertex-connected at minimum possible cost (left). A random permutation of the vertices yields initial vertex ordering (right).

Table 10.1: Initial solution generated by the heuristic algorithm.

3	3	3	3	3	3	3
2	2	3	3	3	3	3
2	2	2	3	3	3	2
2	2	2	2	2	3	2
2	2	2	2	1	2	2
1	2	1	2	1	2	2
1	1	1	1	1	2	2
1	1	1	1	1	1	1
0	1	1	0	1	1	1
0	0	1	0	0	1	1
0	0	0	0	0	0	1
0	−1	0	0	0	0	0

$$
C = \begin{pmatrix}
0 & 20 & 20 & 28 & 31 & 44 & 28 \\
20 & 0 & 28 & 20 & 14 & 28 & 20 \\
20 & 28 & 0 & 44 & 42 & 56 & 20 \\
28 & 20 & 44 & 0 & 14 & 20 & 40 \\
31 & 14 & 42 & 14 & 0 & 14 & 31 \\
44 & 28 & 56 & 20 & 14 & 0 & 44 \\
28 & 20 & 20 & 40 & 31 & 44 & 0
\end{pmatrix} .
$$

We use Algorithm 10.4.5 to generate an initial solution. The procedure is summarized in Table 10.1.

(1)   Initially all vertices have the same deficiency and we start from vertex 0, that is, the first vertex to the left. All of the other vertices are candidates, but the two closest are

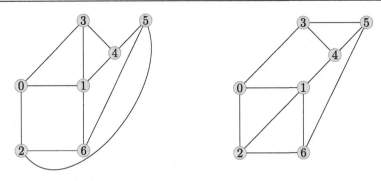

Figure 10.10: An initial feasible solution (left). The optimal 3-connected solution (right).

vertices 1 and 2. Since 1 precedes 2, we choose vertex 1, add the edge $\{0, 1\}$, reduce the deficiency of vertices 0 and 1 by 1, and go on to the next row.

(2) *The leftmost vertex with the highest deficiency is now vertex 2. Among the vertices with the same deficiency, vertex 6 is closest, so we add the edge $\{2, 6\}$, reduce the deficiency of vertices 2 and 6, and proceed to the next row.*

(3) *Starting from vertex 3, which is now the leftmost vertex with the highest deficiency, we see that vertex 4, of the two remaining vertices with deficiency 3, is closest. We add $\{3, 4\}$, reduce the deficiencies, and proceed.*

(4) *The only vertex with deficiency 3 is now vertex 5. Since no other vertex has deficiency 3, we have to select a candidate from vertices with deficiency 2. Of all such candidates, vertex 4 is the closest. We add $\{5, 4\}$, reduce the deficiency of vertices 4 and 5 by 1, and proceed.*

(5) *Again vertex 0 is the leftmost vertex with the highest deficiency. Since there is already an edge $\{0, 1\}$, vertex 1 is forbidden. The closest is then vertex 2, so we add $\{0, 2\}$ and reduce deficiencies.*

(6) *Starting from vertex 1, we see that vertices 3 and 6 are closest, and since 3 precedes 6, we add $\{1, 3\}$.*

(7) *Now only two vertices have deficiency 2: 5 and 6. Since there is no edge between them we add $\{5, 6\}$.*

(8) *Starting from vertex 0 again, the closest of possible vertices is now 3, so we add $\{0, 3\}$.*

(9) *The closest possible candidate to vertex 1 is now vertex 4: we add $\{1, 4\}$.*

(10) *Starting from vertex 2, since there is already an edge $\{2, 6\}$, we have no choice but to connect 2 to 5.*

(11) *Only vertex 6 remains with positive deficiency. The closest possible vertex to connect to is now vertex 1. All deficiencies are now nonpositive so we stop.*

The resulting graph is shown in Fig. 10.10. Next, we need to test its feasibility.

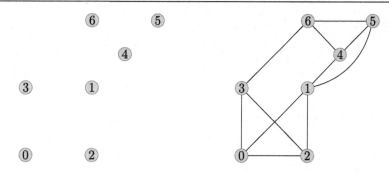

Figure 10.11: Different node enumeration (left). An infeasible initial solution (right).

Table 10.2: Initial solution generated by the heuristic algorithm.

3	3	3	3	3	3	3
2	3	2	3	3	3	3
2	2	2	3	2	3	3
2	2	2	2	2	3	2
2	2	2	2	1	2	2
1	2	2	1	1	2	2
1	1	1	1	1	2	2
1	1	1	1	1	1	1
0	0	1	1	1	1	1
0	0	0	0	1	1	1
0	0	0	0	0	1	0
0	−1	0	0	0	0	0

The initial solution is clearly feasible. We therefore accept it as a starting solution for our search. Its cost, which follows from the cost matrix, is 270.

For the local search, we pick two edges without common vertices and where the vertices belonging to the different edges are not neighbors, and we perform the X-change transformation. Let us say we have picked the edges {1, 3} and {2, 5}. Deleting these edges and creating the edges {1, 2} and {3, 5} produces a new graph. Since the node pairs {1, 3} and {2, 5} still have connectivity 3 after the X-change, the new solution is feasible. Furthermore, it has a cost of 242, so we accept it as a better solution than the initial. (This is, as a matter of fact, the optimal solution [46].) □

Example 10.4.2. *Suppose that we start with the vertices enumerated as in Fig. 10.11. The steps of the algorithm are listed in Table 10.2.*

10.5 A Primal-Dual Algorithm

Survivable network design can be defined differently depending on the resilience property of the network we focus on. To formulate the primal-dual approximation due to Gabow, Goemans, and Williamson [89][90], we use the following definition.

Definition 10.5.1 (The minimum-cost survivable network problem). Given a cost matrix (c_{uv}) and required connectivity matrix $R = (r_{uv})$, the MCSN problem is that of finding a graph $G = (V, E)$ with minimum total cost

$$\sum_{\{u,v\} \in E} c_{uv} \tag{10.4}$$

and vertex connectivity r_{uv} between distinct vertices u, v satisfying

$$s_{uv} \geq r_{uv} \quad \forall u, v, \quad u \neq v. \tag{10.5}$$

●

We are looking for a subset $F \subseteq E$ of edges of minimum total cost such that in the graph (V, F), all pairs u, v of nodes are at least r_{uv}-connected. First, restrict the discussion to the case where all r_{uv} are either 0 or 1. Even in this restricted version, the problem is $\mathcal{NP}$-complete, since it generalizes the Steiner tree problem. The Steiner tree problem can be formulated as follows. We are given a set of terminal nodes $X \subseteq V$, and we want to find a minimum-cost subtree T of G such that all terminals $v \in X$ are connected in T. For that purpose, we may include additional vertices $w \in V$ X into the tree T. The problem is known to be an $\mathcal{NP}$-complete problem, and it can be seen to be a special case of the MCSN design problem by setting $r_{uv} = 1$ for all $u, v \in X$ and $r_{uv} = 0$, whenever $u \notin X$ or $v \notin X$.

We associate a number r_v to each vertex $v \in V$, where $r_v \geq 2$ for sites that need high protection (such as sites belonging to the backbone network), $r_v = 1$ for sites belonging to, for example, an access network site, and $r_v = 0$ for optional sites. Then,

$$r_{uv} = \min\{r_u, r_v\}, \quad u, v \in V.$$

The matrix R is assumed to be symmetric, so that $r_{uv} = r_{vu}$ for all u, v. The goal of the design is to select a subset of edges $E \subset \bar{E}$ so that the nodes u and v are at least r_{uv}-connected, at minimum network cost. First, we note that there are equivalent definitions of r_{uv} connectivity. We have the following result.

Theorem 10.5.1. *The following statements are equivalent:*

(1) The nodes u and v are r_{uv}-connected.

(2) All edge cuts between u and v have at least r_{uv} edges.
(3) The nodes u and v are connected in $(V, \bar{E} \setminus E')$ for any edge set E' with $|E'| < r_{uv}$.
(4) There are r_{uv} edge-disjoint paths between u and v.

We construct a so-called *proper function* $f : 2^E \rightarrow \mathbb{N}$ such that:

(1) $f(\oslash) = 0$,
(2) $f(S) = f(V - S)$, for all $S \subseteq V$ (symmetry),
(3) $f(A \cup B) \leq \max\{f(A), f(B)\}$ (maximality), for A, B disjoint.

We may view the symmetry property of the function as the number of edges between the subgraph formed by the set S and the subgraph formed by the set $V - S$. These two numbers must be equal since the edges must have one endpoint in each set. The maximality principle says that the maximum of the function $f(\cdot)$ of two disjoint sets, for example, such as formed by dividing $G = (V, E)$ into two subgraphs, is always greater than for the union of the set, or the whole graph G in this case. We note that the strength of Section 10.3 is larger in the components of the graph than in the original graph. This follows from the maximality principle. It can be interpreted as follows: there is always a "weakest cut" in a graph G for which, if connecting A and B, the function value of the union $f(A \cup B)$ is lower than for the components A and B.

Now we can formulate the following integer program for the design problem by introducing a decision variable $x_{uv} \in \{0, 1\}$, representing the inclusion or exclusion of link (u, v):

$$\min \sum_{(u,v) \in E} c_{uv} x_{uv}, \tag{10.6}$$

$$x(\delta(S)) \geq f(S), \quad S \subseteq V,$$
$$x_{uv} \in \{0, 1\}, \quad (u, v) \in E.$$

Here, $\delta(S)$ denotes the set of edges with exactly one endpoint in S, called the coboundary of the set S. For the survivable network design problem, we define the proper function $f(S)$ as

$$f(S) = \max_{u \in S, v \notin S} r_{uv}.$$

To obtain a primal-dual algorithm, we adopt a pricing scheme by assigning prices p_S to sets with $f(S) > 0$, which is directly related to the connectivity requirements. The algorithm is building the network in n phases and each phase is ending with a "cleanup" phase. The stopping condition is when the constructed graph fulfills all the connectivity requirements r_{uv}. Then, we have

$$f_{max} = \max_S f(S).$$

The prices p_S should satisfy the following conditions:

(1) $\sum_{S:(u,v)\in\delta(S)} p_S \leq c_{uv}$ for all edges (u, v), so that no edge is "overpaid",
(2) if $(u, v) \in F$, then $\sum_{S:(u,v)\in\delta(S)} p_S = c_{uv}$, implying that if an edge is included in the solution, then it is fully paid for,
(3) if $p_S > 0$, then $|F \cap \delta(S)| = 1$, corresponding to the special case that $r_{uv} \in \{0, 1\}$ for all u, v.

If F and p satisfy all conditions (1)–(3), we say that they are at equilibrium. Since the problem is $\mathcal{NP}$-complete, we cannot determine the dual p in polynomial time. We therefore relax condition (3) to make the problem more tractable, and we write

(3') if $p_S > 0$, then $|F \cap \delta(S)| \leq 2$.

For a solution with p and F satisfying conditions (1), (2), and (3'), we obtain a 2-approximation to the optimum network.

Lemma 10.5.2. *If F and p satisfy constraints (1), (2), and (3') and F^* is an optimal solution to the network design problem, then $c_F \leq 2 \cdot c_{F^*}$.*

Proof. An optimal solution F^* satisfies

$$\sum_{e\in F^*} c_e \geq \sum_{e\in F^*} \sum_{S:e\in\delta(S)} p_S$$

$$= \sum_S p_S \cdot |F^* \cap \delta(S)|$$

$$\geq \sum_S p_S.$$

The last inequality holds because $|F^* \cap \delta(S)| \neq 0$ for all sets S with $p_S > 0$. On the other hand, F satisfies (2) and (3'), so

$$\sum_{e\in F} c_e = \sum_{e\in F} \sum_{S:e\in\delta(S)} p_S$$

$$= \sum_S p_S \cdot |F \cap \delta(S)|$$

$$\leq \sum_S 2 \cdot p_S,$$

which we know to be at most $2 \cdot c_{F^*}$, completing the proof. $\qquad\square$

The following version of the primal-dual algorithm does not quite maintain condition (3')
from above, but only an average of p_S, which suffices to yield a 2-approximation guarantee.
The algorithm consists of two phases, a growing phase and a cleanup phase.

In the growing phase, we maintain a set F of edges, and in each iteration we increase the
prices p_S for *active* sets, that is, sets which have at least one connectivity requirement that has
not yet been met. Once condition (1) is satisfied with equality for some edge (u, v), that is,

$$\sum_{S:e\in\delta(S)} p_S = c_{u,v},$$

we call e *tight* and add it to F.

Let F^k denote the set of edges that are selected at the end of iteration k and $C^{(k+1)}$ the set of
connected components induced by $F^{(k)}$ on V. Let $\mathcal{M}^{(k+1)}$ be those components in $C^{(k+1)}$
which have at least one unsatisfied requirement. That is, $S \in \mathcal{M}^{(k+1)}$ if there is some $u \in S$
and $v \notin S$ with $r_{uv} = 1$. We call the components $s \in \mathcal{M}^{(k+1)}$ *active* components. The prices
assigned to components S after iteration k are denoted by $p_S^{(k)}$. The algorithm maintains conditions (1) and (2) explicitly.

Algorithm 10.5.3 (Survivable network design 1).

Set $p_S^{(0)} \leftarrow 0$ for all S, $F^{(0)} \leftarrow \emptyset$; Let $k \leftarrow 1$.

STEP 1: k
      while $\mathcal{M}^{(k)} \neq \emptyset$ (there are active components) **do**
          Let $\Delta^{(k)}$ be the minimum Δ such that increasing $p_S^{(k)}$ by Δ for all active
          sets $S \in \mathcal{M}^{(k)}$ will make at least one additional edge tight.
          Let $p_S^{(k)} \leftarrow p_S^{(k-1)} + \Delta^{(k)}$ for all active sets $S \in \mathcal{M}^{(k)}$.
          Let $F^{(k)} \leftarrow \{(u, v) \in E | \sum_{S:(u,v)\in\delta(S)} p_S^{(k)} = c_{uv}\}$.
      end (while)

Output design $F^{(k)}$.   ●

In the special case that $r_{st} = 1$ for only two specific nodes $s, t \in V$ and $r_{ij} = 0$ for $\{i, j\} \neq$
$\{s, t\}$, the network design problem becomes equivalent to the shortest-path problem. In that
case, the algorithm will increase prices only for components containing either s or t, and
hence it becomes Dijkstra's shortest-path algorithm, run from the source and sink simultaneously. However, the result from the growing phase may contain many unnecessary edges.
Therefore, the algorithm has a second – cleanup – phase.

Lemma 10.5.4. *Let F be the final output after the cleanup phase and $\mathcal{M}^{(k)}$ the set of active components in some iteration k. Then,*

$$\sum_{S \in \mathcal{M}^{(k)}} |\delta(S) \cap F| \leq 2 \cdot |\mathcal{M}^{(k)}|.$$

Proof. First, we want to show that if we contract all $S \in \mathcal{C}^{(k)}$ to single nodes, the graph induced by F on these nodes is acyclic. By definition, all components $S \in \mathcal{C}^{(k)}$ were connected before any edges between different components are added to F. Hence, if any cycles had been formed at later stages, the cleanup phase would have deleted those edges that formed a cycle (since the connectivity requirements are satisfied without them). The contracted graph is actually a forest.

Next, we claim that in this forest, all inactive components $S \notin \mathcal{M}^{(k)}$ have at least 2 crossing edges, i.e., $|\delta(S) \cap F| \geq 2$, for if an inactive component had degree 1, the edge incident with it could have been deleted in the cleanup phase without violating the connectivity requirements. Therefore, all leaves in the forest resulting from the contraction are active. Because the sum over all degrees in a forest is at most twice the number of nodes, we obtain $\sum_{S \in \mathcal{C}^{(k)}} |\delta(S) \cap F| \leq 2 \cdot |\mathcal{C}^{(k)}|$. Therefore,

$$\sum_{S \in \mathcal{M}^{(k)}} |\delta(S) \cap F| \leq 2|\mathcal{C}^{(k)}| - \sum_{S \in \mathcal{C}^{(k)}/\mathcal{M}^{(k)}} |\delta(S) \cap F|$$

$$\leq 2|\mathcal{C}^{(k)}| - 2|\mathcal{C}^{(k)}/\mathcal{M}^{(k)}|$$

$$= 2|\mathcal{M}^{(k)}|.$$

$\square$

Theorem 10.5.5. *The above primal-dual algorithm is a 2-approximation for the network design problem with requirements $r_{uv} \in \{0, 1\}$.*

Proof. The lower bound on the cost of an optimal solution F^* is unchanged from the proof of Lemma 10.5.2, and hence, we know that $\sum_{e \in F^*} c_e \geq \sum_S p_S$. For an upper bound on the cost of the solution F returned by the algorithm, we divide the costs into the contributions made by different phases of the growing process. More specifically, for any set S, we know that $p_S = \sum_{k:S \in \mathcal{M}^{(k)}} \Delta^{(k)}$. We thus obtain

$$\sum_{(u,v) \in F} c_{uv} = \sum_{(u,v) \in F} \sum_{S:(u,v) \in \delta(S)} p_S$$

$$= \sum_S p_S \cdot |F \cap \delta(S)|$$

$$= \sum_k \Delta^{(k)} \cdot \sum_{S \in \mathcal{M}^{(k)}} |F \cap \delta(S)|$$

$$\leq \sum_k \Delta^{(k)} \cdot 2 \cdot |\mathcal{M}^{(k)}|$$

$$= 2 \cdot \sum_S p_S$$

$$\leq 2 \cdot \text{OPT}.$$

$\square$

Recalling Definition 10.5.1 for connectivity greater than 1, we have a graph $G = (V, E)$ with edge costs $c_{uv} \geq 0$, $(u, v) \in E$, and connectivity requirements $r_{uv} \geq 0$ for each pair of vertices $u, v \in V$, still assuming symmetric requirements $r_{uv} = r_{vu}$. The design problem amounts to finding a subgraph of G of minimum cost such that for all vertices $u, v \in V$ and cuts S with $u \in S$ and $v \notin S$, the number of edges in $\delta(S) \geq r_{uv}$.

For any cut S, define the function $f(S)$ as follows:

$$f(S) = \max_{u \in S, v \notin S} r_{uv}.$$

The constraint of the network design problem is to have $f(S)$ edges leaving S. The problem can again be expressed as an integer program (10.6). The approximation algorithm, however, consists of successive augmentation of an initial solution.

Suppose therefore that we have a set of edges $E_0 \subseteq E$ included in the initial solution where there still are some sets for which we do not have sufficient connectivity. Let $\mathcal{S}$ be a subset of such sets with too few edges, i.e.,

$$\mathcal{S} \subseteq \{S : \delta_{E_0}(S) < f(S)\}.$$

The algorithm aims at finding the minimum-cost subset of edges $F \subseteq E \setminus E_0$ with $\delta_F(S) \geq 1$ for all $S \in \mathcal{S}$. By iteratively adding edges at lowest cost, we can greedily increase the connectivity of the sets $S \in \mathcal{S}$. The challenge is to find the subset $S \in \mathcal{S}$ with the largest connectivity deficiency. To do this, define

$$k = \max_S f(S) - \delta_{E_0}(S),$$

where $S \subseteq E$ including the empty set $\emptyset$. A set S is called a largest deficiency set if $f(S) - \delta_{E_0}(S) = k$. We let $\mathcal{S}_k$ be the set of all largest deficiency sets, that is,

$$\mathcal{S}_k = \{S : f(S) - \delta_{E_0}(S) = k\},$$

and we augment the network if $k > 0$.

To find such sets $\mathcal{S}_k$, for each $u, v \in V$ find the minimum cuts in V, E_0, where u and v are on different sides of the cut, and compute k as

$$k = \max_{u,v}\{r_{uv} - C(u, v)\},$$

where $C(u, v)$ is a minimum u, v-cut. Such minimum u, v-cuts can be found by solving n^2 maximum flows or by using Karger's algorithm.

Theorem 10.5.6. *Suppose that the minimum cost of the original network design problem is OPT and that the deficiency is $k = \max_S f(S) - \delta_{E_0}(S)$. Then we can find $F \subseteq E \backslash E_0$ by augmentation of $\mathcal{S}_k$ with cost $C' \leq \frac{2}{k}OPT$.*

Corollary 10.5.7. *The general survivable network design problem can be solved to an approximation $\left(\frac{2}{R} + \frac{2}{R-1} + \cdots \frac{2}{1}\right) OPT \approx 2\ln(R)OPT$, where $R = \max_{u,v} r_{uv}$.*

Algorithm 10.5.8 (Survivable network design 2).

Set $p_S^{(0)} \leftarrow 0$ for all S, $F^{(0)} \leftarrow \emptyset$; Let $k \leftarrow 1$.

STEP 1: n
      while there are minimal sets in $\mathcal{S}_k$ which are not yet augmented, that is, $\delta_F(S) = 0$,
          Uniformly increase p_S on all such sets.
          If (u, v) is paid for, include it in F.
      end (while)
CLEANUP
      Consider $(u, v) \in F$ in the reverse order of being added and delete (u, v) if not needed for the augmentation.

Output design $F^{(k)}$. ●

In the algorithm, let $\mathcal{M}_i$ denote the minimal sets at each iteration and Δ_i the increase in prices. The stopping condition of the loop is when $\mathcal{M}_i = \emptyset$. Step 1 requires a subroutine that finds minimal sets, that is, a minimum-cut algorithm for every pair u, v of vertices.

Example 10.5.1. *We assume that we have the network in Fig. 10.12. The figure shows the edges usable in the design, and we assume that we have a symmetric requirement matrix $r_{uv} = 2$ for all $u, v \in V$, from which we easily construct the function $f(S) = \max_{u \in S, v \notin S} r_{uv} = (2, 2, 2, 2, 2, 2)$. We also have a matrix with edge costs*

$$c_{uv} = \begin{pmatrix} - & 23 & 20 & 32 & 45 & 51 \\ 23 & - & 10 & 23 & 23 & 37 \\ 20 & 10 & - & 15 & 29 & 32 \\ 32 & 23 & 15 & - & 32 & 20 \\ 45 & 23 & 29 & 32 & - & 32 \\ 51 & 37 & 32 & 20 & 32 & - \end{pmatrix},$$

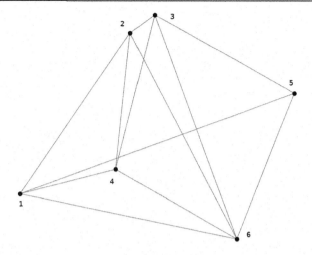

Figure 10.12: Initial graph $G(V, E)$ with feasible edges.

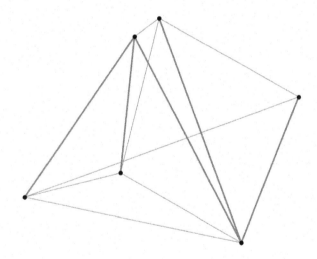

Figure 10.13: First phase adds edges for connectivity requirement $r_{ij} \in \{0, 1\}$.

and we set $E_0 = \emptyset$ and the set of active sets $\mathcal{C} = \{\{1\}, \{2\}, \{3\}, \{4\}, \{5\}, \{6\}\}$, that is, all individual vertices are forming the components of the active set, since no connection does yet exist and all requirements are unfulfilled.

To produce an initial solution we run Algorithm 10.5.3. As the algorithm progresses, the minimum cost $\Delta = \min_{u \neq v} c_{uv}/2$ that has to be paid in order to add an edge is calculated. This step generates $E_0 = 5$ edges as shown highlighted in Fig. 10.13. Next, we iterate to satisfy the connectivity of the deficiency nodes. As shown in Fig. 10.14, this adds four edges to F (drawn

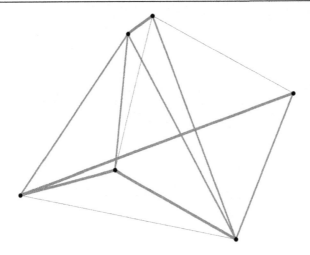

Figure 10.14: Second phase adds edges for requirement $r_{ij} > 1$.

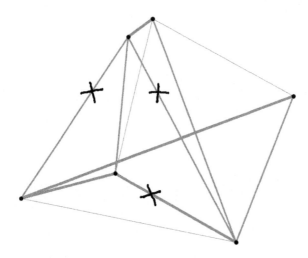

Figure 10.15: Third cleanup phase removes unnecessary edges.

wider than the rest). In the cleanup phase, we can remove marked links in Fig. 10.15 without violating the connectivity requirements. Note that a minimum of six links are necessary to achieve 2-connectivity.

CHAPTER 11

Traffic Engineering

Different traffic types require advanced traffic engineering techniques to ensure quality of service (QoS) and a fair resource allocation in networks. Specifically, we distinguish between real-time traffic, often most sensitive to delay and jitter, and nonreal-time traffic, using protocols like TCP, subjected to error correction to provide uncompromised data transfer.

For real-time traffic, we commonly use tunneling, for example by setting up a virtual private network (VPN). Such (connection-oriented) techniques use bandwidth reservation on specific paths to lower delay and deliver packets to the destination in the same order they are sent. The discussion is somewhat geared towards multipath label switching (MPLS), although the optimization methods are applicable in relation to other technologies as well.

We assume that we have a well-designed network allowing for path diversity, and we discuss how paths should be set up to optimize resource allocation and operational resilience.

11.1 Resilient Routing

For nondisruptive service, MPLS is set up on an active path with a protective path in the case failure occurs. It is crucial that the switchover from the failing active path is fast, and we may want that the protective path is using a completely different set of links to maximize its robustness.

We usually start out with an end-to-end load matrix that defines the resource requirements for active paths. The first planning step is to find the set of possible end-to-end paths and the dependence between these.

The K-Shortest Path

We describe an algorithm for finding K-shortest paths by Yen [91], but note that other algorithm exist as well. The algorithm requires paths to be loopless, which is usually what we are looking for in traffic engineering. By K-shortest paths we mean the shortest path between two nodes s and t, the second shortest path, and so on, up to the Kth shortest path.

As we can find the shortest path by the algorithms by Dijkstra or Bellman–Ford, we can in principle block any of the links along the shortest path and look for the shortest path in the so-restricted graph. In the algorithm by Yen, which as a matter of fact uses a shortest-path algorithm to find partial results, this search for paths is made more structured. We use Dijkstra's algorithm as a subroutine of Yen's algorithm and consequently require all link costs to be non-negative.

The algorithm finds the K-shortest paths iteratively and begins by finding the global shortest path. It uses two containers, A^k, holding the paths found up to level $k = 2, \ldots, K$, and B^k, holding potential shortest paths in step k. The algorithm tries to construct new paths from the shortest path(s), and we call a path a *root path* if the links in this path coincide with a previously found shortest path, and we construct a deviation starting from a *spur node*.

Initially, we determine A^1 by Dijkstra's algorithm. If there are K or more paths with the same cost, we are done. If the number of paths is greater than one and less than K, any of these paths is stored in A^l; the rest of these paths are stored in B^1, the container of candidates for $(k + 1)$-level shortest paths. Otherwise, if we have only one such path, it is in A^l.

In iteration $k = 2, \ldots, K$, we determine A^k as follows. We inspect the subpath consisting of the first i nodes in A^{k-1} with subpaths in previous steps A^j, $j = 1, 2, \ldots, k - 1$. If so, we set $d_{i,i+1} = \infty$, effectively removing the edge from the graph. The subpath is now the root path and node i is the spur node. The spur path is found as the shortest path from i to t, and a k-shortest path candidate is obtained by joining the root path and spur path. All such candidates are stored in B^k. As the kth shortest path we choose the path with shortest length from B^k and this is added to A^k. When we have found K shortest paths, the algorithm terminates.

The algorithm can be summarized as follows.

Algorithm 11.1.1 (Yen's algorithm).
Given an (undirected) graph $G = (V, E)$ with n nodes, nonnegative edge costs d_{ij}, and two vertices s, t.

STEP 0:
> Initiate an empty set of paths B

STEP 1:
> Determine A^1 by Dijkstra's algorithm

STEP 2 to k:
> **for** all deviations A_i^j, $j = 2 \ldots, k - 1, i = 2, \ldots, n$:
> fix a routepath R_i^k and find a spurpath S_i^k
> by Dijkstra's algorithm from v_i to v_n and $d_{ij} = \infty$,
> add the results to B
> Let A^k be the path in B with minimum cost
> **end**

Output: The K-shortest paths. ●

Static and Dynamic Routes

The rerouting strategy can be static, where the protective path is defined at service initiation, or dynamic, where an alternative path is determined at the time of failure of the active path.

To evaluate the reliability of static paths we note that all links need to be in operation for the path to work. Thus, letting p_e be the failure probability of a link indexed by e, we must have $p_a = 1 - \prod_e (1 - p_e)$.

A dynamic strategy potentially selects a path from all possible end-to-end paths; the resilience is given by the pairwise minimum cuts, which can be determined by Karger's algorithm. In most cases, we can assume that the cut is the dominating factor with failure probability $p_d = \prod_e p_e$, where e denotes a link index and p_e is the failure probability associated with that link.

Given a set of alternative feasible routes, we may want to choose two routes (active and protective) so that they are as independent as possible. By resilient routing, we mean imposing such a condition at some stage in the design process. After finding K-shortest paths, we may select maximum resilient paths which are maximum edge-disjoint. Choosing nonoverlapping paths is also important for capacity and load balancing reasons.

11.2 Multiprotocol Label Switching (MPLS)

In MPLS route planning follows a few straightforward principles. For a given node pair, an active tunnel – or label-switched path (LSP) – with a specified capacity is set up. In addition, a backup path is also defined according to one of several possible methods; see [92]. The backup path can be specified statically, or offline, and a failure on the active path triggers a fast switchover to the protective path already defined. Alternatively, the protective path can be defined at the time of failure of the active path. This procedure is known as dynamic route assignment, and it is based on an online routing method.

The obvious advantage of dynamic route assignment is that the protective path is set up on the link having the best conditions, both in terms of resources and load. From a resilience perspective, the dynamic route assignment is more secure, since the backup path can be selected from a large number of possibilities and according to the circumstances prevailing when the failure occurs. Regardless whether the backup route is determined statically or dynamically, it should be defined so that relevant QoS requirements are satisfied, translated into a realistic bandwidth requirement.

The MPLS recovery process is initiated when a fault on the working path is detected, which triggers identification of the protective path, and switchover of on-going traffic onto the backup path. The establishment of backup paths can be triggered and selected in different ways. Here, we will primarily focus on path resilience aspects related to the network topology.

Routing algorithms can be categorized as static or dynamic, depending on the type of routing information used for computing LSP routes. Static algorithms only use network information that does not change with time; dynamic algorithms use the current state of the network, such as link load and blocking probability. On the other hand, routing algorithms can be executed either online (on demand) or offline (precomputed) depending on when this computation is applied. In online routing algorithms path requests are attended to one by one, while of-fline routing does not allow new path route computation. This chapter is focused on dynamic online routing. QoS routing and particular capabilities of MPLS networks are introduced next.

The challenge in MPLS traffic engineering is the proper sizing of LSPs and the consequent efficient allocation of resources. The resource mapping must consider best effort traffic (not transported over LSPs) and ensure minimum disruption in the case of failure and activation of a protective path. A path definition is called dynamic when based on instantaneous network information, and static otherwise. In addition, the routing algorithm may be run online (on de-mand) or offline (scheduled). The calculation of paths subject to QoS requirements is known as QoS routing.

Routing algorithms deploy two main principles for route selection: shortest paths to find the path using the least resources and assignment of (dynamic) routes along least-loaded paths. The two principles aim at achieving routing efficiency and load balancing, respectively – two objectives which often are difficult to combine.

A first approach to the LSP planning problem between two given nodes is to find a set of paths, ordered by their distances. In fiber networks, the number of hops (intermediate nodes) in a path is used to represent distance rather than the actual sum of the cable lengths along the path. The reason for using the number of hops is that nodes better represent operational ex-penditure and delay along the path. Another consequence of using hops is that we are likely to obtain sets of alternative paths having the same number of hops. This serves our purpose, since we can choose the path with the largest available bandwidth from a set of paths with equal hop count.

The problem we study here is the route selection based on network information (in the form of link state metrics) and traffic flow requirements. We tacitly assume that QoS requirements for a certain traffic flow are known and expressed in bandwidth. The actual mapping of QoS to network resources is nontrivial and is usually determined using a queueing theoretic ap-proach; see [93] for a more detailed discussion on probabilistic traffic engineering methods. The focus here is on finding a resilient routing strategy satisfying given bandwidth criteria.

To be able to select an optimal route in a network, some information associated with the route needs to be available. The following metrics are assumed to be known:

- Available bandwidth: Given the required bandwidth of a flow, the link's available bandwidth determines the feasibility of using that particular link to route the flow. Clearly, we require that the available bandwidth is equal to or exceeds the required bandwidth.
- Hop count: As indicated above, the hop count measures (together with the bandwidth) the cost of routing in some general sense. The fewer the hops, the fewer network resources are occupied by the flow.
- Policy: Additional policies are used to filter out links that are unsuitable for the flow for some reason, performance, or other characteristics.

A straightforward per-flow path selection algorithm chooses paths with the minimum number of hops possible and, among those, the path where the available bandwidth is maximal. The available bandwidth is the smallest available bandwidth of the links making up the path, so the metric is a max-min principle, that is, $b_{\max} = \max_{p \in \mathcal{P}} \{\min_{l \in p} b_l\}\}$, where b_l is the available bandwidth on link l and p is a path among the set of minimum-hop paths $\mathcal{P}$. We note that this route selection maximizes the likelihood of satisfying the associated QoS requirements.

In [94], the authors propose a path selection method based on the Bellman–Ford shortest-path algorithm, which is adapted to find the paths of maximum available bandwidth for every number of hops. In other words, in iteration h it finds the path with maximum bandwidth among all paths having at most h hops.

Alternatively, the method can be based on Dijkstra's shortest-path algorithm. Links with insufficient available bandwidth to accommodate the flow can be filtered out in a preprocessing step, or the feasible bandwidth can be checked within the algorithm.

The optimization takes into consideration multiple objectives. In general this leads to intractable problems. Ways around this include sequential optimization steps and using a utility function with weights for each of the targets.

A number of factors complicate matters. Firstly, networks are usually organized hierarchically, so that a number of links constitute a hop. This becomes particularly important when routing resilience is concerned. Hops can therefore refer to physical links of logical links on different levels in the network hierarchy.

Mapping routes onto resources to obtain a global optimum in terms of performance, resilience, and cost is an extended multicommodity flow problem (see [93], for example), which is $\mathcal{NP}$-hard for unsplittable flows, as is the case for VPN tunnels. We may also want to guarantee some capacity for best effort traffic.

Route Assignment and Capacity Allocation

The overall route planning includes definition of active and protective routes and mapping these onto available network paths in a cost efficient manner. It requires a simultaneous assignment of all routes to find a global optimum rather than a per-flow assignment, which is known as a multicommodity flow problem.

The multicommodity flow problem with indivisible flows is $\mathcal{NP}$-complete. If flows are allowed to be split, linear programming can be used to find a solution in polynomial time. However, as the size of a problem instance grows rapidly with the number of vertices, the number of edges, and the number of commodities, linear programming becomes impractical for solving all but small problems.

A more efficient approach is to use an approximation algorithm. It is in general much faster than the linear programming approach and can in theory achieve a solution with arbitrary precision. The running time increases linearly with the number of commodities and inversely to the square of the chosen precision. The execution time is dominated by a number of minimum cost problems that must be solved in each iteration.

Problem Formulation

Let $G = (V, E)$ be an undirected graph with n nodes and m edges. Associated with each edge $(i, j) \in E$ there is a capacity limit u_{ij} and possibly a cost c_{ij} (which, for example, may be proportional to the physical distance). Suppose there is a total of $K > 1$ commodities in the network. Each commodity k is specified by its origin s_k, destination t_k, and demand d_k. Initially, we assume that for each commodity, the flows may be split onto different paths. When the path flows of different commodities interact, the sum of the flows on any single edge is referred to as the *edge flow*. A flow of commodity k carried on edge (i, j) is denoted f_{ij}^k.

Only undirected graphs are considered here. An important consequence of having several commodities is that flow antisymmetry does not hold in general, that is, $f_{ij} \neq -f_{ji}$. It is true only for flows belonging to the same commodity $f_{ij}^k = -f_{ji}^k$. It is also not likely that flows of the same commodity flow in opposite directions in undirected networks in reality. Therefore, we will usually relax this condition here.

The multicommodity flow problem can be expressed as a linear program as follows:

$$\max \Psi$$

$$\sum_{(i,j)\in E} (f_{ij}^k - f_{ji}^k) = \begin{cases} d_k, & i = s_k, \\ 0, & i \neq s_k, t_k, \\ -d_k, & i = t_k, \end{cases}$$

$$0 \leq |f_{ij}^k| \leq u_{ij}^k,$$

$$0 \leq \sum_k |f_{ij}^k| \leq u_{ij}.$$

There are several variants of the multicommodity flow problem. The objective function Ψ above depends on the quantity we wish to optimize. In the *minimum-cost multicommodity flow problem*, there is a cost $c_{ij} f_{ij}$ of sending flow on edge (i, j). In the linear program formulation we then maximize

$$\Psi = - \sum_{(i,j) \in E} c_{ij} \sum_{k=1}^{K} f_{ij}^k.$$

The minimum-cost variant is a generalization of the minimum-cost flow problem. In the *maximum multicommodity flow problem*, there are no restrictions on individual commodities, but the total flow is maximized. The sum of the outflows from each origin s_k for all commodities k gives the total network flow, so that

$$\Psi = \sum_{k=1}^{K} \sum_{j:(s_k,j) \in E} f_{s_k,j}^k.$$

The objective may be to find a *feasible* solution, that is, a multicommodity flow which satisfies all the demands and obeys the capacity constraints. More generally, however, we might want to know the maximum number z such that at least z percent of each demand can be transported without violating the capacity constraints. This problem is known as the *concurrent flow problem*. An equivalent problem is determining the minimum factor by which the capacities must be increased in order to transport all demands. In the *maximum concurrent flow problem*, the commodity with the poorest flow relative to its demand (referred to as its *throughput*) is maximized, i.e.,

$$\Psi = \min_{k \leq K} \frac{\sum_{j \in V} f_{(s_k,j)}^k}{d_k}.$$

It is this variant of multicommodity flow we will discuss here.

An Approximation Algorithm

The algorithm presented here, proposed by [95], is an $(1 + \epsilon)$-approximation algorithm for the maximum concurrent multicommodity flow problem. The algorithm as originally presented is rather complex and we will therefore allow for some simplifications in line with the ones suggested in [97].

Leighton et al. describe an algorithm proposed for solving the general concurrent flow problem with arbitrary capacities and demands. It is initiated with an arbitrarily routed flow which is gradually improved by rerouting individual commodities from highly congested edges to lightly congested ones. Flows are successively rerouted along minimum-cost flows computed in specially constructed auxiliary graphs.

The discussion focuses on the operation aspects of the algorithm and much of the underlying theoretical foundation has been simplified. Some of these modifications are justified by computational aspects – we want to avoid too large or too small numbers in the algorithm that may lead to numerical problems like overflow, underflow, or cancellations.

We note that, using the presented algorithm, approximately computing a k-commodity concurrent flow is about as difficult as computing k single commodity maximum flows.

To formulate the algorithm, we define the problem parameters as follows. Let a network be described by an undirected graph $G = (V, E)$ with positive upper flow limits (or capacities) u_{ij} assigned to each edge $(i, j) \in E$. We assume that G is connected and that it has no parallel edges. To simplify the discussion, we consider flow in only one direction, chosen arbitrarily, for each commodity. This can be justified by the fact that when we have a point-to-point flow in telecommunications, it is likely to be a duplex system. The role of origins and destinations in undirected networks can therefore be considered as symmetric. For any flow, some resources are reserved between the source and the sink, and from a planning point of view it does not matter in which direction the information flows as long as the flow reflects the total demand between the origin–destination pair.

Let n, m, and K be the number of vertices, edges, and commodities, respectively. We assume that the demands and the capacities have integer values. A commodity k is defined by its origin–destination pair $s_k, t_k \in V$ and the corresponding demand $d_k > 0$. This can be expressed in the triple of integers (s_k, t_k, d_k).

In principle, due to the minimum-cost problems that we need to solve – and these are usually solved exactly – we would assume integer demand and capacity values to guarantee convergence. Hence, when the capacities are modified in the course of the algorithm, these values would need to be rounded to the nearest integer values. In practice, however, the outcomes of the minimum-cost flow subproblems that only yield alternative paths for rerouting and approximate solutions are therefore acceptable. We assume that we have a suitable termination condition for minimum-cost flows, perhaps by properly rounding the input values. In any case, this rounding will not be considered as part of the core algorithm.

The optimization variables are the *flows* in the network. We distinguish between *edge flows* – the flow through an edge – and *path flows* – the fraction of a commodity flow following a specific path. Let $\mathcal{P}_k$ denote a collection of routes (or paths) from s_k to t_k in G and let $f_p^k \geq 0$

for every route $p \in \mathcal{P}_k$. Then the value of the flow is the sum of all *route flows* along $p \in \mathcal{P}_k$, i.e.,

$$f^k = \sum_{p \in \mathcal{P}_k} f_p^k.$$

The *edge flow* f_{ij}^k through an edge (i, j) is similarly defined as

$$f_{ij}^k = \sum_r \left\{ f_p^k : p \in \mathcal{P}_k \text{ and } (i, j) \in r \right\}.$$

The total flow on edge (i, j) is the sum of flows from all commodities using edge (i, j), i.e.,

$$f_{ij} = \sum_k f_{ij}^k.$$

In order to describe the optimization problem, we introduce a network scaling parameter $0 < z < 1$, called the *throughput*, whose value is such that at least z percent of each demand can be transported over the network without violating the capacity constraints.

The throughput, z, should be maximized in the maximum concurrent multicommodity flow problem. Since the algorithm is an $(1 + \epsilon)$-approximation, the best result we can expect is a solution for which $z \geq (1 - \epsilon)\hat{z}$, where $\hat{z}$ is the maximum possible throughput. However, rather than maximizing z, we will minimize the parameter $\lambda = 1/z$, called the *congestion*. The congestion on any edge is defined as

$$\lambda_{ij} = f_{ij}/u_{ij}, \quad \text{for any edge } (i, j) \in E.$$

Let the maximum edge congestion be $\lambda = \max_{(i,j) \in E} \lambda_{ij}$ in the network, and denote the smallest possible achievable congestion by $\hat{\lambda} = \min\{\lambda\}$. Without loss of generality, we assume that the multicommodity problem is *simple*, where each commodity has a single source and a single sink. Otherwise, a commodity can be decomposed into a simple multicommodity problem.

The problem of minimizing λ is equivalent to maximizing z, and $z \geq (1 - \epsilon)\hat{z}$ implies $\lambda \leq (1 + \epsilon)\hat{\lambda}$, where $\hat{\lambda}$ is the minimum possible λ. This is given by the fact that

$$\frac{1}{1 + \epsilon} \leq 1 - \epsilon,$$

by the first term of the Maclaurin expansion of the fraction. The reason for using λ in the formulation is that congestion can be defined for any flow $0 \leq f < \infty$, which is convenient when

forming derived quantities for solving the problem (since we want to scale the network resources rather than the demands). With this approach, a solution can be found to a problem
for which no feasible solution exists without scaling.

The parameter ϵ is an input parameter to the algorithm. We shall assume implicitly throughout
that $0 < \epsilon < 1/9$. It should be noted, however, that as ϵ decreases, the running time increases
inversely proportional to its square. In practice, it should be possible to find a value of ϵ which
is of the magnitude of the measurement errors of other input parameters.

Next, we define the *weight* of each edge as

$$w_{ij} = e^{\alpha \lambda_{ij}},$$

where α is a parameter defined as $\alpha = c \cdot s / \lambda$. The product of the two constants c and s scales
the exponent and can be adjusted to increase the performance of the algorithm. In [97] the
values

$$c = 19.1 - \ln m$$

and

$$s = 0.25$$

are used. The weights are used as cost parameters in minimum-cost flow problems that have
to be solved in each iteration of the algorithm. With this choice of α, the same congestion
level gives the same weight for any edge. Note that the value of α must be chosen carefully.
A too small value does not guarantee any improvement, whereas a too large value leads to
very slow progress.

In addition, a potential Φ, defined as

$$\Phi_{ij} = u_{ij} \cdot w_{ij},$$
$$\Phi = \sum_{(i,j) \in E} \Phi_{ij},$$

serves as a measure of the convergence of the algorithm. By rerouting an amount of flow onto
a less congested path results in a significant decrease in Φ. The potential is closely related to
the overall cost of the solution with respect to the length w_{ij}, that is,

$$c_{ij} = f_{ij} \cdot w_{ij} = \lambda_{ij} \cdot \Phi_{ij},$$
$$c = \sum_{(i,j) \in E} c_{ij}, \tag{11.1}$$

where f_{ij} are the edge flows. The cost improvement is used as a termination condition for the algorithm. Finally, the optimal amount of path flow to reroute is described by the parameter $0 < \sigma \leq 1$, whose value is found by minimizing Φ with respect to σ. This step ensures rapid convergence of the algorithm.

The algorithm terminates when the difference in costs between the old path r and the new path q satisfies the bound

$$c_p^k - c_q^k \leq \epsilon(c_p^k + \lambda \cdot \Phi/K) \qquad (11.2)$$

for all path flows f_p^k.

The main principles of the algorithm are as follows. The routine identifies an edge with high congestion λ_{ij} and selects one of the paths flowing through this edge for rerouting. The fictitious weight w_{ij}, based on the congestion and assigned to each edge, is penalizing edges with high congestion and promoting edges with low congestion. To compute the possible benefits of rerouting a path flow, all of the selected path flow is used as required demand in a minimum-cost flow problem. The solution to this problem is an alternative route for the path flow with a cost that can be compared to the cost of the existing path. Usually, only a fraction of the selected flow may need to be rerouted to achieve the optimum. Therefore, an optimization problem is solved, where the minimization of the potential – again proportional to the edge weights – gives the fraction σ of flow to be rerouted. The new flow and its path are thereby completely specified, the rerouting is performed, and the congestion and other parameters are recomputed. The procedure is repeated until the cost improvement goes below the limit (11.2).

The algorithm starts with flows that satisfy all the demands, but not necessarily the capacity constraints. The algorithm then reroutes flow from heavily congested edges to edges with lower congestion in order to decrease the value of λ. In order to do so, it selects a heavily congested edge. This can be done by selecting the edge with the largest λ_{ij}, or, if there are more than one with the same congestion, any of these edges. We can formulate a lower bound on $\hat{\lambda}$.

Lemma 11.2.1. *Suppose that there exists a multicommodity flow satisfying capacities $\lambda \cdot u_{ij}$. Then for any weight function w_{ij} the value $\sum_{k=1}^{K} \hat{c}^k(\lambda)/(\sum_{(i,j) \in E} w_{ij} u_{ij})$ is a lower bound on $\hat{\lambda}$.*

The goal of the algorithm is to find a multicommodity flow f and a weight function w such that this lower bound is within a $(1 + \epsilon)$ factor of optimal, that is,

$$\lambda \le (1+\epsilon) \sum_{k=1}^{K} \hat{c}^k(\lambda) / (\sum_{(i,j)\in E} w_{ij} u_{ij}).$$

In this case, we say that f and w are ϵ-*optimal*. Note that we are using the term ϵ-optimal to refer both to a flow itself and a flow and weight function pair.

Selecting commodity to reroute

Starting with any flows satisfying the demands but not necessarily the capacity constraints, calculate λ_{ij} for each edge. Let $\lambda = \max_{(i,j)} \lambda_{ij}$ and $\alpha = cs/\lambda$. Next, the edge weights w_{ij} are computed. For each commodity, solve a minimum-cost flow problem with the edge weights as costs, the scaled capacities $\bar{u} = \lambda u_{ij}$, and demands d_k. The result is a set of alternative paths for each commodity. The path flow cost c^k per commodity can now be calculated by using (11.1), summing over the edges that build up the route $p \in \mathcal{P}_k$ used by commodity k, that is,

$$c_p^k = \sum_{(i,j)\in p:p\in\mathcal{P}_k} f_{ij}^k w_{ij}.$$

By comparing the costs of using the old paths and the new paths just determined, we have an indication of which commodity to reroute – typically the commodity that gives the highest reduction in cost. Thus, if p is the old path and q the new path, the condition

$$c_p^k - c_q^k > \epsilon(c_p^k + \lambda \cdot \Phi/K)$$

shows that rerouting a proportion of commodity k from p to q would decrease the cost and the network congestion. The costs are computed for routing the entire path flow f_p^k of commodity k on routes p and q, respectively.

The algorithm selects the route with the largest flow through the most congested edge *for which a new route has been found*, or else, a route with the largest flow for which a route already exists. The selected route is the target route for rerouting flow. Compute costs for using the old and the new routes for routing the demand, respectively. The difference in cost shows the potential gain of rerouting. By inspecting the cost gain, a set of flows that can be rerouted can be identified. From this set, a flow can be selected deterministically according to the cost gain, or randomly.

Computing fraction to reroute

In order to calculate σ, the fraction of flow to reroute, we solve an optimization problem, minimizing the difference in potential $\Delta\Phi$ between using the old path and the new path. It is actually sufficient to consider the difference in potential for these two paths only, since the

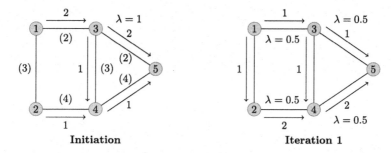

Figure 11.1: A multicommodity flow problem: the initial solution (left) and final flow after rerouting (right).

Table 11.1: Flow demands for Example 11.2.1.

k	s_k	t_k	d_k
1	1	5	2
2	2	5	1
3	3	4	1

rest of the terms in Φ remain unchanged. A simple way to implement this step is to calculate Φ_r and Φ_q for $0 < \sigma \le 1$ with step size, ϵ, say, and find the minimum of $\Phi_p - \Phi_q$, where p is the old path and q is the new path. This gives an approximate value for σ. It seems to be necessary to limit the value σ by a constant $\bar{\sigma}$, with $0.5 \le \bar{\sigma} < 1$, to avoid oscillating behavior of the algorithm. Once σ has been determined, we reroute the flow accordingly, leaving $(1 - \sigma)f_p$ on the old path and σf_p on the new path. Also, to avoid wasting time routing small amounts of flow, we reroute a commodity only if σ_f is at least as large as $O(\varepsilon/\alpha\lambda)$.

Stopping conditions

The stopping condition (11.2) measures the gain in cost achievable by rerouting a flow. In addition, it may be instructive to monitor how λ decreases. If λ does not decrease, either an approximately optimal solution has been found, or the path flow selected for rerouting does not improve the value. Also the potential function can be used to monitor the optimality of the flow; the algorithm yields an ϵ-optimal flow when the potential function becomes sufficiently small.

Example 11.2.1. *Consider the network in Fig. 11.1. Suppose we are given the demands in Table 11.1. Choose the constants $c = 19.1 - \ln m$ with $m = 6$, $s = 0.25$, and $\varepsilon = 0.1$. Also, for simplicity, let $\sigma = 0.5$ in this example. To find initial flows, use the shortest paths for each commodity in the network. The resulting flows are also shown in Fig. 11.1. Note that the flows actually are feasible. Next, calculate the congestion λ_{ij}, weight w_{ij}, potential Φ_{ij}, and cost c_{ij} for each edge. These values are shown in Table 11.2. The maximum congestion is $\lambda = 1$.*

Table 11.2: Initial flow parameters in Example 11.2.1.

(i, j)	u_{ij}	f_{ij}	λ_{ij}	w_{ij}	Φ	c_{ij}
$(1, 2)$	3	0	0.00	1.0	3.0	0
$(1, 3)$	2	2	1.00	75.7	151.4	151.4
$(2, 4)$	4	1	0.25	3.0	11.8	3.0
$(3, 4)$	3	1	0.33	4.2	12.7	4.2
$(3, 5)$	2	2	1.00	75.7	151.4	151.4
$(4, 5)$	4	1	0.25	3.0	11.8	3.0

Table 11.3: Flow parameters after the first iteration in Example 11.2.1.

(i, j)	u_{ij}	f_{ij}	λ_{ij}	w_{ij}	Φ	c_{ij}
$(1, 2)$	3	1	0.33	4.2	12.7	4.2
$(1, 3)$	2	1	0.50	8.7	17.4	8.7
$(2, 4)$	4	2	0.50	8.7	34.8	8.7
$(3, 4)$	3	1	0.33	4.2	12.7	4.2
$(3, 5)$	2	1	0.50	8.7	17.4	8.7
$(4, 5)$	4	2	0.50	8.7	34.8	8.7

We may suspect that this is not an optimal flow, since other congestion values are much lower. The high congestion is caused by commodity 1, so we scale the capacities to $\lambda \cdot u_{ij}$ and solve a minimum-cost flow problem with the scaled capacities and the weights w_{ij} as edge costs. The result is an alternative path for commodity 1, path 1–2–4–5. At this point we may compute a proportion of the demand to reroute, represented by σ. Here we have just set $\sigma = 0.5$, so we reroute half of the demand onto the alternative path. This is illustrated in Fig. 11.1 and the parameters with respect to this flow are tabulated in Table 11.3. Note, however, how the sum of potentials drops from 342 to 130 by this reassignment. Thus, the new multicommodity flow is much closer to the optimum.

Algorithm 11.2.2 (Approximation algorithm for the maximum concurrent multicommodity flow problem).

Given a graph $G = (V, E)$, nonnegative edge costs $\{c_{ij}\}$, edge capacities u_{ij}, and K commodities (s_k, t_k, d_k). Set the precision ϵ.

STEP 0:

Let **w** be the vector of $w_{ij} = c_{ij}$ and $\bar{\mathbf{u}}$ the vector of $\bar{u}_{ij} = \infty$.
for all $k = 1, ..., K$ **do** solve a minimum cost problem
MINCOST($\bar{\mathbf{u}}$, **w**, s_k, t_k, d_k) and assign initial flows f_{ik}^k.

Table 11.4: The specification of the commodities in Example 11.2.2.

k	s_k	t_k	d_k
1	1	7	5
2	3	5	2
3	2	6	2

STEP 1 to N:

    for $i, j = 1, \ldots, n$ **do**

        Compute λ_{ij}, $\lambda = \max_{(i,j)\in E} \lambda_{ij}$ and α

        Compute weights w_{ij}, modified capacities $\bar{u}_{ij} = \lambda u_{ij}$

        and potential $\Phi = \sum_{(i,j)\in E} \Phi_{ij}$

        for all $k = 1, \ldots, K$ **do** solve a minimum cost problem

        MINCOST($\bar{u}, w, s_k, t_k, d_k$), giving a set $\mathcal{Q}_k$ of alternative paths.

        Compute cost of path flows $c_p^k = \sum_{(i,j)\text{in}p:p\in\mathcal{P}_k} c_{ij}^k$ and

        $c_q^k = \sum_{(i,j)\text{in}q:q\in\mathcal{Q}_k} c_{ij}^k$ for each commodity.

        if $c_p^k - c_q^k < \epsilon(c_p^k + \lambda\Phi/K)$ for all k **then** STOP:

        flows are approximately optimal **else**

         Select a flow f_p^k such that $c_p^k - c_q^k \geq \epsilon(c_p^k + \lambda\Phi/K)$

         (typically the flow for which the difference is largest). Fix k.

         Calculate σ: let σ be such that $\Phi_p^k(\sigma) - \Phi_q^k(\sigma)$ is minimum,

         where $\Phi_p^k(\sigma)$ is the potential along $p \in \mathcal{P}_k$ with flow $(1-\sigma)f_p^k$

         and $\Phi_q^k(\sigma)$ is the potential along $q \in \mathcal{Q}_k$ with flow σf_p^k.

         Reroute flow so that $f_q^k := \sigma f_p^k$ and $f_p^k := (1-\sigma)f_p^k$

         Repeat.

Output all path flows f_p^k, $k = 1, \ldots, K$, and $p \in \mathcal{P}_k$.   ●

Example 11.2.2. *Consider the network depicted in Fig. 11.2 and three commodities (to keep the example easy to follow). The commodities are given in Table 11.4.*

Let the parameters $m = |E| = 10$, $c = 19.1 - \ln(m)$, $s = 0.25$ be given as in the description of the algorithm and choose the shortest paths for the initial flows of each commodity. We want to find an approximate solution to this multicommodity flow problem with tolerance $\epsilon = 1/10$. We calculate, in order, edge flows f_{ij}, congestion λ_{ij}, maximum congestion λ, α, edge weights w_{ij}, potentials Φ_{ij}, and costs c_{ij}. The iterations of the algorithm are shown in Fig. 11.2.

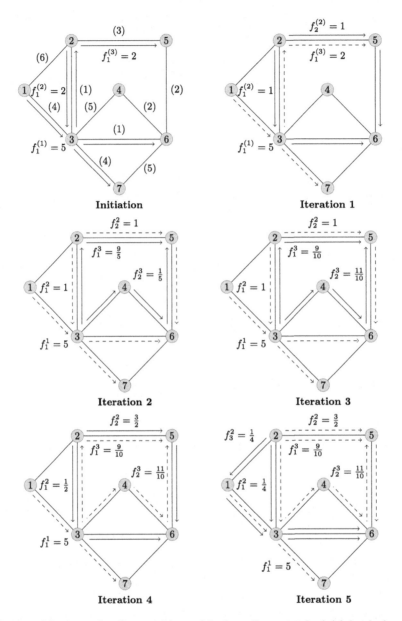

Figure 11.2: A multicommodity flow problem with three flows. In the initial solution (Initiation) shortest paths are used to route the three flows. In iteration 1, flow f^2 is split into two paths. In iterations 2 and 3, flow f^3 is split and rerouted. Iteration 4 shifts flow from f_1^2 to f_2^2 and in iteration 5, a fraction of flow f_1^2 is rerouted onto the third path, f_3^2.

11.3 Wavelength Assignment

Optical networks using wavelength-division multiplexing (WDM) are often considered the transport medium of choice in telecommunications, since they allow for capacity expansion without the need to deploy new optical fiber.

These optical fibers typically run in ducts (or trenches) buried in the ground. When more fibers are needed, fiber is inserted into the duct, commonly by means of jetting, a process combining injection of pressurized air in order to lower friction, and mechanical pushing.

There are different channel divisions in WDM. A typical dense wavelength-division multiplexing (DWDM) system uses 40 channels at 100 GHz separation, 80 channels with 50 GHz separation, or even 320 channels at 12.5 GHz separation (sometimes called ultra dense WDM). Development into higher bandwidth and more flexible channel division is known as elastic optical networks.

There are different WDM types, sometimes denoted xWDM. As this text focuses on general techniques, we simply refer to such networks as WDM networks. At the end points of the fiber, a multiplexer joins different signals onto the high-speed optical carrier, and a demultiplexer at the receiver end is used to split them again. Other equipment, such as wavelength converters and amplifiers may also be part of the system.

Two adjacent nodes are typically connected by one or more full duplex fibers or fiber pairs. On each fiber different wavelengths are used to take advantage of the large bandwidth of the fiber.

A transmission path between two nodes on a certain wavelength is called a lightpath. To establish a lightpath, two different methods can be used. In path multiplexing (PM), the same wavelength is assigned to all links along the path. It follows that this particular wavelength cannot be used for other connections using these links. The second method is link multiplexing, where different wavelengths can be used on different links. This method requires wavelength converters at the nodes along the path, and it is therefore more expensive than path multiplexing.

Planning of WDM networks involves routing and wavelength assignment. This should take into account traffic and performance requirements and can be formulated as an optimization problem with respect to some metric, often the number of wavelengths used. The routes are often chosen along the shortest paths (in terms of number of hops), but one may have to use other paths to reach a global optimum. Resilience aspects can also affect the choice of paths.

Graph Coloring

The assignment of wavelengths in optical WDM networks can conveniently be done using *graph coloring*. To use graph coloring for wavelength assignment, we need to formulate a graph based on the demands and their associated paths. We can obviously have different paths, but we assume here that only one of these is primary; typically the shortest paths. In a demand graph, a node represents a path, and there is an edge between any two such nodes whenever two paths share at least one edge in the original graph.

We approach this design task in a few steps. Graph coloring usually refers to vertex coloring, whereas wavelength assignment is related to paths. We formulate an approximation to the graph coloring problem based on Douglas–Rachford splitting and apply it to a new graph derived from the original network.

Consider an undirected graph $G = G(V, E)$, where V is its vertex set and E its edge set. A k-coloring of the graph G is an assignment of one of k possible colors to each vertex of G such that no two adjacent vertices—that is, vertices having an edge between them—are assigned the same color.

More formally, given the set of colors $K = \{1, \ldots, k\}$, a k-coloring of G is a mapping $f : V \mapsto K$, assigning a color to each vertex. We say that f is proper if

$$f(i) \neq f(j) \quad \text{for all } \{i, j\} \in E \tag{11.3}$$

We are interested in finding the minimum number of colors k such that Eq. (11.3) holds. The problem is known to be $\mathcal{NP}$-complete [99].

For each pair of nodes in a given demand, we can find candidate paths using Yen's algorithm (or Dijkstra's, if only shortest paths are of interest). From these candidate paths we select exactly one path as primary. We construct a new graph with the candidate paths defining the new node set. For any two such nodes we check whether the corresponding paths share any edge in the original graph. If that is the case, the two paths must be assigned different wavelengths, and we create an edge between the nodes in the new graph. Applying graph coloring on this new graph gives a minimal color assignment to the paths.

The Douglas–Rachford Algorithm

The Douglas–Rachford algorithm is an iterative scheme to minimize functionals of the form

$$\min_x F(x) + G(x) \tag{11.4}$$

where F and G are convex functions for which one is able to compute the proximal mappings $\text{prox}_{\gamma F}$ and $\text{prox}_{\gamma G}$, which are defined as

$$\text{prox}_{\gamma F}(x) = \arg\min_y \frac{1}{2}||x - y||^2 + \gamma F(y) \tag{11.5}$$

and similarly for G.

The important point is that F and G do not need to be smooth. They only need to be "proximable" [100].

Starting from a solution $Z^{(0)}$, the Douglas–Rachford iteration can be written, indexed by r,

$$X^{(r)} = \text{prox}_{\gamma G}(Z^{(r-1)}), \tag{11.6}$$

$$Y^{(r)} = \text{prox}_{\gamma F}(2X^{(r)} - Z^{(r-1)}), \tag{11.7}$$

$$Z^{(r)} = Z^{(r-1)} + Y^{(r)} - X^{(r)}. \tag{11.8}$$

We now reformulate this following [99]. Given k colors and n nodes, that is, $n = |V|$, and let $l = |E|$ be the number of edges. We define the sets $K = \{1, \ldots, k\}$, $I = \{1, \ldots, n\}$, $P = \{n+1, \ldots, n+l\}$, and $E = \{e_1, \ldots, e_l\}$. We can then define [99] the coloring problem as finding the solution $Z = (z_{it}) \in \mathbb{R}^{(n+l) \times k}$ satisfying

$$Z = C_1 \cap C_2 \cap C_3 \cap C_4$$

where

$$C_4 := \{z_{1,1}\},$$

$$C_3 := \{\sum_i^n z_{it} \geq 1, \forall t \in K\},$$

$$C_2 := \{z_{it} + z_{jt} - z_{pt} = 0, e_{p-n} = \{i, j\} \in E \forall p, t\},$$

$$C_1 := \{\sum_{t=1}^k z_{it} = 1, z_{it} \in \{0, 1\}\}.$$

We maintain a matrix A and define four projections P_{C_1}, P_{C_2}, P_{C_3}, and P_{C_4}. The matrix $A = (a_{pq}) \in \mathbb{R}^{l \times (n+1)}$ is defined by

$$a_{pq} = \begin{cases} 1 & \text{if } e_p = \{i, j\} \text{ and } q \in \{i, j\}, \\ -1 & \text{if } q = n + p, \\ 0 & \text{otherwise}, \end{cases}$$

for each $p = 1, \ldots, l$ and $q \in I \cup P$.

The projections of any $Z \in \mathbb{R}^{(n+l) \times k}$ onto C_1, C_3, and C_4 are given, pointwise, by

$$(P_{C_1}(Z))[i, t] = \begin{cases} 1 & \text{if } i \in I, t = \arg\max\{z_{i1}, z_{i2}, \ldots, z_{ik}\}, \\ z_{it} & \text{if } i \in P, \\ 0 & \text{otherwise}, \end{cases} \qquad (11.9)$$

$$(P_{C_3}(Z))[i, t] = \begin{cases} 1 & \text{if } i = \arg\max\{z_{1t}, z_{2t}, \ldots, z_{nt}\}, \\ \min\{1, \max\{0, \text{round}(z_{it})\}\} & \text{otherwise}, \end{cases} \qquad (11.10)$$

$$(P_{C_4}(Z))[i, t] = \begin{cases} 1 & \text{if } i = t = 1, \\ z_{it} & \text{otherwise}, \end{cases} \qquad (11.11)$$

for each $i \in I \cup P$ and $t \in K$, where the lowest index is chosen in argmax (the projections onto C_1 and C_3 need not be unique). Since A is full row rank, the projection onto C_2 is given by

$$P_{C_2}(Z) = (I_{n+l} - A^T (AA^T)^{-1} A) Z. \qquad (11.12)$$

We can improve the performance of the algorithm by taking into consideration maximal cliques, since such a clique with the number of n_q vertices needs exactly n_q colors. We decompose the edge set into subsets of those in a maximal clique and those which are not.

Let $Q \subset 2^V$ be a nonempty subset of maximal cliques of the graph $G = (V, E)$ and let $\hat{E} := E \cup Q$. Let $Q = \{e_{l+1}, \ldots, e_r\}$, with $r \geq l + 1$. Thus, $\hat{E} = \{e_1, \ldots, e_l, e_{l+1}, \ldots, e_r\}$.

The matrix $\hat{A} = (\hat{a}_{pq}) \in \mathbb{R}^{r \times (n+r)}$ is defined by

$$\hat{a}_{pq} = \begin{cases} 1 & \text{if } q \in e_p, \\ -1 & \text{if } q = n + p, \\ 0 & \text{otherwise}, \end{cases} \qquad (11.13)$$

for each $p = 1, \ldots, r$ and $q \in \{1, \ldots, n + r\}$. The maximal cliques are found using the Bron–Kerbosch algorithm described next.

The Bron–Kerbosch Algorithm

The Bron–Kerbosch algorithm is a method for finding maximal cliques in an undirected graph.

It produces all subsets of vertices with the two properties that each pair of vertices in one of the listed subsets is connected by an edge, and furthermore the returned subset cannot have any additional vertices added to it preserving its complete connectivity. The recursive algorithm can be stated as follows.

Algorithm 11.3.1 (Bron–Kerbosch).

Let a graph $G = (V, E)$ be given. Set $R = \emptyset$, $P = V$, $X = \emptyset$ and let $N(v)$ be the neighbors of vertex v.

STEP 0:
     if $P = X = \emptyset$, **then** R is a maximal clique.
STEP 1:
     for each vertex v in P **do** Bron–Kerbosch($R \cup \{v\}$, $P \cap N(v)$, $X \cap N(v)$)
     $P \leftarrow P/v$
     $X \leftarrow X \cup v$

Output R.   •

The basic form of the Bron–Kerbosch algorithm is a recursive backtracking algorithm that searches for all maximal cliques in a given graph G.

Given three sets of vertices R, P, and X, it finds the maximal cliques that include all of the vertices in R, possibly some of the vertices in P, and none of the vertices in X.

In each call to the algorithm, P and X are disjoint sets whose union consists of those vertices that form cliques when added to R.

When P and X are both empty there are no further elements that can be added to R, so R is a maximal clique and the algorithm returns R.

Initially $R = \emptyset$, $X = \emptyset$ and $P = V$. With each recursive call, the algorithm tests the vertices in P one by one. If there are no vertices in P, it either returns R as a maximal clique provided X is empty, or it backtracks.

For each vertex v in P, it makes a recursive call in which v is added to R and P and X are restricted to the set of neighbors $N(v)$ of v, and it returns all cliques R extended by v. In the next step, it moves v from P to X in which it is excluded from further consideration, and it continues with the next vertex in P [98].

It appears that the coloring algorithm is not very efficient for graphs that are almost complete. However, for an all end-to-end demand matrix, the resulting graph is likely to be almost complete.

By taking every second path, we obtain a sparser graph on which the algorithm works better. Consider the example network in Fig. 11.3. The network has $n = 7$ nodes and with 12 out of the $n(n - 1)/2 = 24$ end-to-end demands, we obtain the coloring shown in Fig. 11.4. The coloring uses six colors. Since the largest clique has four nodes, we know that we need at least four and at most twelve colors to cater for these end-to-end flows. Suppose the second 'half'

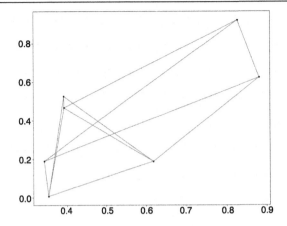

Figure 11.3: **Example network topology of order** $n = 7$ **and size** $m = 11$.

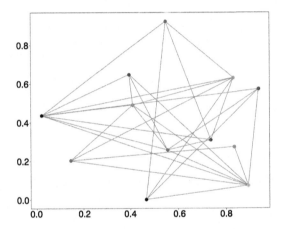

Figure 11.4: **Vertex coloring of half of the paths of the demand graph. This associated graph is noticeably larger than the topological graph.**

of the flows can be colored by six colors as well; then with a new set of colors, or a suitable permutation of the colors already used, the two solutions can be superimposed on each other using no more than twelve colors.

11.4 Preplanned Cycle Protection

We now outline a fault-resilient routing strategy on a network topology based on so-called preplanned or preconfigured cycle protection, *p-cycles* in short. We consider rather dense networks, such as the aggregation layer in 5G.

Network resilience can be seen as a combination of topological properties of a network and flow assignment onto the given topology. In high-speed networks, any failure potentially causes loss of huge amounts of data, and cost effective protective measures are therefore of fundamental importance.

Nodes are generally relative inexpensive compared to links, although the failure of a well-connected node has a larger impact on the network's operation than the failure of a single link.

Flow assignments are traditionally done either by creating logical rings or on a logical mesh, each with advantages and disadvantages. The concept of p-cycles offers a combination of these two planning principles taking advantages of both.

The cost of resilience consists in new links in the case the network topology is modified, and spare capacity on existing links in the case of flow assignment. Here, we consider the topology as given.

The difference between assignment of flows according to a logical mesh and a logical ring topology is that the mesh is more efficient in terms of spare capacity, whereas the rings allow for a faster restoration—in the order of milliseconds—by requiring rerouting of the flow in only two nodes. In a mesh topology, routing can utilize spare capacity more efficiently following shortest paths, but restoration can be in the order of seconds.

The principle of p-cycles provides a hybrid between mesh and ring: they allow fast restoration and routing independently of other parts of the network, and are at the same time capacity efficient.

The idea behind p-cycles is to expand the ring structure to include straddling links, that is, links with end nodes lying on the cycle, not being any of the link segments making up the cycle. The p-cycles are configured with spare capacity to cater for any single link failure protected by the p-cycle. With each link (i, j) in a network, we associate a (total) capacity c_{ij} and a working capacity w_{ij} that reflects the traffic that the link carries. Clearly, we always have $w_{ij} \leq c_{ij}$. The spare capacity s_{ij} is the difference between the two quantities, $s_{ij} = c_{ij} - w_{ij}$. p-Cycles use this spare capacity for protection of the primary path working capacity. The operation of a p-cycle is as follows. Recall that it contains on-cycle links and straddling spans. In the case an on-cycle link (i, j) fails, any path including this link is rerouted in the reverse direction along the p-cycle. Thus, for each on-cycle link there is one protection path. In the case a straddling link fails, the traffic is rerouted along the p-cycle in either direction, thereby providing two protection paths. The capacity efficiency of p-cycles follows from their ability to protect straddling spans, which is provided "free-of-charge".

Figure 11.5: **Example topology with four** *p*-cycles marked.

Finding Cycles in Graphs

A cycle (or circuit) in a graph G is a path in which the first and last vertices are identical. (See Fig. 11.5.) A path is called simple (or elementary) if no vertex appears twice. A cycle is called simple if no vertex but the first and last appears twice. Two simple cycles are distinct if one is not a cyclic permutation of the other. We will assume that all paths are simple. The algorithm by Johnson [101] is one of the best-known methods to find all cycles in a directed or undirected graph.

We note that if there exists an optimal algorithm to list all st-paths in a graph G, there exists an optimal algorithm to list the cycles in G: for any cycle q_{vv}, there exists a vertex u such that q_{vv} consists of a path p_{uv} followed by the edge (u, v).

Similar to shortest paths, cycles are constructed from a root vertex s in the subgraph induced by s and vertices "larger than s" in some ordering of the vertices. To avoid duplicate cycles, a vertex v is blocked when it is added to a path beginning in s.

Algorithm 11.4.1 (Johnson).

Input: A graph G represented as a list of edges, a set of cycles $C = \emptyset$.

STEP 1 TO N

      for each edge in G **do:**

        for each v in edge **do:**

          $C = \text{findNewCycles}(\{v\}, C)$

        end

      end

Output C.   •

Algorithm 11.4.2 (findNewCycles).

Input: a path p and the current set of cycles, C.

STEP 0:

Set $s = p[0], t = \emptyset, sub = \emptyset$.

STEP 1 TO N:

    for each edge in G **do:**

      $[u, v]$ = edge

      if s in edge **then:**

        if $v1 == s$ **then:**

          $t = v2$

        else

          $t = v1$

        end

      end

      if t not in p **then:**

        $sub = \{t\}$

        $sub \rightarrow sub \cup p$

        C = findNewCycles(sub, C)

      end

      else if $|p| > 2$ and $t == p[-1]$ **then:**

        if p is not in C **then:**

          $C \rightarrow C \cup p$

        end

      end

    end

Output C.

                                                                    ●

p-Cycle Design

The design of p-cycles clearly depends on the underlying topology. In principle, the design can be modeled as an integer program; we note that flows typically are treated as non-splittable, which gives the problem its integer character.

A common objective is to minimize the spare capacity for 100% restorability, or maximizing restorability for a given spare capacity. We define the set of parameters detailed in Table 11.5 [102].

Maximizing p-cycle restorability for a given spare capacity is equivalent to minimizing the total number of unrestorable spans u_i in the network. The integer program for this problem reads

Table 11.5: Parameters used in *p*-cycle design.

Parameter	Description
x_{ij}	The number of feasible paths a *p*-cycle j provides in the case span i fails
S	The set of network spans
P	The set of elementary cycles in a graph
δ_{ij}	The binary variable that equals unity if cycle i includes span j and zero otherwise
w_j, s_j	The number of working channels and spare channels on span j
p_j	The number of available protection paths in the design in excess of those required for span j (defined as a slack variable)
u_i	The number of unrestorable working channels on span i
n_j	The number of unit capacity copies of a cycle j in the design

$$\min_{i \in S} \quad u_i$$

$$s_k \geq \sum_{j \in P} \delta_{kj} n_j, \quad \text{for all } k \in S,$$

$$u_i + \sum_{j \in P} x_{ij} n_j = w_i + r_i, \quad \text{for all } i \in S,$$

$$0 \leq u_i \leq w_i, \quad \text{for all } i \in S,$$

$$u_i \geq 0, r_i \geq 0, \quad \text{for all } i, j \in S,$$

and the integer program for minimization of spare capacity with 100% restorability is

$$\min_{k \in S} \quad c_k s_k$$

$$s_k \geq \sum_{j \in P} \delta_{kj} n_j, \quad \text{for all } k \in S,$$

$$w_i \leq \sum_{j \in P} x_{ij} n_j, \quad \text{for all } i \in S,$$

$$n_j \geq 0, s_k \geq 0, \quad \text{for all } j, k \in S,$$

where c_k is the cost of using link k.

Rather than using the integer program approach, which tends to be very large for realistic networks, we use an efficient heuristic. To formulate this algorithm, we need to characterize the efficiency of *p*-cycles. Three efficiency measures proposed in the literature are topological score (TS), a priori efficiency (AE) and efficiency ratio (ER).

Definition 11.4.1 (Topological score). *The* topological score *of a candidate p-cycle j is defined as*

$$TS(j) = \sum_{i \in S} x_{ij}, \tag{11.14}$$

where x_{ij} takes the values 0 or 1 or 2 corresponding to the cases where the link does not have end points on the cycle, the link is part of the cycle, or the link is a straddling link.

The topological score measures the total number of protection paths a cycle provides in the network, but it does not reflect the cost of constructing the cycle in terms spare capacity. A slightly more sophisticated measure is the a priori efficiency (AE). It reflects the cost by comparing the number of protection paths with the actual capacity.

Definition 11.4.2 (A priori efficiency). *The a priori efficiency is defined as*

$$AE(j) = \sum_{i \in S} x_{ij} / \sum_{k \in S} \delta_{kj} c_k, \tag{11.15}$$

where $\delta_{kj} = \{0, 1\}$ depending on whether link k is part of cycle j or not.

We note that the more straddling links a cycle has, the higher its AE, because the contribution for such links is x_{ij}, whereas the cost is zero.

Finally, the efficiency ratio (ER) takes into account the actual demand on link i in the efficiency estimation. Specifically, if a straddling link does not carry any traffic, its potential protection should not contribute to the efficiency of the p-cycle.

Definition 11.4.3. *The efficiency ratio (ER) is defined as*

$$ER(j) = \sum_{i \in S} \min\{x_{ij}, w_i\} / \sum_{k \in S} \delta_{kj} c_k. \tag{11.16}$$

We can now formulate a heuristic based on the ER metric, presented by Zhang et al. [103]. It is based on finding all cycles in a network, and each cycle is then evaluated with respect to its efficiency. To reduce the inherent hardness of finding candidate cycles, we can introduce a limit on the lengths of admissible cycles in terms of hops.

The algorithm works with directed traffic, so that working and protection capacity between two nodes need not be symmetric. In [103], a capacity unit is defined as one wavelength in xWDM, and so all capacities are expressed as integer multiples of such wavelengths. A p-cycle design is consequently expressed as a collection of unity p-cycles (representing one wavelength).

The efficiency ratio is calculated for each candidate cycle, where a higher ER indicates greater efficiency. By using as efficient p-cycles as possible, the total spare capacity is reduced.

Algorithm 11.4.3 (p-Cycle design).
Input: A given network topology $G = (V, E)$ and traffic demand D.

STEP 1 TO N:

(1) Find all candidate cycles (Johnson's algorithm), and determine the working capacity on each span (typically determined by shortest-path routing). Candidate cycles may be subject to certain constraints, such as the maximum cycle length or hop count.

(2) For each candidate cycle, calculate the ER of its unity p-cycle.

(3) Select a unity p-cycle with the maximum ER. If multiple unity p-cycles have the same maximum ER, then randomly select one.

(4) Update the working capacity by removing those working units that can be protected by the selected unity p-cycle.

(5) Return to Step 2 until the working capacity on every span becomes 0.

Output: collection of p-cycles.   ●

There are several variations on p-cycle design and related algorithms. Some designs are built on preselected p-cycles, where the set of candidate cycles are chosen to be those having an a priori efficiency above a certain threshold.

For large networks and given an upper limit on the number of the number of distinct cycles, the candidates can be selected [104] as

- Select 50% from the cycles with highest AE rank.
- Add 20% from the cycles ranked highest in terms of maximum straddling links.
- Add 20% from the longest cycles.
- Add 10% from randomly selected cycles.

Duplicate cycles should be removed from the set.

The Straddling Span Approach

It is also possible to build a design starting from a presumptive straddling span. At any node with degree at least three, we try to find two disjoint paths connecting the end nodes of the straddling span. The p-cycles generated in this way are guaranteed to have at least one straddling span. If the node degree is two, we can possibly find a different path back to the start node of the span. p-Cycles so constructed may lack any straddling span, since the original span now is part of the p-cycle itself.

An iterative improvement method can be constructed by replacing any span on the cycle by a disjoint shortest path between the end nodes of the span, if it exists. The original span then becomes a straddling span and the resulting cycle has a higher efficiency than the original one.

Node Failures

Node failures in a network are usually protected against using node redundancy. When this is not sufficient, *p*-cycles can be used for node protection as well. This may be desirable for routers, for example.

In the case the node fails, all connections passing through that node effectively become straddling spans of the cycle, which then are protected. However, it may not be possible to use simple cycles for node encircling in such a case [105]. We need a *p*-cycle that includes all nodes adjacent to the protected one, but not the node itself. The *p*-cycle then protects all the flows passing through the protected node. The links connected to the node effectively become straddling spans. However, traffic originating or terminating at the protected node cannot be recovered in the case of failure. A network with a number of nodes *n* requires *n* *p*-cycles to protect all the nodes with node encircling *p*-cycles.

CHAPTER 12

Methods of Big Data Analytics

With big data we usually mean data produced and delivered at high frequency and from different heterogeneous sources. It is often described by the popular *3V model*: volume, velocity, and variety [106]. Big data present a number of challenges; traditional data management and analysis methods are insufficient and/or unsuitable to process such volumes in a timely fashon. We use the term *streaming data* for real-time, high frequency data. Due to the sheer volume of data generated, we assume that it would be too costly or even impossible to store it.

In high-speed networks, efficient network supervision and management is challenged by the cost and speed of analyzing the huge amount of data generated. From a Big Data viewpoint, we want to deploy algorithms that are resource efficient while yielding acceptable accuracy.

In optical network, packets arrive at the nodes at high speed, and measurements should therefore be performed at high frequencies for accuracy. Some of the quantities we wish to derive from a high frequency data stream for the purpose of design, traffic engineering, billing or anomality detection are

- The number of distinct flows and average flow size
- Elephant flows (very large or high-frequent flows)
- Pathwise entropy of the traffic
- Flow size distributions
- Traffic matrix

12.1 Discretization

In many data processing algorithms we assume integer data values. In general, we can view any data as a multivariate time series and perform discretization by performing clustering so that the loss of information is minimized using an information-theoretic approach [150]. This method finds the number of levels that best preserve the characteristics of the data, such as correlation between variables and dynamics inherent in the time series.

The Discretization Problem

Firstly, we formally define a discretization for real-valued vector $\mathbf{v} = (v_1, \ldots, v_N)$.

5G Networks. https://doi.org/10.1016/B978-0-12-812707-0.00017-6

Definition 12.1.1. A discretization of a real-valued vector $\mathbf{v} = (v_1, \ldots, v_N)$ is an integer-valued vector $\mathbf{d} = (d_1, \ldots, d_N)$ with the following properties:

(1) Each element of $\mathbf{d}$ is in the set $\{0, 1, \ldots, D - 1\}$ for some (usually small) positive integer D, called the degree of the discretization.
(2) For all $1 \leq i, j \leq N$, we have $d_i \leq d_j$ if $v_i \leq v_j$.

A discretization $\mathbf{d}$ is called a *spanning discretization* of degree D if the smallest element of $\mathbf{d}$ is equal to 0 and the largest element of $\mathbf{d}$ is equal to $D - 1$.

The close link to clustering suggests that we can use a standard clustering algorithm, for example k-means, to perform discretization. Here, clearly k corresponds to D. In the algorithm, however, we would prefer not having to specify D explicitly.

Top-down clustering algorithms start from the entire data set and iteratively split it up until either the degree of similarity reaches a certain threshold or every group consists of one object only. For the purpose of data analysis, it is impractical to let the clustering algorithm produce clusters containing only one real value. The iteration at which the algorithm is terminated is crucial since it determines the degree of the discretization, and one of the most important features of such a discretization method is the definition of the termination criteria.

We assume that the data can be represented in one or more real-valued vectors, without making any assumptions about the distribution or discretization thresholds. We describe an algorithm that uses a graph based on the data points where the edge weights are the Euclidean distances between the points.

The method arranges the data points into clusters according to their relative distance between each other and the information content of the result. When more than one vector has to be discretized, the algorithm discretizes each vector independently.

If the vector contains m distinct entries, a complete weighted graph on m vertices is constructed, where a vertex represents an entry and an edge weight is the Euclidean distance between its endpoints. The discretization process starts by deleting the edge(s) of highest weight until the graph gets disconnected. If there is more than one edge labeled with the current highest weight, then all of the edges with this weight are deleted. The order in which the edges are removed leads to components, in which the distance between any two vertices is smaller than the distance between any two components. We define the distance between two components G and H to be $\min\{|g - h| \, g \in G, h \in H\}$. The algorithm produces a discretization of the vector in which each cluster corresponds to a discrete state and the vector entries that belong to one component are discretized into the same state. We illustrate the algorithm by an example borrowed from [150].

Suppose that vector $\mathbf{v} = (1, 2, 7, 9, 10, 11)$ is to be discretized. We start with constructing the complete weighted graph based on $\mathbf{v}$ which corresponds to iteration 0.

Eight edges with weights 10, 9, 9, 8, 8, 7, 6, and 5, respectively, have to be deleted to disconnect the graph into two components. This is the first iteration with the discretization where one level contains vertices 1 and 2 and the another vertices 7, 9, 10, and 11. Having disconnected the graph, the next step is to determine if the obtained degree of discretization is sufficient; if not, the components are further disconnected in a similar manner to obtain a finer discretization. A component is further disconnected if one of the following four conditions is satisfied ("disconnect further" criterion):

(1) The average edge weight of the component is greater than half the average edge weight of the complete graph.
(2) The distance between its smallest and largest vertices is greater than or equal to half this distance in the complete graph. For the complete graph, the distance is the graph's largest weight.
(3) The minimum vertex degree of the component is less than the number of its vertices minus 1. The contrary implies that the component is a complete graph by itself, that is, the distance between its minimum and maximum vertices is smaller than the distance between the component and any other component.
(4) Finally, if the above conditions fail, a fourth condition is applied: disconnect the component if it leads to a substantial increase in the information content carried by the discretized vector.

The discretization method can be seen as a clustering problem, where we use entropy as an optimization parameter. The information measure criterion for discretizing the entries of data into a finite number of states reduces the information carried by the discrete vector, measured by its entropy. For a set of n possible events whose probabilities of occurrence are known to be $p_1, p_2, \ldots, p_n$, we recall the entropy

$$H = -\sum_{i=1}^{n} p_i \log p_i. \tag{12.1}$$

The base 2 of the logarithm is commonly used so that the result can be related to bits. Empirically, Eq. (12.1) is written

$$H = -\sum_{i=1}^{n} \frac{w_i}{n} \log\left(\frac{n}{w_i}\right),$$

where w_i is the number of entries discretized into state i (assuming a spanning discretization). An increase in the number of states leads to an increase in entropy, having an upper bound of $\log_2 n$.

However, we want the number of states to be small. That is why it is important to notice that H increases by a different amount depending on which state is split and the size of the resulting new states. For example, if a state containing the most entries is split into two new states of equal size, H will increase more than if a state of fewer entries is split or if we split the larger state into two states of different sizes.

Suppose that we split state 0 into two states containing m and $w_0 - m$ entries, respectively, where $0 < m < w_0$. This will change only the first term of the right-hand side of the above entropy expression and leave the rest of the summation unchanged. It is easy to verify that $h(w_0) = \frac{w_0}{n} \log\left(\frac{n}{w_0}\right)$ achieves its maximum value over $0 < m < w_0$ at $m = \frac{w_0}{2}$. Thus, splitting a state into two states of equal size maximizes the entropy increase.

The information measure criterion is applied to a component only after the component has failed the other three conditions. When this happens, we consider splitting it further only if doing so would provide a very significant increase of the entropy, that is, if the component corresponds to a "large" collection of entries. As a rule, we split a component further only if it contains at least half the vector entries. Unlike the other criteria, when a component is split under the information condition, the corresponding sorted entries are split into two (almost) equal parts rather than divided between the two most distant entries. This is to guarantee a maximum increase of the information measure.

In initial example, the two components that were obtained by removing the edges of heaviest weight both fail the "disconnect further" conditions 1–3. If the discretization process stops at this iteration, then the vector $\mathbf{d} = (0, 0, 1, 1, 1, 1)$ has entropy 0.78631. Having most of the entries discretized into the same state, 1, reduces the information content of $\mathbf{d}$.

Suppose discretization of $\mathbf{d}$ continues without enforcing the fourth condition of "disconnect further". The next step is to remove the edges of highest weight until a component gets disconnected. This yields the removal of the four edges of weights 4, 3, 2, and 2, respectively, giving the discretization d = $(0, 0, 1, 2, 2, 2)$. The entropy of the new discretization of v is 1.43534. Still half of the entries of v remain at the same discrete level, now 2, which does not allow for a maximal increase in the information content of $\mathbf{d}$. If instead discretization proceeds by applying the information criterion to the larger component, the resulting discretization becomes $\mathbf{d} = (0, 0, 1, 1, 2, 2)$ with entropy 1.58631, higher than the previous entropy of 1.43534.

The discretization algorithm produces a discretization which is consistent with the definition given above, keeps the number of discrete states small, and maximizes information content. We summarize the algorithm as follows.

Algorithm 12.1.1 (Discretization).

Input: set $S_r = \{\mathbf{v}_i | i = 1, \ldots, m\}$ where each $\mathbf{v}_i = (v_{i1}, \ldots, v_{iN})$ is a real-valued vector of length N to be discretized. Set $S_d = \{\mathbf{d}_i | i = 1, \ldots, m\}$ where each $\mathbf{d}_i = (d_{i1}, \ldots, d_{iN})$ is the discretization of $\mathbf{v}_i$ for all $i = 1, \ldots, m$.

(1) For each $i = 1, \ldots, m$, construct a complete weighted graph G_i where each vertex represents a distinct v_{ij} and the weight of each edge is the Euclidean distance between the incident vertices.

(2) Remove the edge(s) of the highest weight.

(3) If G_i is disconnected into components $C_{i1}^{G_i}, \ldots, C_{iM_i}^{G_i}$, go to 4. Else, go to 2.

(4) For each $C_{ik}^{G_i}$, $k = 1, \ldots, M_i$, apply "disconnect further" criteria 1–3. If any of the three criteria holds, set $G_i = C_{ik}^{G_i}$ and go to 2. Else, go to 5.

(5) Apply "disconnect further" 4. If criterion 4 is satisfied, go to 6. Else, go to 7.

(6) Sort the vertex values of $C_{ik}^{G_i}$ and split them into two sets: if $|V(C_{ik}^{G_i})|$ is even, split the first $|V(C_{ik}^{G_i})|/2$ sorted vertex values of $C_{ik}^{G_i}$ into one set and the rest into another. If $|V(C_{ik}^{G_i})|$ is odd, split the first $|V(C_{ik}^{G_i})|/2 + 1$ sorted vertex values of $C_{ik}^{G_i}$ into one set and the rest into another.

(7) Sort the components $C_{ik}^{G_i}$, $k = 1, \ldots, M_i$, by the smallest vertex value in each $C_{ik}^{G_i}$ and enumerate them $0, \ldots, D_i - 1$, where D_i is the number of components into which G_i got disconnected. For each $j = 1, \ldots, N$, d_{ij} is equal to the label of the component in which v_{ij} is a vertex.

Output discretization S_d.   •

Given M variables, with N time points each, we compute $N(N - 1)/2$ distances to construct the distance matrix so the complexity of this step is $O(N^2)$. The distance matrix is used to create the edge and vertex sets of the complete distance graph, containing $N(N - 1)/2$ edges. This can also be accomplished in $O(N^2)$ time. These edges are then sorted into decreasing order, so that the largest edges are removed first. A standard sorting algorithm, such as merge sort, has complexity $O(N \log N)$. As each edge is removed, the check for graph disconnection involves testing for the existence of a path between the two vertices of the edge. This test for graph disconnection can be accomplished with a breadth-first search, which has order $O(E + V)$, where E is the number of edges and V the number of vertices of the component. This has complexity $O(N^2)$. Edge removal is typically performed for a large percentage of the $N(N - 1)/2$ edges, so this step has overall complexity $O(N^4)$. The edge removal step dominates the complexity so that the overall complexity is $O(MN^4)$ to discretize all M variables. While this is the theoretical worst-case performance, the typical performance should be significantly better.

While any discretization results in information loss, a good multivariate discretization should preserve some important features of the data, such as correlation between the variables. The authors of [150] show that the discretization algorithm to a large degree preserves the correlation structure.

12.2 Data Sketches

Analytics for big data require methods adapted for streaming data. This means that such methods need to be fast and using a small memory space. In return, we can allow relaxation of the precision in our results – after all, a streaming data source does not have a well-defined beginning or end. The algorithms are based on *data sketches*, that is, a significant subset of data values, from which approximate statistical properties can be derived within a given ratio from the exact value. We outline the methodology to calculate statistical properties from data streams.

Data Stream Models

Data sketches are based on basic assumptions on the basic principles governing the arrival of data, called data stream models. Let the data stream be values $i_1, i_2, \ldots$ arriving sequentially. The stream can be viewed either as an infinite sequence of inputs, or as a finite sequence of n inputs. In general, data points are not stored and so the algorithms can only access a data value once. This is known as a single pass over the data. In some situations, however, we need to allow multiple passes.

There are two common data stream models, the *cash register model* and the *turnstile model*. We envisage data arriving as tuples $i_t = (s_j, c_t)$, containing a data element, s_j, and a value, c_t, for $j \in \{1, 2, \ldots, m\}, t \in \{1, 2, \ldots, n\}$. The difference between the models can be illustrated by the way we can create an update, $v_j = v_j + c_t$. In the cash register model, the value is required to be positive, $c_t \geq 0$. The cash register model handles data as counters, which are only allowed to increase with time.

In the turnstile model, the value c_t in the update can be both positive and negative, and it is therefore interpreted as a gauge. There are two variants: in the strict turnstile model the update can only be strictly nonnegative, whereas in the general turnstile model, the update can be negative as well. In general, we obtain algorithms with different properties, depending on the data stream model we choose. Important aspects of data stream algorithms are the processing time and storage requirements, which are commonly expressed in the number of unique elements, m, and the input size, n.

In the data stream model, we envisage a situation where we continuously receive a stream of data and use it in some computations in order to answer some queries about the data stream.

We can think of the data stream as a ticker tape passing through before us. We can only access the data in this order and typically these data can no longer be accessed after having passed by. The challenge is to find meaningful answers with some accuracy about the properties of the data given these limitations.

A difference between deterministic and randomized algorithms is that the former only work in the cash register model, where only increments are allowed. Some of the randomized algorithms work in the turnstile model as well. In the data stream model, the values of m and n are usually magnitudes larger than the available main memory and trivial algorithms would therefore have to store their counters in the much slower external memory. In such cases the slowdown from the need of external memory is so significant that the algorithm will not be practical.

Hash Functions

An important tool in randomized algorithms are *hash functions*. Generally speaking, a hash function h is a function that maps data of arbitrary size onto some data set of fixed size. When a large space is mapped onto a much smaller space, we need to be aware of the probability of collision, that two different elements in the original space are mapped onto the same element after hashing. Formally, we map a universe $\mathcal{U}$ of arbitrary size into m bins, so that $h : \mathcal{U} \rightarrow \{0, 1, \ldots, m\}$. We often like to bound the probability of a collision, and this concern along with other properties can be used to define different classes of hash functions.

A *c-universal family* of hash functions are functions h_i chosen uniformly at random from the family $H = \{h_i : \mathcal{U} \rightarrow \{0, 1, \ldots, m\}\}$, with probability of collision bounded by

$$x, y \in \mathcal{U}, x \neq y : \mathop{\mathbf{P}}_{h_i \in H} (h_i(x) = h_i(y)) \leq \frac{c}{m}.$$

In other words, in selecting h_i randomly from H, we are guaranteed that the probability of a collision between two distinct elements in $\mathcal{U}$ is no more than $\frac{m}{c}$, for some constant c. Such hash functions can easily be generated, which is important in applications. A widely used family of hash functions is designed by Carter and Wegman [107], where

$$H = \{h(x) = (((ax + b) \mod p) \mod m)\}$$

for $a \in \{1, 2, \ldots, p - 1\}$, $b \in \{0, 1, \ldots, p\}$, and with $p \geq m$ a prime. To generate primes we may use the recurrence relation

$$a_n = a_{n-1} + \gcd(n, a_{n-1}), \quad a_1 = 7,$$

where $\gcd(x, y)$ denotes the greatest common divisor of x and y.

Approximate Counting

Approximate counting is a method monitoring a sequence of events and at request returning an estimate of the number of events thus far in the sequence. Rather than simply counting events as they pass, we use a data sketch and maintain an approximate number of events, requiring much less memory.

Suppose we want to monitor a sequence of events and upon request retrieve (an estimate of) the number of events that has occurred thus far (see Morris [108]). Let the number of events be represented by a single integer n and assume we require the following operations:

- init(); sets $n \leftarrow 0$,
- update(); increments $n \leftarrow n + 1$,
- query(); outputs n or an estimate of n.

Initially, we assume that $n = 0$. A trivial algorithm uses a simple counter, storing n in binary format as a sequence of $\lceil \log_2 n \rceil = O(\log_2 n)$ bits. If query(); is required to return the exact value of n, this representation is necessary, and we cannot use fewer bits. Suppose that for some algorithm, we use $f(n)$ bits to store the integer n. Then there are $2^{f(n)}$ possible configurations for these bits. If we require the algorithm to store the exact value of all integers up to n, the number of possible configurations must be greater than or equal to the number n, that is,

$$2^{f(n)} \geq n \Rightarrow f(n) \geq \log n.$$

Suppose we would like to use much more than $O(\log n)$ and allow the output from query(); to be some estimate $\tilde{n}$ of n. We want $\tilde{n}$ to be a "good" estimate of n, satisfying

$$\mathbf{P}(|\tilde{n} - n| > \epsilon n) < \delta,$$

for some constants $\epsilon > 0$ and $0 < \delta < 1$ that are specified initially. The parameter ϵ is called the approximation factor and δ the failure probability.

The algorithm by Morris provides such an estimator for some ϵ, δ. The algorithm works as follows:

- init();: set $X \leftarrow 0$,
- update();: increment X with probability 2^{-X},
- query();: output $\tilde{n} = 2^X - 1$.

In the algorithm, the variable X is storing a value that is approximately $\log_2 n$. Denote by X_n the value of X after n updates.

Proposition 12.2.1. *In Morris's algorithm,* $\mathbf{E}\, 2^{X_n} = n + 1$.

Proof. We will prove the claim by induction. When $n = 0$, we set $X \leftarrow 0$ and no update has yet been made. Thus, $X_n = 0$ and $\mathbf{E}\, 2^{X_n} = n + 1$. Now suppose that $\mathbf{E}\, 2^{X_n} = n + 1$ for some fixed n. Then,

$$
\begin{aligned}
\mathbf{E}\, 2^{X_{n+1}} &= \sum_{j=0}^{\infty} \mathbf{P}(X_n = j) \cdot \mathbf{E}(2^{X_{n+1}} | X_n = j) \\
&= \sum_{j=0}^{\infty} \mathbf{P}(X_n = j) \cdot \left(2^j \left(1 - \frac{1}{2^j} \right) + \frac{1}{2^j} \cdot 2^{j+1} \right) \\
&= \sum_{j=0}^{\infty} \mathbf{P}(X_n = j) 2^j + \sum_{j} \mathbf{P}(X_n = j) \\
&= \mathbf{E}\, 2^{X_n} + 1 \\
&= (n + 1) + 1.
\end{aligned}
$$

$\square$

This shows that the estimate of n returned is given by $\tilde{n} = 2^X - 1$, which is an unbiased estimator of n. To verify the probability limits, we also need to control the variance of the estimator. By Chebyshev's inequality,

$$
\mathbf{P}(|\tilde{n} - n| > \epsilon n) < \frac{1}{\epsilon^2 n^2} \cdot \mathbf{E}(\tilde{n} - n)^2 = \frac{1}{\epsilon^2 n^2} \cdot \mathbf{E}(2^X - 1 - n)^2.
$$

Proposition 12.2.2. *In Morris's algorithm, we have*

$$
\mathbf{E}\, 2^{2X_n} = \frac{3}{2}n^2 + \frac{3}{2}n + 1.
$$

Proof. By induction, we have

$$
\begin{aligned}
\mathbf{E}\, 2^{2X_{n+1}} &= \sum_{j=0}^{\infty} \mathbf{P}(2^{X_n} = j) \cdot \mathbf{E}(2^{2X_{n+1}} | 2^{X_n} = j) \\
&= \sum_{j=0}^{\infty} \mathbf{P}(2^{X_n} = j) \cdot \left(\frac{1}{j} \cdot 4j^2 + \left(1 - \frac{1}{j} \right) \cdot j^2 \right) \\
&= \sum_{j=0}^{\infty} \mathbf{P}(2^{X_n} = j) \cdot (j^2 + 3j) \\
&= \mathbf{E}\, 2^{2X_n} + 3\, \mathbf{E}\, 2^{X_n}
\end{aligned}
$$

$$= \left(\frac{3}{2}n^2 + \frac{3}{2}n + 1\right) + (3n + 3)$$

$$= \frac{3}{2}(n + 1)^2 + \frac{3}{2}(n + 1) + 1.$$

Since the variance $\text{Var}(Z)$ can be written $\mathbf{E}\,Z^2 - (\mathbf{E}\,Z)^2$, we have

$$\mathbf{E}(\tilde{n} - n)^2 = \text{Var}(2^{X_n} - 1) = (1/2)n^2 - (1/2)n - 1 < (1/2)n^2$$

and thus

$$\mathbf{P}(|\tilde{n} - n| > \epsilon n) < \frac{1}{\epsilon^2 n^2} \cdot \frac{n^2}{2} = \frac{1}{2\epsilon^2}.$$

$\square$

This result is not very good, since the algorithm only succeeds with a probability better than $1/2$ for an $\epsilon \geq 1$-approximation. The estimator therefore has a nonneglibible probability of always being zero.

We can improve the basic algorithm by making s independent copies of the estimator and take their average. Let $\tilde{n}_1, \ldots, \tilde{n}_s$ be s independent copies from s instantiations of Morris's algorithm and let the query return

$$\tilde{n} = \frac{1}{s}\sum_{i=1}^{s}\tilde{n}_i.$$

Since each $\tilde{n}_i$ is an unbiased estimator of n, so is their average. The variance of the average now leads to

$$\mathbf{P}(|\tilde{n} - n| > \epsilon n) < \frac{1}{2s\epsilon^2} < \delta$$

for $s > 1/(2\epsilon^2\delta) = \Phi(1/(\epsilon^2\delta))$. This technique is often used to reduce the variance of an estimator. We refer to this modified algorithm as the Morris+ algorithm.

We can actually improve the estimate further, reducing the failure probability from $\mathcal{O}(1/\delta)$ to $\mathcal{O}(\log(1/\delta))$. The technique is as follows. We can run t instantiations of the Morris+ algorithm, each with $\delta = \frac{1}{3}$, each with $s = \Phi(1/\epsilon^2)$ copies. The expected number of successful Morris+ instantiations is at least $2t/3$. Now take the median of the t instantiations. For this estimate to be bad, the number of successful estimations has to deviate by at least $t/6$. Defining the variable

$$T_i = \begin{cases} 1, & \text{if the } i\text{th Morris+ instantiation successful,} \\ 0, & \text{otherwise,} \end{cases}$$

we have by the Chernoff bound,

$$\mathbf{P}\left(\sum_{i=1}^{t} Y_i \leq \frac{t}{2}\right) \leq \mathbf{P}\left(\left|\sum_{i=1}^{t} Y_i - \mathbf{E}\sum_{i=1}^{t} Y_i\right| \geq \frac{t}{6}\right) \leq 2e^{-t/3} < \delta$$

for $t \in \Phi(\log(1/\delta))$. The memory requirement for the algorithm is actually a random variable and can be shown to be, with probability $1 - \delta$, at most

$$O(\epsilon^{-2}\log(1/\delta)(\log\log(n/(\epsilon\delta))))$$

bits. For constant ϵ, δ (say each 1/100), we need $O(\log(\log n))$ space with constant probability that can be compared with $\log n$ space by storing a counter.

Note that the space is a random variable, and the total space complexity is, with probability $1 - \delta$, at most

$$O(\epsilon^{-2}\log(1/\delta)(\log\log(n/(\epsilon\delta))))$$

bits. In particular, for constant ϵ, δ (say each 1/100), the total space complexity is $O(\log\log n)$ with constant probability. This is exponentially better than the $\log n$ space achieved by storing a counter.

Counting Distinct Elements

Consider a stream of elements, that is, integer values $i_1, i_2, i_3, \ldots, i_m \in \{1, \ldots, n\}$, and suppose we want to estimate the number of distinct elements in the stream.

Similarly to the case of counting events, we could define a bit vector of length n, setting the bit in position i whenever this element is encountered. Alternatively, we can store the elements using $\min\{m, n\} \log_2 n$.

Denoting the number of elements by A, we would like to design an algorithm that returns an estimate $\tilde{A}$ such that $\mathbf{P}(|\tilde{A} - A| > \epsilon \cdot A) < \delta$.

The algorithm published by Flajolet and Martin [109] for counting elements is based on the following steps:

(1) Choose a random hash function $h : [n] \to [0, 1]$.
(2) Update the smallest hash encountered so far; $X = \min_i h(i)$.
(3) Output $1/X - 1$.

Suppose t is the number of distinct elements. We can interpret the counting as a partitioning of the interval $[0, 1]$ into bins of size $1/(t + 1)$.

Proposition 12.2.3. *We have* $E(X) = \frac{1}{t+1}$.

Proof. We have

$$
\begin{aligned}
E(X) &= \int_0^\infty P(X > \lambda)d\lambda \\
&= \int_0^\infty \prod_{i \in \text{stream}} P(h(i) > \lambda)d\lambda \\
&= \int_0^1 (1 - \lambda)^t d\lambda \\
&\quad \frac{1}{t+1}.
\end{aligned}
$$

$\square$

Similarly, we get the second moment

$$
\begin{aligned}
E(X^2) &= \int_0^\infty P(X^2 > \lambda)d\lambda \\
&= \int_0^\infty P(X > \sqrt{\lambda})d\lambda \\
&= \int_0^1 (1 - \sqrt{\lambda})^t d\lambda \\
&\quad \frac{2}{(t+1)(t+2)}
\end{aligned}
$$

and the variance

$$
\text{Var}(X) = \frac{2}{(t+1)(t+2)} - \frac{1}{(t+1)^2} = \frac{t}{(t+1)^2(t+2)} < (E(X))^2.
$$

We refer to this algorithm as the FM algorithm. Using the same method as earlier, we can improve the algorithm by running q independent copies of FM. It can be shown that with $q = \frac{1}{\epsilon^2 \eta}$ copies generating $X_1, \ldots, X_q$, query() should return

$$
\frac{1}{\frac{1}{q}\sum_{i=1}^q X_i} - 1.
$$

Proposition 12.2.4. *We have*

$$
P\left(\left|\frac{1}{q}\sum_{i=1}^q X - i - \frac{1}{t+1}\right|\right) < \eta.
$$

Proof. Chebyshev's inequality yields

$$
\mathbf{P}\left(\left|\frac{1}{q}\sum_{i=1}^{q}X_i - \frac{1}{t+1}\right| > \frac{\epsilon}{t+1}\right) < \frac{\mathrm{Var}(\frac{1}{q}\sum_i X_i)}{\frac{\epsilon^2}{(t+1)^2}} < \frac{1}{\epsilon^2 q} = \eta.
$$

$\square$

We call the improved version of the algorithm FM+.

Again, we can improve the result further by running $t = \Phi(\log\frac{1}{\delta})$ independent copies of FM+ with $\eta = 1/3$ and let `query()` return the median of all FM+ estimates.

The new space requirement for the improved algorithm is $\mathcal{O}\left(\frac{1}{\epsilon^2}\log\frac{1}{\delta}\right)$.

Estimation of Vector Norms

It is very useful to interpret data streams in terms of vectors and their algorithms in terms of vector norms. We recall the characteristics of a vector norm, considering real numbers only. A *vector norm* is a function $f : V \rightarrow \mathbb{R}$ over a vector space V that for $\mathbf{x}, \mathbf{y} \in V$ and a scalar $a \in \mathbb{R}$ has the following properties:

(1)

$$
f(a\mathbf{x}) = |a|f(\mathbf{x}),
$$

(2)

$$
f(\mathbf{x} + \mathbf{y}) \leq f(\mathbf{x}) + f(\mathbf{y}),
$$

(3)

$$
f(\mathbf{x}) \geq 0,
$$

(4)   if $f(\mathbf{x}) = 0$ then $\mathbf{0} = \mathbf{0}$, that is, the zero vector.

The ℓ_p vector norm is defined as

$$
||\mathbf{x}||_p = \left(\sum_{i=1}^{n}|x_i|^p\right)^{1/p}, \quad p \geq 1.
$$

From this definition, we have for $p = 1$ the *taxicab* (or Manhattan grid) norm

$$
||\mathbf{x}||_1 = \sum_{i=1}^{n}|x_i| \tag{12.2}
$$

and for $p = 2$ the Euclidean norm

$$||\mathbf{x}||_2 = \sqrt{x_1^2 + x_2^2 + \cdots + x_n^2}.$$

We also have the maximum norm for $p = \infty$,

$$||\mathbf{x}||_\infty = \max\{|x_1|, |x_2|, \ldots, |x_n|\}.$$

Note that

$$\sum_{i=1}^{n} x_i$$

is not a norm, since it can be negative in general. This is the reason for using the cash register model rather than the strict turnstile model, where it is equivalent to (12.2) following the assumption that $x_i >= 0$ for all i.

We also introduce the notation $||\mathbf{x}||_0$ for the count of distinct elements in $\mathbf{x}$, that is,

$$||\mathbf{x}||_0 = \text{Number of } (i \,|\, x_i \neq 0).$$

Norms are important as they link the vector interpretation of the stream with statistical properties of the data. We can interpret the vector $\mathbf{x}$ as the empirical distribution of the data in the stream, and then $||x||_2$ corresponds to the second-order moment of x, closely related to its variance, that is, a measure of dispersion of x. We notice for each $||x||_1 = n$ fixed, the Euclidean norm $||x||_2$ has its minimum value when all $x_i = n/m$, which corresponds to a uniform distribution.

We discuss two algorithms for estimating the ℓ_2 norm. The first, published by Alon, Matias, and Szegedy [110], we refer to as the AMS algorithm. The second, by Johnson and Lindenstrauss [111], we call the JL algorithm.

The AMS Algorithm

The aim of the AMS algorithm is to obtain an unbiased estimator of $||x||_2^2$ by using a *linear sketch* and proving its precision by a variance argument of the estimator.

The algorithm works as follows. Choose $r_1, r_2, \ldots, r_m \in \{-1, 1\}$ as independent and identically distributed random variables such that $\mathbf{P}(r_i = -1) = \mathbf{P}(r_i = 1) = 0.5$ for every i. Construct the variable $Z = (r, x) = \sum r_i x_i$. Since Z is linear in x, whenever an update of x by (i, a) is necessary, we only need to increment Z by $r_i a$. The returned estimator of $||x||_2^2$ is Z^2.

Being a randomized algorithm, we verify the performance of the AMS algorithm by showing $\mathbf{E}(Z^2) = ||x||_2^2$ and the boundedness of the variance. Thus,

$$\mathbf{E}(Z^2) = \mathbf{E}(x^T r r^T x) = x^T \mathbf{E}(r r^T)x = x^T x = ||x||_2^2,$$

using the fact that for every i, $\mathbf{E}(r_i^2) = 1$ and for every $i \neq j$, $\mathbf{E}(r_i r_j) = \mathbf{E}(r_i)\mathbf{E}(r_j) = 0$. For the variance $\text{Var}(Z^2)$ is $O(||x||_2^4)$, we have

$$\text{Var}(Z^2) = \mathbf{E}(Z^4) - \mathbf{E}(Z^2)^2 = \mathbf{E}(Z^4) - ||x||_2^4.$$

We decompose $\mathbf{E}(Z^4)$ as

$$\mathbf{E}(Z^4) = \sum_{i,j,k,l} \mathbf{E}(x_i x_j x_k x_l r_i r_j r_k r_l) = \sum_{i,j,k,l} x_i x_j x_k x_l \, \mathbf{E}(r_i r_j r_k r_l).$$

We note that $\mathbf{E}(r_i r_j r_k r_l)$ is 0 if there is one index only appearing once in i, j, k, l, so we only need to consider the case where every distinct index appears at least twice. Then there are two cases such that $\mathbf{E}(r_i r_j r_k r_l) = 1$:

- There are two pairs of distinct values of indices (i, j, k, l).
- All indices i, j, k, l take identical values.

We have

$$\mathbf{E}(Z^4) = \frac{1}{2}C_4^2 \sum_{i \neq j} x_i^2 x_j^2 + \sum_i x_i^4 = 3 \sum_{i \neq j} x_i^2 x_j^2 + \sum_i x_i^4,$$

and $\mathbf{E}(Z^4)$ can be bounded by

$$\mathbf{E}(Z^4) = 3 \sum_{i \neq j} x_i^2 x_j^2 + \sum_i x_i^4 \leq 3 \sum_{i \neq j} x_i^2 x_j^2 + 3 \sum_i x_i^4 \leq 3||x||_2^4,$$

which gives $\text{Var}(Z^4) \leq 2||x||_2^4$. By the Chebyshev inequality,

$$\mathbf{P}(|\mathbf{E}(Z^2) - ||x||_2^2| \leq \sqrt{2}c||x||_2^2) \leq 1/c^2.$$

This bound is rather loose for the approximation of $||x||_2^2$. We can improve this by running k independent copies of the AMS algorithm and taking the average. This does not affect the expectation, but reduces the variance by a factor k.

We therefore generate $Z_1, Z_2, \ldots, Z_k$ where for every j, $Z_j = \sum_i r_{ji} x_i$ are independent and identically distributed random variables. Then the estimator for $||x||_2^2$ is modified to $(\sum_j Z_j^2)/k$. Let $Y = (\sum_j Z_j^2)/k$, so that $\mathbf{E}(Y) = (\sum_j \mathbf{E}(Z_j^2))/k = ||x||_2^2$ and $\text{Var}(Y) = (\sum_j \text{Var}(Z_j^2))/k^2 \leq 2||x||_2^4/k$. The Chebyshev inequality gives

$$\mathbf{P}(|\mathbf{E}(Z^2) - ||x||_2^2| \leq c\sqrt{2/k}||x||_2^2) \leq 1/c^2.$$

With $c = \Phi(1)$ and $k = \Phi(1/\epsilon^2)$, we get a $(1 \pm \epsilon)$-approximation with constant probability. The space requirement of the algorithm is dominated by Z_j for all j. The maximum possible value for Z_j is mn so the space requirement is $\log(mn)$, ignoring the space to generate r_j. With $\mathcal{O}(1/\epsilon^2)$ copies of Z_j we need $\mathcal{O}(\log(mn)/\epsilon^2)$ bits.

We can also formulate the AMS algorithm in vector form. Let R be a $k \times m$ matrix where $R_{ij} = r_{ij}$. Then we compute the vector $\mathbf{Z} = R\mathbf{x}$ return $||R\mathbf{x}||_2^2/k$; $R\mathbf{x}$ is called the *linear sketch* of $\mathbf{x}$. It can be seen as a compression of $\mathbf{x}$ which greatly reduces the dimension of the vector but still contains enough information to give a good estimate for the property in question, that is, the ℓ_2 norm.

The linearity of the sketch makes it convenient for data stream concatenation, which is common in applications. Suppose we have two data streams $\mathbf{x}$ and $\mathbf{y}$ from two different sources, and we wish to estimate $||x + y||_2^2$; it suffices to compute $R\mathbf{x}$ and $R\mathbf{y}$ and return $||R\mathbf{x} + R\mathbf{y}||_2^2/k$.

The Johnson–Lindenstrauss Algorithm

We can improve the error bound in the AMS algorithm by choosing a different sketch. The sketch proposed by Johnson and Lindenstrauss [111] uses Gaussian random weights rather than the two-point distribution in the AMS algorithm. The advantage of using the normal distribution is that it is straightforward to analyze and that it is closed under linear operations.

We refer to such a sketch as a Johnson–Lindenstrauss sketch, or JL sketch for short. The sketch matrix R is again a $k \times m$ matrix, where entry r_{ij} is an independent and identically distributed random variable drawn from $\mathcal{N}(0, 1)$. The JL algorithm is apart from the sketch the same as in the AMS algorithm.

In vector notation, we compute the vector $R\mathbf{x}$ and return the estimate of $||\mathbf{x}||_2^2$ as $||R\mathbf{x}||_2^2/k$. To analyze the JL algorithm, we note that we still have $\mathbf{E}(||R\mathbf{x}||_2^2/k) = ||\mathbf{x}||_2^2$, since

$$\mathbf{E}(||R\mathbf{x}||_2^2/k) = \frac{1}{k}\mathbf{E}(\mathbf{x}^T R^T R\mathbf{x}) = \frac{1}{k}\mathbf{x}^T \mathbf{E}(R^T R)\mathbf{x} = ||\mathbf{x}||_2^2,$$

where $\mathbf{E}(R^T R)$ is a diagonal matrix with all diagonal entries equal to k. For every $i \in \{1, \ldots, k\}$, the ith diagonal entry in the matrix is $\mathbf{E}(\sum_k R_{ki}^2) = \sum_k \mathbf{E}(R_{ki}^2) = k$. Similarly, for every entry in position (i, j) with $i \neq j$, we have $\mathbf{E}(\sum_k R_{ki} R_{kj}) = \sum_k \mathbf{E}(R_{ki} R_{kj}) = 0$.

Proposition 12.2.5. *We have*

$$\mathbf{P}(|\,||R\mathbf{x}||_2^2 - k||\mathbf{x}||_2^2| \geq \epsilon k||\mathbf{x}||_2^2) \leq \exp(-C\epsilon^2 k).$$

Proof. Suppose $||\mathbf{x}||_2^2 = 1$ and let $Z = R\mathbf{x}$. We want to show that

$$\mathbf{P}(|\,||Z||_2^2 - k| \geq \epsilon k) \leq \exp(-C\epsilon^2 k).$$

Consider the one-sided inequality

$$\mathbf{P}(||Z||_2^2 - (1+\epsilon)k) \leq \exp(-\epsilon^2 k + O(k\epsilon^3)).$$

Let $Y = ||Z||_2^2$ and $\alpha = k(1+\epsilon)^2$. Then we have for $s > 0$

$$\mathbf{P}(Y > \alpha) = \mathbf{P}(\exp(sY) > \exp(s\alpha)) \leq \exp(-s\alpha)\,\mathbf{E}(\exp(sY))$$

from the Markov inequality. We can decompose $\mathbf{E}(\exp(sY))$ by the independence of the Z_i, that is,

$$\mathbf{E}(\exp(sY)) = \prod_i \mathbf{E}(\exp(sZ_i^2)).$$

By the closure property of the normal distribution, Z_i are also normal. Since, for every i,

$$\mathbf{E}(Z_i) = \sum_j \mathbf{E}(r_{ij}x_j) = 0,$$

$$\mathrm{Var}(Z_i) = \mathbf{E}(Z_i^2) = \sum_j \mathbf{E}(r_{ij}^2 x_j^2) = ||x||_2^2 = 1,$$

we see that $Z_i \sim \mathcal{N}(0, 1)$, so we can calculate $\mathbf{E}(\exp(sZ_i^2))$ analytically as

$$\mathbf{E}(\exp(sZ_i^2)) = \frac{1}{\sqrt{2\pi}} \int \exp(sT^2)\exp(-t^2/2)\mathrm{d}t = \frac{1}{\sqrt{1-2s}},$$

leading to

$$\mathbf{P}(Y \geq \alpha) = \exp(-s\alpha)(1-2s)^{-k/2}.$$

Now, choosing $\alpha = k(1+\epsilon)^2$ and plugging it into the equation, we find

$$\mathbf{P}(Y \geq \alpha) = \exp(-\epsilon k - \epsilon^2 k/2 + k\ln(1+\epsilon)) = \exp(-k\epsilon^2 + kO(\epsilon^3)),$$

using the Taylor expansion $\ln(1+x) = x - x^2/2 + O(x^3)$. From this tail bound, we see that the estimation is correct, and with a narrower error bound. Letting $\exp(-C\epsilon^2 k) = \delta$, we conclude that we need $k = \mathcal{O}(1/\epsilon^2 \log(1/\delta)$ to have a $1 \pm \epsilon)$-approximation with probability $1 - \delta$. $\qquad\square$

The time complexity of the JL algorithm is $\mathcal{O}(k)$ to update the sketch.

The Median Algorithm

We present a third algorithm for ℓ_2 norm estimation that can easily be generalized to ℓ_p with $0 < p \leq 2$. This algorithm can be used to estimate the ℓ_1 norm, for example. However, the algorithm only works for positive updates ($a \geq 0$).

The algorithm uses sampling instead of sketches and requires $O(k\frac{m^{1-1/k}}{\epsilon^2})$ space to obtain a $(1 \pm \epsilon)$-approximation with constant probability.

The first step of the algorithm is identical to the previous algorithm, using a linear sketch $R\mathbf{x} = [Z_1 \cdots Z_k]$, with each entry of R drawn from a normal distribution $\mathcal{N}(0, 1)$, $k = O(1/\epsilon^2)$. Therefore, each of Z_i has an $N(0, 1)$ distribution with variance $\sum_i x_i^2 = ||\mathbf{x}||_2^2$. We can view the Z_i as weighted samples of $||\mathbf{x}||_2^2$, so that $Z_i = ||\mathbf{x}||_2 G_i$, where G_i is drawn from $\mathcal{N}(0, 1)$.

In the previous two algorithms, we used $Y = (Z_1^2 + \cdots + Z_k^2)/k$ to estimate $||\mathbf{x}||_2^2$, but it is possible to use other estimators as well. A candidate estimator of $||\mathbf{x}||_2$ from $Z_1, Z_2, \ldots, Zk$ is

$$Y = \frac{\text{median}\{|Z_1|, \ldots, |Z_k|\}}{\text{median}\{|G|\}},$$

where G is drawn from $\mathcal{N}(0, 1)$. Here, the median of a sequence is the usual midpoint number of the sorted sequence, whereas the median of a random variable U is the value such that $\mathbf{P}(U \leq \text{median}) = 0.5$. The intuition here is that $\text{median}\{|Z1|, \ldots, |Zk|\} = ||x||_2\text{median}\{|G_1|, \ldots, |G_k|\}$. For large enough k, $\text{median}\{|G_1|, \ldots, |G_k|\}$ becomes close to $\text{median}\{|G|\}$.

Lemma 12.2.6. *Let $U_1, \ldots, U_k$ be independent and identically distributed random variables chosen from a distribution with continuous cumulative distribution function (CDF) F and median M, so that $F(t) = \mathbf{P}(U_i \leq t)$ and $F(M) = 1/2$. Define $U = median\{U_1, \ldots, U_k\}$. Then, for some constant $C > 0$,*

$$\mathbf{P}\left(F(U) \in \left(\frac{1}{2} - \epsilon, \frac{1}{2} + \epsilon\right)\right) \geq 1 - e^{-C\epsilon^2 k}.$$

Proof. For simplicity, we assume that k is odd so that the median is exactly the midpoint. Consider the one-sided events $E_i : F(U_i) < \frac{1}{2} - \epsilon$. We have $p = \mathbf{P}(E_i) = \frac{1}{2} - \epsilon$. We see that $F(U) < \frac{1}{2} - \epsilon$ if and only if at least $k/2$ of these events hold. By the Chernoff bound, the probability that at least $k/2$ of the events hold is at most $e^{-C\epsilon^2 k}$. Therefore, $\mathbf{P}(F(u) < \frac{1}{2} - \epsilon) \leq e^{-C\epsilon^2 k}$. The other side is treated analogously. $\square$

Lemma 12.2.7. *Let F be the CDF of a random variable $|G|$, where G is drawn from $\mathcal{N}(0, 1)$. There exists a $C' > 0$ such that for some z we have*

$$F(z) \in \left(\frac{1}{2} - \epsilon, \frac{1}{2} + \epsilon \right),$$

and then

$$z = \text{median}(g) \pm C'\epsilon.$$

Combining Lemmas 12.2.6 and 12.2.7, we have the following theorem.

Theorem 12.2.8. *By using the median estimator*

$$Y = \frac{\text{median}\{|Z_1|, \ldots, |Z_k|\}}{\text{median}\{|g|\}},$$

where $Z_j = \sum_i r_{ij} x_i$ and r_{ij} are independent and identically distributed random variables drawn from $\mathcal{N}(0, 1)$, we have

$$Y = \frac{\text{median}\{g\} \pm C'\epsilon}{\text{median}\{|g|\}} = ||\mathbf{x}||_2 (1 \pm C''\epsilon)$$

with probability

$$1 - e^{-C\epsilon^2 k}.$$

To extend the algorithm to $0 < p \le p$, we note a key property of the normal distribution. If $U_1, \ldots, U_k$ are independent normal random variables, then $x_1 U_1 + \cdots + x_m U_m$ is distributed as

$$(|x_1|^p + \cdots + |x_m|^p)^{1/p} U$$

with $p = 2$. Distributions having this property are known as "p-stable" and exist for $p \in (0, 2]$. For $p = 1$, we have the Cauchy distribution with probability density function

$$f(x) = \frac{1}{\pi (1 + x^2)}$$

and cumulative distribution function

$$F(z) = \arctan(z)/\pi + \frac{1}{2}.$$

This means that for 1-stability, we know that $x_1 U_1 + \cdots + x_m U_m$ is distributed as $(|x_1| + \cdots + |x_m|)U$.

The Cauchy distribution does not have a first or second moment. However, the arguments in showing the validity of the estimator are still valid. Thus, we can generate Cauchy variables from uniformly distributed variables $u \in [0, 1]$ and computing $F^{-1}(u)$, and then estimate ℓ_1 norm. Similarly, for the $\ell_{1/2}$ norm, we use the Levy distribution.

We can now formulate an algorithm for computing the ℓ_p-norm, which only works for a stream of elements $i_1, \ldots, i_n$, with the ith update as $x_i = x_i + 1$. The algorithm does not work for negative updates. The space requirement is $O(m^{1-1/k}/\epsilon^2)$ for a $(1 \pm \epsilon)$-approximation with constant probability. Let the frequency moment of a stream $i_1, \ldots, i_n$ be $F_k = \sum_{i=1}^{m} x_i^k = ||\mathbf{x}||_k^k$ and consider the following algorithm.

(1)   Choose an element $i = i_j$ uniformly at random.
(2)   Update $x_i \leftarrow x_i + 1$.

Return the estimator $Y = nx^{k-1}$.

The expectation of the estimator is $\mathbf{E}(Y) = \sum_i \frac{x_i}{n} nx_i^{k-1} = \sum_i x_i^k = F_k$. We obtain the second moment from

$$\mathbf{E}(Y^2) = \sum_i \frac{x_i}{n} n^2 x_i^{2k-2} = n \sum_i x_i 2k - 1 = n F_{2k-1}.$$

The variance can be bounded so that $n F_{2k-1} \leq m^{1-1/k}(K_k)^2$. This can be seen from

$$n F_{2k-1} = n||x||_{2k-1}^{2k-1} \leq ||x||_1 ||x||_k^{2k-1}$$
$$\leq m^{1-1/k} ||x||_k ||x||_k^{2k-1} = m^{1-1/k} ||x||_k^{2k} = m^{1-1/k} F_k^2.$$

Therefore, by averaging over $O(m^{1-1/k}/\epsilon^2)$ samples and using the Chebyshev inequality, we see that the algorithm indeed is an $(1 \pm \epsilon)$-approximation. However, the algorithm has to pass over the data twice: first to pick a random element and then to make the updates.

We improve the algorithm by first choosing an element $i = i_j$ uniformly at random from the stream and then compute r, the number of occurrences of i in the remainder of the stream $i_j, \ldots, i_n$. We can use r in place of x_i in the estimator. By construction, $r \leq x_i$, but it is clear that $E(r) = \frac{x_i+1}{2}$, so by proper scaling, the algorithm should work well. Alternatively, we can use the estimator $Y' = n(r^k - (r-1)^k)$, which has the expectation

$$\mathbf{E}(Y') = n \mathbf{E}(r^k - (r-1)^k) = n \frac{1}{n} \sum_i \sum_{j=1}^{x_i} \left(j^k - (j-1)^k \right) = \sum_i x_i^k.$$

For the second moment, we observe that $Y' = n(r^k - (r-1)^k) \leq nkr^{k-1} \leq kY$ and so,

$$\mathbf{E}(Y'^2) \leq k^2 \mathbf{E}[Y^2] \leq k^2 m^{1-1/k} F_k^2.$$

These two results verify the algorithm. The space requirement is $O(k^2 m^{1-1/k}/\epsilon^2)$ for a $(1 \pm \epsilon)$-approximation.

The Count-Min Sketch

The fixed sketch matrix R used in the linear transformation would require larger space than that required to store $\mathbf{x}$. To circumvent this, many sketches use hash functions to generate a linear transformation having smaller memory requirement.

The count-min sketch by Cormode and Muthukrishnan [112] is a frequency-based sketch for data streams $i_1, \ldots, i_n$. Algorithms based on the count-min sketch take the parameters ϵ and δ, where ϵ is the approximation factor and δ the corresponding failure probability. When queried, it returns the approximate frequency $\hat{x}_i$, such that for $i \in \{1, 2, \ldots, m\}$, $\hat{v}_i$ is no more than $\epsilon ||\mathbf{x}||_1$ larger than x_i with probability $1 - \delta$. Here $||\mathbf{x}||_1 = \sum_{i=1}^{m} x_i$ is the ℓ_1-norm of the actual frequency vector $\mathbf{x}$.

The sketch uses a matrix $V \in \mathbb{R}^{d \times w}$ where the amount of rows, d, will be called the depth and the amount of columns, w, will be called the width. Initially all entries are zeroed and the dimensions of the matrix are defined as $w = \lfloor b/\epsilon \rfloor$ and $d = \lfloor \log_b \delta^{-1} \rfloor$, where the base of the logarithm, b, can be chosen freely for all $b > 1$.

For the linear transformation, we choose k hash functions $h_1, \ldots, h_k$ independently, from a family of hash functions $h_i : \{1, 2, \ldots, m\} \to \{1, 2, \ldots, w\}$ for all i.

In the original algorithm [112], the hash functions are chosen from a pairwise independent family. However, as shown by Hovmand and Nygaard [113], this requirement can be relaxed to choosing hash functions from a c-universal family for some constant c, by appropriately scaling w. Each row in V is associated with a hash function. When an update is performed, the matrix is updated row by row in V, that is, if element x_i is updated with value a at time i, all entries are updated by $V(j, h_j(s_i)) \leftarrow V(j, h_j(s_i)) + a$ for all $j \in \{1, 2, \ldots, k\}$. These updates are independent following the independence of the hash functions.

A point query $Q(s_i)$ to the count-min sketch returns the estimated frequency $\hat{x}_i$ of element s_i. The closeness of the estimated frequency to the true frequency is given by the approximation factor ϵ and the failure probability δ. In the point query for an element s_i, we inspect all buckets $\hat{x}_{i,j} = V[j, hj(s_i)]$ for $j \in \{1, 2, \ldots, k\}$. Since the hash functions h_j are different, we expect different estimates for each j. Furthermore, using the strict turnstile model, so that updates take $a > 0$, the error of each estimate is one-sided, so that $\hat{x}_{i,j} \leq x_i$ for all j. Therefore, taking the minimum of all buckets $\min_j \hat{x}_{i,j}$ gives the closest estimate to x_i. For the performance of a point query to the count-min sketch, we have the following theorem.

Theorem 12.2.9. *A point query to the count-min sketch returns an estimated frequency $\hat{x}_i$ such that*

(1) $x_i \leq \hat{x}_i$ *and*
(2) $\hat{x}_i \leq x_i + \epsilon ||\mathbf{x}||_1$ *with probability* $1 - \delta$.

Proof. Let $I_{i,j,k}$ indicate if a collision occurs when two distinct elements are applied to the same hash function, for which

$$I_{i,j,k} = \begin{cases} 1, & \text{if } i \neq k \text{ and } h_j(s_i) = h_j(s_k), \\ 0, & \text{otherwise} \end{cases}$$

for all i, j, k, where $i, k \in [m]_1$ and $j \in [d]_1$. The expected amount of collisions can be derived from the choice of the family of hash functions. The probability of collision for a c-universal is by definition $\mathbf{P}(h_j(s_i) = h_j(s_k)) \leq \frac{c}{w}$ for $i \leq k$. The expectation of $I_{i,j,k}$ then becomes

$$\mathbf{E}(I_{i,j,k}) = \mathbf{P}(h_j(s_i) = h_j(s_k)) \leq \frac{c}{w} \leq c/\lceil\frac{b}{\epsilon}\rceil \leq \frac{\epsilon c}{b}.$$

Let $X_{i,j}$ be the random variable $X_{i,j} = \sum_{k=1}^{m} I_{i,j,k} v_k$ for $i \in [m]_1$, $j \in [d]_1$, from the independent choices of the hash functions; $X_{i,j}$ then expresses all the additional mass contributed by other elements as a consequence of hash function collisions. Since v_i is nonnegative by the definition of the strict turnstile model, $X_{i,j}$ must also be nonnegative. By the construction of the array V of the count-min sketch data structure, an entry in the array is then $V[j, h_j(s_i)] = v_i + X_{i,j}$. This implies that item 1 is true, since $\hat{v}_i = \min_j V[j, h_j(s_i)] \geq v_i$. To prove 2, observe that the expected collision mass for $i \in [m]_1$, $j \in [d]_1$, and the constant c can be defined as

$$\mathbf{E}[X_{i,j}] = \mathbf{E}\left[\sum_{k=1}^{m} I_{i,j,k} v_k\right] = \sum_{k=1}^{m} v_k \mathbf{E}(I_{i,j,k}) \leq \frac{\epsilon c}{b} \sum_{k=1}^{m} v_k = \frac{\epsilon c}{b} ||v||_1.$$

Using this, we can calculate the probability that $\hat{v}_i > v_i + \epsilon ||v||_1$, that is, the probability that the estimate of the frequency is larger than the expected error bound introduced by approximation. This is reasonably easy to do, since taking the minimum of all the estimated frequencies for an element s_i from each of the rows gives us the closest estimate to the true frequency. For this estimate to fail ($\hat{v}_i > v_i + \epsilon ||v||_1$) it must by the definition of the minimum have failed for all rows. This is what is expressed in the following:

$$\mathbf{P}(\hat{v}_i > v_i + \epsilon ||v||_1) = \prod_j \mathbf{P}(V(j, h_j(s_i)) > v_i + \epsilon ||v||_1$$

$$= \prod_j \mathbf{P}(v_i + X_{i,j} > v_i + \epsilon ||v||_1) \quad = \prod_j \mathbf{P}\left(X_{i,j} > \frac{b}{c} \mathbf{E}(X_{i,j})\right) < \prod_j \frac{c}{b}$$

$$= c^d b^{-d} \leq q c^d \delta.$$

Note that the product of the probabilities follows since the hash functions h_j are all independent of each other. If we rescale w with the constant c, that is, $w = \lceil bc/\epsilon \rceil = O(\epsilon^{-1})$, and redo the analysis in this theorem we end up removing the c constant throughout the proof, giving us $\mathbf{P}(\hat{v}_i \leq v_i + \epsilon ||v||_1) = 1 - \delta$, which proves item 2. $\qquad\square$

The Count-Median Sketch

The count-median sketch by Charikar et al. [114] allows updates according to the generalized turnstile model. We let ϵ be the approximation ratio and δ the failure probability. The sketch is based on a matrix $R \in \mathbb{R}^{k \times w}$, initially with all entries set to zero. The dimensions are given by $w = \lceil q/\epsilon^2 \rceil$ and

$$k = \left\lceil \frac{\ln(\delta^{-1})}{\frac{1}{6} - \frac{1}{3q}} \right\rceil$$

for some constant $q > 2$ determining the error probability of a specific bucket.

A point query $Q(s_i)$ to the count-median sketch returns an estimate of the frequency $\hat{x}_i$ of an element s_i, such that the absolute error of $\hat{x}_i$ is bounded by $\epsilon ||\mathbf{x}||_2$ from the true value x_i with probability $1 - \delta$. Here $||\mathbf{x}||_2 = \sqrt{\sum_{i=1}^{m} v_i^2}$ is the ℓ_2-norm of the data.

We associate a pair of hash functions (h_j, g_j) with each row j in R. The hash functions $h_1, \ldots, h_k$ and $g_1, \ldots, g_k$ linearly transform the input vector $\mathbf{x}$ as $h_j : \{1, 2, \ldots, m\} \rightarrow \{1, 2, \ldots, w\}$ and $g_j\{1, 2, \ldots, m\} \rightarrow \{-1, 1\}$.

It turns out that it suffices to draw the hash functions h_j from a c-universal family as long as w is scaled by the constant c so that $w = \lceil qc/\epsilon^2 \rceil$. For the g_j hash functions, a tighter guarantee is required by choosing g_j from a pairwise independent family.

At an update with respect to element s_i by a value a, each row of R is updated according to $R(j, h_j(s_i)) \leftarrow R(j, h_j(s_i)) + g_j(s_i)a$ for all $j \in \{1, 2, \ldots, k\}$. The updates are independent due to the independence of the hash functions h_j and g_j.

The space requirement is

$$|V| = wk = \left\lceil \frac{q}{\epsilon^2} \right\rceil \frac{\ln(\delta^{-1})}{\left(\frac{1}{6} - \frac{1}{3q} \right)} \sim O\left(\epsilon^{-2} \ln \delta^{-1} \right).$$

Since the sketch uses signed updates we cannot take the minimum value over the rows of R. Instead, we return the median of the k estimated frequencies, which can be shown to satisfy the performance guarantees. For the error, we have the following lemma.

Lemma 12.2.10. *The expected error from colliding elements in a bucket $R(j, h_j(s_i))$ is zero, and the variance is*

$$\frac{c||\mathbf{x}||_2^2}{w}.$$

Proof. Let $I_{i,j,k}$ be the indicator variable indicating if a collision occurs for two distinct elements when applied to the same hash function, defined as

$$I_{i,j,k} = \begin{cases} 1, & \text{if } i \neq k \text{ and } h_j(s_i) = h_j(s_k), \\ 0, & \text{otherwise} \end{cases}$$

for $i, k \in [m]_1$, $j \in [d]_1$. The expected amount of collisions can then be derived since h_j is chosen from a c-universal family, which by definition has probability $\leq c/w$ for two distinct elements to collide, that is,

$$\mathbf{E}(I_{i,j,k}) = \mathbf{P}(h_j(s_j) = h_j(s_k)) \leq \frac{c}{w}.$$

Let $X_{i,j} = \sum_{k=1}^{m} I_{i,j,k} v_k g_j(s_k)$ be the random variable describing the mass of all elements that collide with s_i for hash function h_j. Hence, we can associate $X_{i,j}$ with the content of a bucket, since we can rewrite $V[j, h_j(s_i)] = X_{i,j} + g_j(s_i)v_i$. If we expect over $X_{i,j}$ we get the expected collision mass introduced in each bucket of the sketch,

$$\mathbf{E}(X_{i,j}) = \mathbf{E}\left(\sum_{k=1}^{m} I_{i,j,k} v_k g_j(s_k)\right)$$

$$= \sum_{k=1}^{m} \left(\mathbf{E}(I_{i,j,k}) v_k \, \mathbf{E}(g_j(s_k))\right) = \frac{c}{w} \sum_{k=1}^{m} \left(v_k \, \mathbf{E}(g_j(s_k))\right) = 0.$$

In our expectations, the error of each bucket is then canceled out due to the use of the g_j hash functions. The result follows from that the hash functions g_j and h_j are independent of each other, and g_j are chosen from a pairwise independent family of hash functions.

Furthermore, one can estimate the variance of $X_{i,j}$, to see how much the estimated error will variate from the expected value of $X_{i,j}$. We have

$$\text{Var}(X_{i,j}) = \mathbf{E}(X_{i,j}^2) - \mathbf{E}(X_{i,j})^2 = \mathbf{E}(X_{i,j}^2) = \mathbf{E}\left(\left(\sum_{k=1}^{m} I_{i,j,k} v_k g_j(s_k)\right)^2\right)$$

$$= \mathbf{E}\left(\sum_{k=1}^{m} (I_{i,j,k} v_k g_j(s_k))^2 + \sum_{k' \neq k} I_{i,j,k} v_k g_j(s_k) I_{i,j,k'} v_{k'} g_j(s_{k'})\right)$$

$$= \mathbf{E}\left(\sum_{k=1}^{m} (I_{i,j,k} v_k g_j(s_k))^2\right) + \sum_{k' \neq k} \mathbf{E}(I_{i,j,k} I_{i,j,k'}) v_k \, \mathbf{E}(g_j(s_k)) v_{k'} \, \mathbf{E}(g_j(s_{k'}))$$

$$= \mathbf{E}\left(\left(\sum_{k=1}^{m} I_{i,j,k}^2 v_k^2 g_j^2(s_k)\right)\right) = \mathbf{E}\left(\left(\sum_{k=1}^{m} I_{i,j,k}^2 v_k^2\right)\right) = \mathbf{E}\left(\left(\sum_{k=1}^{m} I_{i,j,k} v_k^2\right)\right)$$

$$= \sum_{k=1}^{m} \left(\mathbf{E}(I_{i,j,k}) v_k^2\right) = \frac{c}{w} \sum_{k=1}^{m} v_k^2 = \frac{c\|v\|_2^2}{w},$$

which follows from the linearity of expectation, the fact that h_j and g_j are chosen independently of each other, and that g_j are pairwise independent. $\qquad\square$

Theorem 12.2.11. *The count-median sketch returns the estimated frequency $\hat{x}_i$ with probability $1 - \delta$, $|\hat{x}_i - x_i| \le \epsilon ||\mathbf{x}||_2$ with $k = \frac{\ln(\delta^{-1})}{(\frac{1}{6} - \frac{1}{3q})}$, and $w = \frac{cq}{\epsilon^2} = O(\epsilon^{-2})$.*

Proof. We denote the output of the algorithm $\hat{v}_i = \text{median}_j V[j, h_j(s_i)] g_j(s_i)$ and further denote a specific bucket $\hat{v}_{i,j} = V[j, h_j(s_i)] g_j(s_i)$. First, let us compute the expectation of the output of each bucket. We have

$$
\mathbf{E}(\hat{v}_{i,j}) = \mathbf{E}(V[j, h_j(s_i)] g_j(s_i)) = \mathbf{E}((X_{i,j} + v_i g_j(s_i)) g_j(s_i))
$$
$$
= \mathbf{E}(X_{i,j} g_j(s_i) + v_i) = \mathbf{E}(X_{i,j}) \mathbf{E}(g_j(s_i)) + v_i = v_i,
$$

which follows from the fact that $g_j(s_i)$ is independent from the hash function $h_j(s_i)$ from the $X_{i,j}$ variable and furthermore because it is independent of the choices of the other sign hashes $g_j(s_k)$ likewise defined in the $X_{i,j}$ variable. Next, the variance of each bucket can be calculated as follows:

$$
\begin{aligned}
\text{Var}(\hat{v}_{i,j}) &= \mathbf{E}(\hat{v}_{i,j}^2) - \mathbf{E}(\hat{v}_{i,j})^2 = \mathbf{E}(\hat{v}_{i,j}^2) - v_i^2 \\
&= \mathbf{E}\left((V[j, h_j(s_i)] g_j(s_i))^2\right) - v_i^2 = \mathbf{E}\left(((X_{i,j} + v_i g_j(s_i)) g_j(s_i))^2\right) - v_i^2 \\
&= \mathbf{E}\left((X_{i,j} + v_i g_j(s_i)) g_j)^2\right) - v_i^2 = \mathbf{E}\left((X_{i,j} g_j(s_i) + v_i)^2\right) - v_i^2 \\
&= \mathbf{E}\left((X_{i,j} g_j(s_i))^2 + 2(X_{i,j} g_j(s_i) v_i) + v_i^2\right) - v_i^2 \\
&= \mathbf{E}\left(X_{i,j}^2 + 2(X_{i,j} g_j(s_i) v_i) + v_i^2\right) - v_i^2 \\
&= \mathbf{E}(X_{i,j}^2) + 2(\mathbf{E}(X_{i,j}) \mathbf{E}(g_j(s_i)) v_i) + v_i^2 - v_i^2 = \mathbf{E}(X_{i,j}^2) = \frac{c||v||_2^2}{w}.
\end{aligned}
$$

The final step is to ensure that large deviations from the mean in any bucket do not occur very frequently. This can be done using Chebyshev's inequality

$$
\mathbf{P}\left(|\hat{v}_{i,j} - v_i| \ge \epsilon' \sqrt{\frac{c||v||_2^2}{w}}\right) \le \frac{1}{\epsilon'^2} \Rightarrow \mathbf{P}\left(|\hat{v}_{i,j} - v_i| \ge \epsilon' \frac{\sqrt{c}||v||_2}{\sqrt{w}}\right) \le \frac{1}{\epsilon'^2}
$$

$$
\Rightarrow \mathbf{P}(|\hat{v}_{i,j} - v_i| \ge \epsilon ||v||_2) \le \frac{c}{w\epsilon^2}, \quad \epsilon' = \frac{\sqrt{w}\epsilon}{\sqrt{c}},
$$

and choosing $w = \frac{ck}{\epsilon^2}$ gives that a bucket deviates from the mean with more than the variance, with probability at most k^{-1}. Having bounded the error on each bucket of the sketch, we are able to bound the whole sketch by bounding the probability over the median of all the buckets that are associated with a frequency estimate $\hat{v}_i$. The median is the middle element in a sorted set; hence for the median to have a large deviation from the mean, $d/2$ of the buckets must deviate with at least as much. Let $Y_{i,j}$ be an indicator variable indicating if a bucket estimate $\hat{v}_{i,j}$ is off by more than allowed, defined as

$$Y_{i,j} = \begin{cases} 1, & \text{if } |\hat{v}_{i,j} - v_i| > \epsilon||v||_2, \\ 0, & \text{otherwise.} \end{cases}$$

Let $Y_i = \sum_{j=1}^{d} Y_{i,j}$ be the total number of failed buckets over all rows, for element s_i. Since we have $j \in [d]_1$ and all of the d rows are independent of each other in the sketch, $E[Y_i] \leq d/k$, using a union bound, which gives us

$$P(|\hat{v}_i - v_i| > \epsilon||v||_2) \leq P\left(Y_i \geq \frac{d}{2}\right) \leq P\left(Y_i \geq \frac{k}{2} E(Y_k)\right)$$

$$\leq P\left(Y_i \geq \left(1 + \left(\frac{k}{2} - 1\right)\right) E(Y_i)\right) \leq e^{-\frac{(k/2-1) E(Y_i)}{3}} \leq e^{-d\left(\frac{1}{6} - \frac{1}{3k}\right)}.$$

Here, we require that $k > 2$. Choosing $d = \frac{\ln(\delta^{-1})}{(1/6 - 1/3k)}$ gives us

$$P(|\hat{v}_i - v_i| > \epsilon||v||_2) \leq \delta,$$

which proves the theorem.   □

Heavy Hitters

Heavy hitters are elements showing with high frequency. Such elements may be IP addresses on the internet, IP packet lengths, or web searches. The problem is usually specified in relation to a norm of the input data in order to define what *frequent* means. We can set a threshold factor ϕ of either the l_1 or the ℓ_2-norm of the input data. This broad specification of the problem leads to a plethora of different algorithms.

We present a solution to the ℓ_1 heavy hitters problem in the strict turnstile model, following [113]. The algorithm allows for both insertion and deletion of elements, while the frequencies always remain nonnegative.

Consider a data stream of size n with m unique elements and updates by (s_i, a) where $i \in \{1, 2, \ldots, m\}$ and $a \in \mathbb{R}$ is an update value. The task is to maintain the set of elements $S = \{s_1, \ldots, s_m\}$ by storing the frequencies of the elements indexed by time t as a vector $\mathbf{x}(t)^{\mathsf{T}} = (x_1(t) = f_t(s_1), \ldots, x_m(t) = f_t(s_m)]$, where $f_t(s_i)$ is a function estimating the frequency element s_i at time t. An update from time $t - 1$ to time t is then performed by changing the index in the vector corresponding to element i, $x_i(t) = x_i(t-1) + a$, while the rest of the indices remain the same, that is, $x_j(t) = x_j(t-1)$, for $j \neq i$, $j \in \{1, 2, \ldots, m\}$. We want to find all elements having a frequency $x_i(t) \geq \phi||\mathbf{x}(t)||$, where $0 < \phi \leq 1$ is a proportion of the norm. Elements which have frequency $x_i(t) \geq \phi||\mathbf{x}(t)||$ are called *heavy hitters*.

An algorithm for finding heavy hitters has the two operations Update(s_i, a) and Query(), not taking any argument. The Update operation will modify the inherent data structure to be able to support the Query operation at any time t, returning all heavy hitters. A trivial algorithm would be storing a counter for each unique element s_i and returning those elements that exceed the $\phi||\mathbf{x}(t)||$ threshold. In practice, however, n and m of the data stream are often of such size that maintaining counters of all elements would exceed the amount of available memory.

To reduce the space requirements, we consider an approximation algorithm which returns all heavy hitters, that is, all elements s_i exceeding frequency $\phi||\mathbf{x}(t)||$, but also limits the number of other less frequent elements to be returned. We limit this error by an approximation factor ϵ. That is, for $\epsilon < \phi$ no elements with a frequency less than $(\phi - \epsilon)||\mathbf{x}(t)||$ is returned with high probability, $1 - \delta$. The approximation factor ϵ then represents the allowable absolute error of the estimated frequency of any specific element s_i and δ the probability of failing to keep this limit.

Comparing the ℓ_1 heavy hitters problem with a point query, we find that:

(1) for the ℓ_1 point query, we have query(s_i) = $x_i \pm \phi||\mathbf{x}||_1$,
(2) for the ℓ_1 heavy hitters, Query() returns a set $L \in \{1, 2, \ldots, n\}$ such that
    (a) $|x_i| > \phi||\mathbf{x}||_1 \Rightarrow i \in L$,
    (b) $|L| = \mathcal{O}(1/\phi)$.

The number of ϕ-heavy hitters, that is, the elements i satisfying $|x_i| > \phi||\mathbf{x}||_1 \rightarrow i \in L$, is less than $1/\phi$. This means that L should not be larger than a constant multiplied by the maximum possible size.

In the turnstile model with both insertions and deletions, updates in one time period T_1 may decrement some frequencies and in some other disjoint time interval T_2 increase frequencies. This implies that $\mathbf{x}$ in two queries will reflect the difference of two frequency vectors. Turnstile heavy hitters algorithms can therefore also be used to detect large changes in frequency. We introduce the notion of point query with a tail guarantee and ϵ-tail heavy hitters.

(1) For the ℓ_1 point query with a tail guarantee we have Query(s_i) = $x_i \pm \epsilon||\mathbf{x}_{[1/\epsilon]}||_1$.
(2) For ℓ_1 tail heavy hitters, Query() returns $L \in [n]$ such that
    (a) $|x_i| > \epsilon||\mathbf{x}_{[1/\epsilon]}||_1 \Rightarrow i \in L$,
    (b) $|L| \sim O(1/\epsilon)$.

Here we denote by $\mathbf{x}_{[\bar{w}]}$ the vector $\mathbf{x}$ obtained after setting its w largest entries to zero. The number of elements i such that $|x_i| > \epsilon ||\mathbf{x}_{[\bar{w}]}||_1$ is at most $w + 1/\epsilon$, since apart from the w zeroed entries in $\mathbf{x}$, the number of other elements i satisfying $x_i > \epsilon \sum_{j \notin S} |x_j|$ must be less than $1/\epsilon$.

To solve the heavy hitter problem, we can perform a ℓ_p point query for each element i and return the $w = 1/\phi$ largest values $\tilde{x}_i$. In the heavy hitter problem, we have the additional complication of identifying the estimated frequencies relative to the threshold. The sketch-based approach uses the concept of *dyadic ranges* to create hierarchical data structures that enables faster heavy hitter queries.

Dyadic ranges are formed by dividing the m data elements into intervals, each having an increasing number of elements. In the case of heavy hitters, intervals are such that the sums of the frequencies of the elements in those ranges are increasing. The dyadic ranges for the elements $\{1, 2, \ldots, m\}$ can be defined as the sets of all ranges from $(z2^y + 1 \ldots (z + 1)2^y)$ for all $y \in \{0, 1, \ldots, \log m\}$, $z \in \mathbb{N}$, and $(z + 1)2^y \leq m$. The idea of the dyadic ranges can also be viewed as a tree of height $\log(m)$ and where y defines the levels and z the nodes.

For the heavy hitters problem a data structure is associated with each of the $\log(m)$ levels of the dyadic ranges. In our case, we can use the count-min sketch or the count-median sketch, each solving the frequency estimation problem. These structures are used to make decisions when querying about heavy hitters. They hold information about the frequencies of the z ranges for each level of the tree. Since the nodes z on each level span a certain range of elements, the data structures provide an estimated range sum of the frequencies of the corresponding elements.

Updates are performed for each of the data structures from the top of the tree down to the leaves. Each sketch is updated with a value $a \in \mathbb{R}$ associated with each dyadic range. The frequencies provided by the data structures for each range z provide a way to compare the sum of the frequencies in the range with the heavy hitter threshold, and if larger than the threshold, that range is investigated on the level below, with a smaller y. Such a search in the dyadic ranges can be done recursively using binary search, or iteratively with the stack-based approach.

12.3 Sample Entropy Estimation

The entropy of traffic distributions is a useful metric in many network management applications; for example traffic pattern classification, fault detection and load prediction. Since packets in an optical network arrive in nanosecond intervals, an algorithm needs to be both fast and memory efficient.

There has been a shift in traffic flow analysis from volume-based to distribution-based analysis. Lall et al. [116] present a streaming-based estimation algorithm for the entropy of the flow distribution.

We identify traffic flows by some identifier based IP address pairs, port numbers, protocol, or a combination thereof. The identifier is assumed to be an integer $\{1, 2, \ldots, n\}$, where n is large enough to accommodate the largest identifier.

The frequency of item $i \in \{1, 2, \ldots, n\}$ is denoted by m_i and the total number of items in the stream by $m = \sum_{i=1}^{n} m_i$. Let the jth element in the stream be denoted by a_j and define n_0 to be the number of distinct items that appear in the stream, since not all n items necessarily are present. The analysis is carried out in terms of m rather than n, since in general $n >> m$.

The entropy of an empirical distribution is defined as

$$H = -\sum_{i=1}^{n} \frac{m_i}{m} \log\left(\frac{m_i}{m}\right)$$

$$= \frac{-1}{m}\left(\sum_i m_i \log(m_i) - \sum_i m_i \log(m)\right) = \log(m) - \frac{1}{m}\sum_i m_i \log(m_i).$$

Normally, the logarithms are to base 2 and we use the definition $0\log(0) = 0$. To estimate the entropy it suffices to compute $S = \sum_i m_i \log(m_i)$, since we can keep a count of m exactly with $\log(m)$ bits.

It can be useful to normalize this number to be able to compare entropy estimates across different measurement epochs. We define the standardized entropy as $H/\log(m)$. Since H can be expressed in terms of $S = \sum_i m_i \log(m_i)$, we focus on an estimation of this quantity. The relative error is defined as $|S - \hat{S}|/S$, where $\hat{S}$ is the estimated value and S the actual value. For practical applications, we require that the relative error be low (say less than 2–3%).

Let $\hat{S}$ be the estimated value of S and $\hat{H}$ the estimated value of H computed from $\hat{S}$, that is, $\hat{H} = \log(m)\hat{S}/m$. Suppose we have an algorithm to compute S with relative error at most ϵ. Then the relative error in estimating H can be bounded as

$$\frac{|H - \hat{H}|}{H} = \frac{|\log m - S/m - \log m + \hat{S}/m|}{H}$$

$$= \frac{S - \hat{S}}{Hm} \le \epsilon \frac{S}{Hm}.$$

The relative error in H actually depends on the ratio S/Hm, which may become large if H is close to zero. If we impose a lower bound for H, we can convert an algorithm that approximates S with a relative error at most ϵ to one that approximates H with relative error $\epsilon' = f(\epsilon)$.

The entropy estimation algorithm consists of an online estimation phase and a postprocessing phase. The estimation phase is based on the Alon–Matias–Szegedy sketch.

Algorithm 12.3.1 (Entropy estimation).

Given a data stream with m different elements and parameters ϵ, δ.

STEP 0:

Let $z = \lceil 32 \log m/\epsilon^2 \rceil$, $g = \lceil 2 \log(1/\delta) \rceil$:

choose $z \cdot g$ locations in the stream at random.

Online stage

for each item a_j in the stream **do**

if a_j already has one or more counters **then**

increment all a_j counters

if a_j is a new randomly chosen location **then**

initialize the counter for a_j with value 1 **end**

STEP 1:

Let $z = \lceil 32 \log m/\epsilon^2 \rceil$, $g = \lceil 2 \log(1/\delta) \rceil$:

choose $z \cdot g$ locations in the stream at random.

Postprocessing stage

for $i = 1 : g$ **do**

for $j = 1 : z$ **do**

$X_{ij} = m(c_{ij} \log(c_{ij}) + (c_{ij} - 1) \log(c_{ij} - 1))$

end

end

for $i = 1 : g$ **do**

Compute $\bar{X}_i$, the average of the X in group i.

end

Output: the median of $\bar{X}_1, \ldots, \bar{X}_g$. ●

Fig. 12.1 shows the exact and the estimated entropy for a Pareto-distributed data sequence ($\alpha = 2.3$) and $n = 1000$.

12.4 Flow Size Distribution

Knowledge of the flow-size distribution is very valuable for design, traffic forecasting and anomality detection. It can be used to identify simultaneously active applications without packet inspection, such as streaming music, streaming video or voice over IP. The flow distribution can also help identify significant changes in the network dynamics, such as a sudden

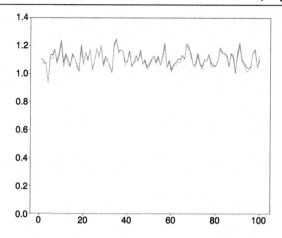

Figure 12.1: Exact and estimated entropy.

increase in the number of large flows (elephants) in a link, which may indicate link failure, route flapping or certain types of attacks.

We define a flow as all packets having the same flow label. This label can be defined depending on the granularity needed and the application. Thus, we may combine any fields from the IP header, for example Source IP, Source Port, Destination IP, Destination Port and Protocol.

Traditionally, flow distributions have been estimated from sampled network traffic. With increasing frequency and data volumes, such sampling either becomes slow and costly, or the accuracy of the results is limited by a low sampling rate.

Kumar et al. [117] propose a fast streaming data algorithm using a resource efficient lossy data structure. Like many other streaming data algorithms it uses hash functions to map readings onto counters, and the data structure is therefore lossy in the sense that different readings may collide onto the same counter. The algorithm uses expectation-maximization to infer the empirical flow-size distribution from the counter values.

We let the set of possible flow sizes be integers between 1 and z, where z is the maximum flow size observed in the data. We denote the total number of flows by n, and the number of flows that have i packets by n_i. Let the fraction of flows containing i packets be $\phi_i = n_i/n$.

We introduce a lossy data structure consisting of an array of counters. For each packet, the algorithm updates one single counter through a hash function. The data structure is lossy, since there is a possibility of collision due to hashing, so the counters can deviate significantly from the actual flow distribution.

The algorithm then uses expectation-maximization (EM) to infer the flow-size distribution from the observed counter values. The authors report a high measurement accuracy, within

2% relative error. In the algorithm, we need to estimate n, the total number of flows with reasonable accuracy ($\pm 50\%$), to have a sufficient number of counters.

The algorithm consists of two parts: online measurement and offline processing. Measurements are performed in epochs, where at the end of each epoch the current counter values, or raw data, are exported from the online module to the processing module and counters are reset to zero. The raw data is processed offline by using Bayesian methods for the best possible estimates.

When a packet arrives at a node, its flow label is hashed and mapped into the array and the corresponding counter is incremented by one. Collisions may occur when two or more labels result in the same counter index. The raw data can be represented as a list of tuples (i, v), where i is the counter index and v its frequency.

Internet traffic may have some very large flows, so the counters need to be large enough to accommodate this. Therefore, the measurement is divided into epochs with a small counter in the online module of, say, 7 bits which is exported to a larger offline counter. When an online counter exceeds a threshold value, say, 64, it increments the offline counter by this value and resets the online counter.

We estimate the number of flows in an array of m uniform hash functions as

$$\hat{n} = m \ln \left(\frac{m}{m_0} \right), \tag{12.3}$$

where m_0 is the number of zero entries in the array after the insertion of n flows.

The first step is to estimate flows of size one. The number of flows consisting of a single packet is important for two reasons: it is a crucial step in ascertaining possible heavy-tailedness of the flow distribution, and some types of attacks show a significant increase in such flows.

Let n_1 be the number of single-packet flows and $\hat{\lambda} = \hat{n}/m$ the estimated load factor, that is, the average number of flows that are mapped to the same index in the array. Then

$$\hat{n}_1 = y_1 e^{\hat{\lambda}}, \tag{12.4}$$

where y_1 is the number of counters with value 1. This simple estimator turns out to be very accurate.

Unfortunately, the above process cannot easily be extended to flows of larger size. Instead, we use expectation-maximization (see Chapter 7) to estimate the posterior distribution $p(\beta|\phi, n, v)$ from an initial estimate of the flow distribution. We assume that both n and m are large so that the binomial distribution can be approximated by a Poisson distribution.

Let λ_j denote the average number of size j flows that are hashed into the array. Let $\lambda_j = n_j/m = n\phi_j/m$ and $\lambda = \sum_{j=1}^{z} \lambda_j$, the average number of flows that are hashed to any array index. For any hash index i with value v, we need to estimate the probability $p(\beta|\phi, n)$ that different flows collide into this index. Let β be the event that f_1 flows of size s_1, f_2 flows of size $s_2, \ldots, f_q$ flows of size s_q collide into this index, where $a \le s_1 < s_2 < \ldots < s_q \le z$. We state two results and refer to [117] for their proofs.

Lemma 12.4.1. *Given ϕ and n, the a priori (before observing the value v) probability that event β happens is*

$$p(\beta|\phi, n) = e^{-\lambda} \prod_{i=1}^{q} \frac{\lambda_{s_i}^{f_i}}{f_i!}. \tag{12.5}$$

After the value v is observed, the probability $p(\beta|\phi, n, v)$ is given by the following.

Theorem 12.4.2. *Let Ω_v be the set of all collision patterns that add up to v. Then*

$$p(\beta|\phi, n, v) = \frac{p(\beta|\phi, n)}{\sum_{\alpha \in \Omega_v} p(\alpha|\phi, n)}, \tag{12.6}$$

where $p(\beta|\phi, n)$ and $p(\alpha|\phi, n)$ can be computed using Lemma 12.4.1.

For large counter values, the number of possible events becomes enormously large. The heavy-tailed nature of flow-size distributions can be used here. To reduce the complexity of enumerating all events that could add up to large counter value (say larger than 300), we ignore the cases involving the collision of 4 or more flows at the corresponding index. Since the number of counters with a value larger than 300 is quite small, and collisions involving 4 or more flows must occur with a low probability, this assumption has very little impact on the overall estimation.

We may also ignore events involving 5 or more collisions for counters having a value between 50 and 300 and those involving 7 or more collisions for other counters. This reduces the asymptotic computation complexity of partitioning a counter's values. Since the number of counters with very large values (say larger than 1000) is small, we can ignore splitting such counter values entirely and instead use the counter value as the size of a single flow.

Algorithm 12.4.3 (Flow estimation by expectation maximization).
Input y_i, the number of counters that have value i, $(1 \le i \le z)$.

STEP 0:

Estimate the total flow count, $\hat{n}^{(\text{old})} = m \ln(m/m_0)$ and an initial flow distribution $\phi^{(\text{old})}$

for $i = 1 : z$ **do**

$\quad \phi^{(old)} = y_i / \hat{n}^{(old)}$.

end

Set $\phi^{(new)} = \phi^{(old)}$, $\hat{n}^{(new)} = \hat{n}^{(old)}$

STEP 1:N

while convergence condition not met **do**

$\quad \phi^{(old)} = \phi^{(new)}$, $\hat{n}^{(old)} = \hat{n}^{(new)}$.

for $i = 1 : z$ **do**

for each $\beta \in \Omega_i$ **do**

Suppose β is that f_1 flows of size s_1, f_2 flows of size s_2, ..., and f_q flows of size s_q collide into a counter of value i, then

for $j = 1 : q$ **do**

$\quad n_{s_j} = n_{s_j} + y_i * f_j * p(\beta | \phi^{(old)}, \hat{n}^{(old)}, V = i)$.

end

end

end

$\hat{n}^{(new)} = \sum_{i=1}^{z} n_i$.

for $i = 1 : z$ **do**

$\quad \phi^{(new)} = n_i / \hat{n}^{(new)}$.

end end

Output ϕ, $\hat{n}$. $\qquad\qquad\qquad\qquad\qquad\qquad\qquad\qquad\qquad\qquad\qquad\qquad\bullet$

Multi-Resolution Estimation

Kumar et al. point out that the accuracy of the estimation of the flow distribution deteriorates when the number of counters fall below about 2/3 of the number of flows n. Since n can be large, this observation leads to a large and costly SRAM memory requirement. To reduce the memory needed, the authors propose a mapping of the counters, say $M = 2^r m$, to $r + 1$ physical arrays of size m each.

Half of the hash space is mapped to array 1, a quarter to array 2, and so on until two blocks of m counters remain, which are mapped onto array r and $(r + 1)$. The total space of the arrays is $m(\log_2(M/m) + 1)$. In other words, the arrays $A_1, A_2, \ldots, A_r, A_{r+1}$ are used, respectively, for the hash ranges $[0, \frac{1}{2}M), [\frac{1}{2}M, \frac{7}{8}M), \ldots, [(1 - \frac{1}{2^{r-1}}M, (1 - \frac{1}{2^r}M), [(1 - \frac{1}{2^r}M, M)$. A hash index k is therefore mapped to array $(k \mod m)$.

In order not to lose too much accuracy, the authors suggest that the arrays should be chosen so that no more than 1.5 flows collide onto the same slot. In this division of the memory block, the base $b = 2$ is used, but the argument can be generalized to any base.

The traffic volume between node pairs in a network, known as the traffic matrix, is essential for resource planning and traffic engineering. Traditionally, the traffic matrix is estimated based on statistical inference and/or packet sampling, because of which a high accuracy is difficult to achieve.

Zhao et al. [118] propose a data streaming algorithm designed to process traffic stream at very high speed and creates traffic "snapshots", or digests. By correlation of the digests collected for a node pair using Bayesian statistics, the sizes of pathwise flows can be accurately determined.

The problem of measuring the traffic matrix TM can be formalized as follows. Assume that there are m ingress nodes and n egress nodes in a network. We denote by TM_{ij} the total traffic volume traversing the network from the ingress node $i \in \{1, 2, \ldots, m\}$ to the egress node $j \in \{1, 2, \ldots, n\}$. The authors outline a method to estimate the TM on a high-speed network in a specified measurement interval.

There are two distinct approaches to estimating the traffic matrix, indirect and direct methods. An indirect method determines the traffic from related information that is readily accessible, such as link counts, and infers the volume from a traffic model. Such methods have the drawback of being dependent on a priori assumptions on the characteristics of traffic. In contrast, direct methods are based entirely on measurements and do not rely on traffic models. Direct methods are in general more accurate than indirect methods, but they require more sophisticated measurements.

To estimate a traffic matrix element TM_{ij}, two sets of bitmaps are collected from the ingress and egress nodes, i and j, respectively, and they are processed to obtain an estimator.

This allows estimation of a submatrix using the minimum amount of information possible, namely, only the bitmaps from the rows and columns of the submatrix. This feature of the scheme is practically important in two aspects. First, in a large ISP network, most applications are often interested in only a portion of elements in the traffic matrix instead of the whole traffic matrix. Second, this feature allows for the incremental deployment of the scheme since the existence of non-participating nodes does not affect the estimation of the traffic submatrix between all participating ingress and egress nodes.

The Bitmap Algorithm

The data structure is based on a bitmap B indexed by a hash function h, initially with all entries set to zero. When a packet p arrives, the invariant portion of the packet, $\phi(p)$, is extracted and hashed by h.

The result of the hashing is an integer which can be viewed as an index into B and the bit count at the corresponding index is set to 1. When the bitmap reaches a size threshold, it is exported to larger and slower memory. This time interval is called the "bitmap epoch". The algorithm is applied to all relevant ingress and egress nodes, using the same hash function h and the same bitmap size b.

The invariant portion of a packet used as the input to the hash function must uniquely represent the packet and should remain the same when it travels from one router to another. At the same time, it is desirable to make its size reasonably small to allow for fast hash processing. In the scheme, the invariant portion of a packet consists of the packet header, where the variant fields (e.g., TTL, ToS, and checksum) are marked by 0s, and the first 8 bytes of the payload if there is any.

When we would like to know TM_{ij} during a certain time interval, the bitmaps corresponding to the bitmap epochs contained or partly contained in that interval will be requested from nodes i and j and shipped to the central server.

Initially, for the simplicity of the discussion, assume that node i and node j produce exactly one bitmap during a measurement interval. Let the set of packets arriving at the ingress node i during the measurement interval be T_i and the resulting bitmap be B_{T_i}. Let U_{T_i} denote the number of bits in B_{T_i} that are zeros. An estimator of $|T_i|$, the number of packets in T_i, is

$$D_{T_i} = b \ln \frac{b}{U_{T_i}}. \tag{12.7}$$

The quantity we wish to estimate, TM_{ij}, is given by $|T_i \cap T_j|$. An estimator for this quantity is

$$\hat{TM}_{ij} = D_{T_i} + D_{T_j} - D_{T_i \cup T_j}. \tag{12.8}$$

Here, $D_{T_i \cup D_j}$ is defined as $b \ln \frac{b}{U_{T_i \cup T_j}}$, where $U_{T_i \cup T_j}$ denotes the number of zeros in $B_{T_i \cup T_j}$—the result of hashing the set of packets $T_i \cup T_j$ into a single bitmap. The bitmap $B_{T_i \cup T_j}$ is computed as the bitwise OR of B_{T_i} and B_{T_j}.

Let t_{T_i}, t_{T_j}, $t_{T_i \cap T_j}$, and $t_{T_i \cup T_j}$ denote $|T_i|/b$, $|T_j|/b$, $|T_i \cap T_j|/b$, and $|T_i \cup T_j|/b$, respectively. These are the load factors of the array when the corresponding set of packets are hashed into the array.

Theorem 12.4.4. *The variance of the TM estimates is*

$$\mathrm{Var}(\hat{TM}_{ij}) = b(2e^{t_{T_i \cap T_j}} + e^{t_{T_i \cup T_j}} - e^{t_{T_i}} - e^{t_{T_j}} - t_{T_i \cap T_j} - 1).$$

The average relative error of the estimator $T\hat{M}_{ij}$ is given by

$$\frac{\sqrt{2e^{tT_i \cap T_j} + e^{tT_i \cup T_j} - e^{tT_i} - e^{tT_j} - t_{T_i \cap T_j} - 1}}{\sqrt{b}t_{T_i \cap T_j}}. \tag{12.9}$$

As the error is scaled by $\sqrt{b}$, the larger the paged bitmap size, the better accuracy is obtained. Zhao et al. suggest paging a bitmap when it reaches a size in the order of 512 KB.

The parameter $T\hat{M}_{ij}$ estimates the number of distinct packets in TM_{ij}. Identical packets are only counted once during a bitmap epoch, as they are hashed onto the same bit. The most common reason for transmission of identical packets is the TCP retransmissions. We can adjust for such a retransmission by maintaining a packet counter C at the ingress node i, counting all packets in an measurement epoch, including retransmitted ones. Since D_{T_i} estimates the number of distinct packets T_i in the ingress node, the quantity $C - D_{T_i}$ estimates the total number of retransmitted packets. We then scale $T\hat{M}_{ij}$ by C/D_{T_i} for an adjusted estimate of TM_{ij}, assuming the same retransmission rate from i to all its destinations j.

For capacity planning and routing decisions, traffic measurements are typically required on much larger time scales (minutes or hours) than the measurement epochs (seconds). We note that due to varying traffic patterns, the bitmaps fill up unequally fast, and therefore the bitmap epochs may not be aligned in time.

To cater for different number of bitmaps, we enumerate these as $1, 2, \ldots, k_1$ and $1, 2, \ldots, k_2$, respectively. Then the traffic matrix element at (i, j) is given by

$$T\hat{M}_{ij} = \sum_{q=1}^{k_1} \sum_{r=1}^{k_2} \hat{N}_{qr} \times I(q, r), \tag{12.10}$$

where $\hat{N}_{qr}$ is the estimate of the traffic between bitmap q at node i and bitmap r at node j, and $I(q, r)$ equals 1 if bitmap q overlaps in time with bitmap r at least partially, and 0 otherwise. To synchronize the bitmap pages, we store the time stamps of the pages separately.

We may sample the packets at a lower frequency, which gives a lower measurement error but higher approximation error. Zhao et al. give the following rule of thumb:

If the expected traffic demand in a bitmap epoch does not make the resulting load factor exceed T^*, no sampling is needed. Otherwise, the sampling rate p^* should be set so that the load factor of the sampled traffic on the bitmap is approximately t^*.

If we let each ingress and egress node use the same load factor t after sampling, the value of t can be optimized for accurate estimates of the traffic matrix.

Denote by α and p_α the ingress bitmap q at node i and its chosen sampling rate, and by β and p_β the bitmap r and its sampling rate at the egress node j.

Using Eq. (12.8) to obtain the first estimate $\hat{N}_{ij}$ from the sampled packets, the traffic matrix estimator is given by $\hat{TM}_{ij}$ by $\hat{N}_{ij}/p$ with $p = \min\{p_\alpha, p_\beta\}$. Based on their empirical investigation, the authors suggest a load factor of $t^* \approx 0.7$.

CHAPTER 13

Dynamic Resource Management

The strict performance requirements for certain applications require a very high degree of flexibility in the utilization of network resources. Such flexibility is offered by software defined networking and network function virtualization, enabling nodes to be configured based on instantaneous network states through a central control function. Such an architecture is especially beneficial for real-time congestion control.

By congestion control we mean a set of functions ensuring service quality for diverse traffic types and demands with high robustness and flexibility. In principle, congestion control can be viewed from two different angles – centralized or distributed control. In the case of a centralized control, we may assume that the controller has access to information about the entire network and can therefore determine network optimal policies. A centralized function also ensures consistency and minimizes control message overhead. The disadvantages of a single controlling function are its vulnerability to failure and communication latency caused by processing load or transmission delay. A distributed control, on the other hand, is very resilient and allows fast decisions for packet processing and routing, but the logic must for the same reason be based on simple policies and limited network status information.

On a node level, congestion control amounts to traffic aggregation and fast routing. Such controls have no or limited possibility to adapt to changing network conditions. It can be viewed as a heuristic, node-centric approach that may result in network states far from the global optimum. On a network level, however, traffic routes can be assigned so that the *network-wide end-to-end performance* is optimized.

We describe a framework for congestion control, taking advantage of both the centralized and the distributed views through functionality separation by means of software-defined networking (SDN) and NFV. The functionality is separated into (i) centrally performed optimization of traffic aggregation strategies and routing policies and (ii) local traffic aggregation and routing. By adopting SDN and NFV, we assume a hierarchical implementation on three levels to ensure both high resilience and resource utilization.

The network optimal congestion control is implemented in three hierarchies (see Fig. 13.1) consisting of:

(1) SDN control, responsible for network monitoring and assignment of orchestrator virtual machines and acting as a direct interface to network management systems;

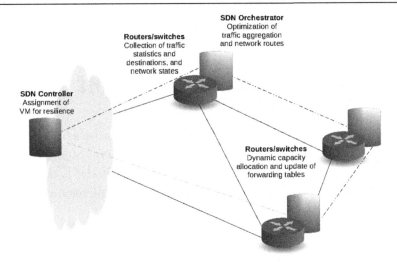

Figure 13.1: Schematic architecture of SDN/NFV-based congestion control.

(2)   SDN orchestration, responsible for global optimization and policy creation, traffic collection, statistical analysis, and network state reporting to the SDN controller, as well as updating policies and coordinating routing tables on a router level;

(3)   aggregation and routing, responsible for real-time traffic aggregation and route selection in accordance with policies.

This function separation allows reduction of control plane communication overhead, resilience on a concentrator level and efficient use of resources. In effect, this framework aims at improving all five main requirements on 5G networks.

We show that congestion control implemented using available resources more flexibly leads to substantially improved QoS. We show by simulation that dynamic traffic aggregation improves resource utilization on an individual router level, whereas optimal routing takes advantage of the router load levels to distribute the traffic throughout the network, improving QoS. The experimental environment includes a simulator generating three representative traffic types with different characteristics, aggregation and resource scaling on the node level, and a network route optimizer. An optimal strategy of control function allocation is also presented.

13.1 Network Traffic

The requirements on service quality and network capabilities are dependent on traffic type. Traditionally, traffic has been classified in real-time and nonreal-time types of services, where real-time services tend to be more delay-sensitive and nonreal-time services are more sensitive to packet loss. A typical distinction in IP are services using UDP and TCP, respectively. In ATM this roughly corresponds to CBR or rt-VBR and nrt-VBR or ABR services.

Characterization of Traffic

To describe the vast range of communication services available today, many having different characteristics, many models have been suggested in the literature. In our simulations we are simulating three traffic types – smooth, bursty, and long range–dependent traffic (see Chapter 4). These roughly correspond to the services telephony (audio), streaming video, and data traffic. These three traffic types can for our purpose be described on a high level by three quantities – intensity, burstiness, and autocorrelation.

Traditional plain voice traffic is often modeled as arriving according to a Poisson process with exponential duration. Such traffic is also known as Markovian, or memoryless. It is well known that this model is inappropriate to model bursty video traffic or data traffic exhibiting a nonnegligible memory. Therefore, we use a Markov additive process (MAP) to model bursty traffic and a fractional Brownian motion (FBM) to represent aggregate data traffic. These models are described below. The main characteristic of traffic is its *intensity*, which is defined as

$$\rho = \lambda/\mu, \tag{13.1}$$

where λ is the number of packet arrivals per time unit and μ^{-1} is the mean processing time of a packet, so that μ is the average number of processed jobs per time unit. The traffic intensity represents the mean number of requests in progress per time unit, or equivalently, the work load in a queue. The intensity is used to describe traffic at any point in the network, that is, offered traffic, carried traffic, and blocked traffic (as a percentage of offered traffic). It should be pointed out, however, that for traffic types with long memory, this quantity is difficult to determine. Bursty traffic can be characterized by its *peakedness factor*, defined as

$$Z = \frac{\mathrm{Var}(X)}{\mathrm{E}(X)}.$$

We have for a Poisson process $\mathrm{E}(X) = \lambda = \mathrm{Var}(X)$, so its peakedness factor is $Z = 1$. In contrast, bursty traffic has a peakedness $Z > 1$.

It is also useful to define the autocorrelation of a process X as

$$\rho(i, j) = \gamma(i, j)/\sigma^2,$$

where

$$\gamma(i, j) = \mathbf{E}((X_i - \lambda)(X_j - \lambda)),$$

where X_i and X_j are observations (number of packets) at times i and j, respectively, on some time scale. The quantity λ is the average arrival rate of the process.

For Poisson (memoryless) traffic, $\gamma(i, j)$ is negligible for $i \neq j$. In fact, this property defines lack of memory. Short-memory processes have nonzero correlation on lags $i \neq j$ which tends to decrease exponentially – at least asymptotically. For long-memory processes, however, the correlation decreases slower than exponentially and can be pronounced on very long time lags. This has the effect of relatively long periods of low or high arrival rates, which has a large impact on network performance.

Entropy

Traffic models have different statistical properties and are described by different sets of parameters. This makes them hard to compare with traditional statistical methods. For example, the mean and variance of long range–dependent traffic is difficult to determine on short time scales.

A different approach is to use the entropy of the traffic as a single parameter characteristic. The entropy measures the "randomness" of an arrival process, so it can be interpreted as a measure of the memory and the behavior of a given traffic trace. The entropy $H(X)$ of a stochastic process X is

$$H(X) = - \sum_{i=1}^{n} p_i \log p_i,$$

where n is the number of states and p_i the probability of the process attaining state i. The states are defined by the number of packets arriving during a time interval corresponding to two consecutive sampling times. The entropy is higher the more random the process is. We therefore expect the entropy of the Poisson process to be the highest, which also turns out to be the case. On average, the MAP traffic is close to the Poisson traffic, but it has higher variance. The FBM traffic has significantly lower entropy on average, but much larger variance. The variability in entropy allows fast characterization of traffic and decision of traffic aggregation so that the entropy of the aggregated traffic has lower variance than the FBM traffic itself. Along with appropriate resource scaling of the router queues, we show that the QoS and/or throughput is improved.

Since entropy is not additive, we cannot use it directly for scaling. We need both an estimation of mean and of high-frequency variability to estimate the scaling parameters. In our investigation, we use the resource scaling factor $1.75 H_a$ (where H_a is the entropy of the traffic aggregate) to the entropy H_t of the remaining traffic.

13.2 Traffic Aggregation

Network traffic is very heterogeneous, exhibiting variation on various time scales. It has been widely accepted that many traffic types are self-similar (long range–dependent), which may lead to starvation of traffic with shorter memory, just as prioritization of short-memory traffic may throttle long range–dependent traffic.

Inasmuch long range–dependent traffic may cause congestion where resources are allocated statically, it might be possible to improve the network performance by allocating resources dynamically based on the characteristics of the traffic. In fact, the long-range dependence of traffic means that the load it induces can be predicted using its autocorrelation structure. The predicted load may serve as a control variable for dynamic traffic aggregation, which must be performed on much longer time scales than, for example, routing table lookups.

Even if the control is assumed to be much slower than the traffic processing speed, the analysis of offered traffic and network load should be parsimonious and fast enough to capture variations on, say, time scales of seconds or fractions of a second. We suggest using entropy of the traffic traces and the resulting aggregation strategies as a single measure of their randomness.

Given the three traffic types and their different characteristics, we are interested in the most efficient way to aggregate the traffic and map it onto available resources. We use as efficiency measure the statistical performance in throughput, packet loss, and delay.

Firstly, it is instructive to look at the scaling properties of these processes (see, for example, [21]). We define the time-aggregated traffic as the average of a time block of size m, so that

$$X_t^{(m)} = \frac{1}{m} (X_{tm-m+1} + \ldots + X_{tm}).$$

When looking at the sample mean $\bar{X}$ of a traffic process X (where $\bar{X}$ is approximating λ in Eq. (13.1) above), a standard result in statistics is that the variance of $\bar{X}$ decreases linearly with sample size. That is, if $X_1, X_2, \ldots, X_n$ represent instantaneous traffic with mean $\lambda = \mathbf{E}(X_i)$ and variance $\sigma^2 = \mathrm{Var}(X_i) = \mathbf{E}((X_i - \lambda)^2)$, then the variance of $\bar{X} = n^{-1} \sum_{i=1}^{n} X_i$ equals

$$\mathrm{Var}(\bar{X}) = \sigma^2 n^{-1}. \tag{13.2}$$

For the sample mean $\bar{X}$, we have for large samples

$$\lambda \in [\bar{X} \pm z_{\alpha/2}s \cdot n^{-1/2}], \tag{13.3}$$

where $z_{\alpha/2}$ is the upper $(1 - \alpha/2)$ quantile of the standard normal distribution and $s^2 = (n - 1)^{-1}\sum_{i=1}^{n}(X_i - \bar{X})^2$ is the sample variance estimating σ^2.

The conditions under which Eqs. (13.2) and (13.3) hold are:

(1)   the process mean $\lambda = \mathbf{E}(X_i)$ exists and is finite,
(2)   the process variance $\sigma^2 = \text{Var}(X_i)$ exists and is finite,
(3)   the observations $X_1, X_2, \ldots, X_n$ are uncorrelated, that is,

$$\rho(i, j) = 0, \quad \text{for} \quad i \neq j.$$

We assume that conditions (1) and (2) always hold, but not necessarily condition (3).

13.3 Congestion Control

The aim of congestion control in communication networks is to provide services to its users that meets certain performance criteria, often represented by a set of QoS parameters. The performance criteria typically differ between services. One of the challenges in meeting QoS requirements is avoiding bandwidth starvation of certain traffic types by others. This may happen when one or more traffic types are given unconditional priorities. Another cause of congestion is traffic long-range dependence.

We study congestion control by using two fundamental techniques for efficient resource utilization – traffic aggregation and dynamic routing. Traffic aggregation refers to the resource utilization in the nodes, that is, processing capacity and buffer size. Dynamic routing, on the other hand, aims at directing traffic so that the overall network performance is maximized. It affects both loads at the nodes and transmission resources. In this study, however, the transmission resources are assumed to have sufficient capacity for any traffic stream and do not induce any significant propagation delay. This assumption is reasonable in optical DWDM backbone networks.

We distinguish between congestion control on the flow level, such as TCP (also referred to as congestion avoidance), and congestion control on network level, being the focus of this investigation. On the network level, we may have static QoS guarantees implemented by resource reservation, or soft QoS guarantees implemented through traffic prioritization and scheduling. As noted in [24], such QoS enforcement may adversely affect traffic not subjected to QoS restrictions and lead to bandwidth starvation.

The described algorithm does not impose any QoS restrictions. It is designed to allocate resources so that they are utilized in the best possible way. The only physical restriction affecting the traffic is the scaling of node resources with respect to traffic entropy.

Letting λ be the arrival rate and γ the throughput, a common type of feedback congestion control on flow level has the form

$$\frac{d\lambda}{dt} = \begin{cases} \epsilon, & \text{if } d\gamma/d\lambda, \\ -\alpha\lambda, & \text{if } d\gamma/d\lambda, \end{cases} \tag{13.4}$$

where $\epsilon, \alpha > 0$ are constants. Thus, if an increase in traffic flow rate results in an increase in throughput ($d\gamma/d\lambda > 0$), then the flow is increased linearly by ϵ, whereas if an increase in flow rate leads to a decrease in throughput ($d\gamma/d\lambda < 0$), then the flow rate is decreased exponentially.

Gerla and Kleinrock [119] describe different effects of congestion and propose mechanisms to mitigate these effects. Most notable are variants of resource reservation, where a single type of traffic is prevented from fully occupying a network resource (for example, a link or a buffer). Setting rules for resource allocation has been shown to be efficient in avoiding congestion and network instability effects.

Rather than using a protocol-based congestion control like Eq. (13.4), we investigate how available network resources can best be utilized and how that affects performance on a network level. We should point out, however, that a combination of congestion control methods on different levels need to be implemented to have a robust network operation, such as admission control and rate control on the flow level (such as TCP-controlled flows).

Congestion Control by Traffic Aggregation

Congestion control by aggregation uses the statistical multiplexing gain by superimposing different traffic types on a single queue. Due to long-range dependence inherent in many traffic types and present on long time scales, this type of control logic is well suited to be located in the SDN orchestrator.

We consider three traffic types – voice, video and long range–dependent traffic, modeled by a Poisson process, an MAP, and FBM, respectively.

We consider a deterministic queue model $G/D/n$ for simplicity, where $X \sim G$ is any traffic aggregate following a general distribution G, s is the deterministic server capacity in packets per time unit related to D, and n is the number of queues. For comparison, we study the delay, packet loss, and throughput in the absence of traffic aggregation, where each traffic type is fed to its own queue. The traffic streams have similar long-term intensity, and each queue can either be allocated capacity statically or dynamically. In the former case, all resources are simply divided into n equal blocks allocated to the queues. In dynamic capacity allocation, a queue is configured for a capacity determined by the entropy of the traffic it is fed with.

By using entropy, we can formulate an aggregation logic based on a single parameter. The entropy of a traffic trace is a measure of its information content or, equivalently, its "randomness". The more similar to a uniform distribution the traffic is, the higher the entropy.

In our scheme, we assume that an instantaneous traffic descriptor is available from every node at all times. It is sufficient to let the traffic be described by the number of packets arriving at each discrete time instant (of some convenient granularity) of each source type.

The statistical multiplexing gain by aggregation of traffic increases the utilization of the queue resources and thereby likely improves the throughput and the QoS for all traffic types. Heuristically, the idea is that short-memory traffic should be allocated resources sufficient to provide services with low (but nonzero) overflow probability. At the same time, the variability of the self-similar traffic can be expected to be limited by the scaled buffer size. The QoS parameters used are the delay and packet loss for each queue during the simulation time frame, from which we determine the total and maximum delay and packet loss of the node.

It is assumed that the orchestrator can measure the traffic intensity of each traffic type so that the control logic can decide which two traffic types to aggregate onto a single server, leaving the third traffic type on a separate server to avoid bandwidth starvation of any of the traffic types.

Congestion Control by Optimal Routing

A natural method for congestion control is using routes involving nodes with low load. Such routing achieves load balancing of the network resources. With a central knowledge of the load and average delay of each node, an end-to-end minimum-delay route can in principle be found by solving a shortest-path problem for each flow.

Simulation of Congestion Control

In this study, we use the end-to-end delay as the main optimization target parameter and performance indicator. Two other system characteristics are also measured: packet loss and throughput. Packet loss is considered less important than delay due to the retransmission capabilities of TCP. Throughput, on the other hand, is a measure of resource utilization rather than service quality. It is therefore of less importance when considering end-to-end connections in a network context. We highlight that we are investigating performance on a network level rather than on a flow level. In this context, throughput is not an additive measure, since large throughput downstream would not be useful when the throughput upstream is lower.

For the performance analysis, both the total delay and the maximum delay of end-to-end connections are determined. In addition to being a measure of service quality, the delay can also be seen as a measure of the load of the individual queues, since it is directly related to the average queue length.

There is also an obvious trade off between delay and packet loss. The probability of packet loss increases with decreasing buffer size, but at the same time, the delay decreases. Indeed, small buffers have been suggested for self-similar traffic [4]. The principle is that for such traffic, it is better to discard packets than causing overload with large delays as a consequence.

We use simulation to show the performance improvements of the proposed traffic aggregation and routing strategies. The use of simulation can be motivated by two facts. Firstly, an analytic treatment of the strategies is #$\mathcal{P}$-complete and additional complexity is introduced by traffic long-range dependence, which consequently affects all routes and performance measures. Secondly, the control logic would be based on statistical network properties, so a simulation approach directly gives a realistic "blueprint" of the algorithm.

To carry out simulation of congestion control strategies, we need to define:

(1)   network topology,
(2)   node capabilities,
(3)   end-to-end traffic distribution,
(4)   simulated load from the different traffic types,
(5)   aggregation and routing logic and QoS evaluation.

In simulating the effect of the congestion control logic, we keep the network topology fixed. A network topology consisting of seven nodes is used. These nodes are assumed to be identical for the sake of clarity of the performance evaluation results.

The end-to-end traffic distribution is randomly sampled to allow for maximum flexibility in routing scenarios. Each node is fed with an individually simulated traffic load, consisting of Poisson, MAP, and FBM traffic. The intensity of the simulated traffic traces is the same for each node, again to simplify the interpretation of the simulation results.

The control logic performs measurements on incoming traffic streams and is assumed to possess information of the queue states of the nodes and the traffic distribution matrix. Based on this information, it configures the nodes depending on the aggregated traffic characteristics and determines end-to-end routes in the network.

Finally, the QoS parameters delay and packet loss in each node are collected and the end-to-end parameters are determined. A comparison of both the total sum of end-to-end delays, packet losses, and throughputs and the sums of maximum end-to-end delays, packet losses, and throughputs are used for performance evaluation.

The logical representation of the congestion-controlled network is shown in Fig. 13.1. It consists of a network of nodes (routers or switches) with high-speed optical links connecting them and an SDN orchestrator (hosting the central logic) that is able to pull traffic statistics and the states of each node from the network, to configure buffer sizes and transmission rates in the nodes and to modify their forwarding tables.

For full flexibility in traffic generation, node configuration, route optimization, and detailed statistics, we developed a simulation framework dedicated to the present study, rather than using SDN development platforms like Mininet or NS-3. The simulator is written in Python and was run on a 64-bit dual-processor HP 6830s Linux Fedora 25 platform. Each simulation consists of generating three traffic traces, each of size 512 (that is, 512 time steps), performing three different aggregations and calculating the resulting entropy for each node in the network. Based on the entropy, the nodes are configured proportionally, and the queues are simulated by feeding them with the computed traffic aggregates. Next, the performance measures are calculated and used to determine optimal routes. The route optimization is iterative and uses a fixed-point equation which converges in a few iterations. In each step, the traffic aggregation at the nodes has to be recalculated. Each iteration on the network runs for approximately 100 s on the platform used.

Each network scenario – route optimization with respect to total end-to-end delay or maximum end-to-end delay – is simulated 100 times to generate statistical gain figures. The simulation is facilitated by using a $G/D/n$ queue, which means that the time steps are equidistant with $t_0 = 0$ and $t_n \in ((n-1)h, nh]$, where h is an arbitrary time unit. The packet loss and delay induced in each node are calculated from the simulated queue length distributions.

Network Topology

The network topology is chosen so that there are sufficiently many routing possibilities to yield interesting results and still small enough to allow for fast simulation. The speed of simulation is imperative in generating a large number of cases, which is necessary due to the self-similar nature of the traffic.

The topology is a 3-connected network, described in [93]. The 3-connectivity ensures a certain level of resilience, which also implies path diversity (see Fig. 13.2).

The network is modeled by a graph, on which shortest paths are determined. Since the network topology is assumed given and the shortest paths are determined with respect to delay rather than distance, only the connectivity matrix is used in most of the simulations.

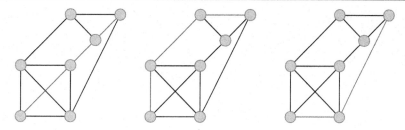

Figure 13.2: The topology used allows path diversity, like the three different paths between bottom left and upper right nodes.

We assume that the links have infinite capacity and induce zero delay. This is considered a good approximation for optical fiber networks carrying moderate data volumes. To be able to compute the shortest paths, however, the links are associated with a fictitious delay parameter deduced from the router delays.

Node Capabilities

The nodes are initially assigned a total processing and buffer capacity that can be configured into n queues, with the resources dynamically distributed among the queues.

The number n of queues is here either 3 or 2, where 3 queues correspond to no traffic aggregation and each traffic type simply is fed onto its own queue, and 2 queues correspond to aggregation of two traffic types. We note that the single-queue case is trivial from the point of view of aggregation strategy, that is, *how* traffic should be aggregated. The traffic separation by queues can in principle be implemented using for example traffic type identification as discussed in Section 12.4.

Whenever traffic aggregation is disabled, the processing and buffer capacity is equally divided between the three queues. When enabled, resources are divided proportionally to the entropy of the aggregates or single traces.

Traffic is represented by some capacity unit, which we may think of as packets. Simulation is carried out for traffic traces of length $n = 512$, following Lindley's recursion

$$Q_{t+1} = \max\{0, Q_t + A_t - S_t\},$$

where Q_t is the buffer content, A_t denotes arriving packets, and S_t denotes serviced packets between time t and $t + 1$. Whenever the number of arriving packets exceeds the free buffer space, so that $A_t > B - Q_t$, loss of $A_t - (B - Q_t)$ packets occurs.

Assuming unit time increments, the queue delay is calculated as the expected queue length, determined for each queue. Letting $p(k)$ denote the probability of finding k packets in the buffer gives the delay

$$\mathbf{E}(d) = \sum_{k=1}^{B} kp(k).$$

Packet loss is measured by counting the number of packets exceeding system capacity, and throughput is the number of forwarded packets.

Traffic Distribution

For each simulation scenario, the end-to-end traffic distribution matrix – expressed as a stochastic matrix – is simulated. For a stochastic matrix, the row sum must equal unity. The minimum assumption on the distribution is a uniform distribution $p_{ij} \sim U(0, 1)$ of end-to-end destination probabilities, ensuring that the row sum is unity. The uniform distribution may not be appropriate for real networks, but in order to obtain as general results as possible, we need to impose a minimum of network-specific assumptions. The same traffic distribution has been used for all three traffic types for simplicity.

Traffic Simulation

The traffic is simulated independently for each type, and new traffic simulations are carried out for each new scenario and node. The three traffic types Poisson, MAP, and FBM are simulated using the same traffic parameters in all cases to make the comparison of aggregated traffic more discernable. The time scale is a generic time step reflecting the packet arrival intensity and processing capacity of the nodes.

To simplify the simulated queue characteristics and traffic aggregation, we express the work load as a number of packets, each assumed to be of equal length. These packets are generated by one of the three traffic processes (Poisson, MAP, or FBM). The servers are assumed to process packets with a constant bitrate, without any priority between traffic types.

Using such an experimental framework, the traffic in a network is mainly determined by

- the intensity of the traffic types,
- the traffic distribution on a network level, given by a traffic matrix,
- the server capacities, and
- traffic aggregation and routing strategies.

To study the performance gain on a network level as free from market-specific assumptions as possible, we keep the offered traffic intensity constant and the node capacities uniform. The traffic distribution is uniform in order to stochastically spread its effect. The important quantity is the quotient of the total mean arrival rate to the server processing capacity (Eq. (13.1)), which should be close to the node capacity in order to obtain performance measures of interest.

Simulation of Poisson processes

The Poisson arrivals are simulated using the following algorithm. Let $\{N(t) : t \geq 0\}$ be the counting process of a Poisson process with rate λ. Then $N(1)$ is Poisson-distributed with mean λ. Let $Y = N(1) + 1$ and $t_n = X_1 + X_2 + \ldots + X_n$ denote the nth arrival of the Poisson process, where the X_i are independent and identically distributed according to an exponential distribution with rate λ. We note that $Y = \min\{n \geq 1 : t_n > 1\} = \min\{n \geq 1 : X_1 + X_2 + \ldots + X_n > 1\}$ is a stopping time. Suppose we use the inverse transform method to generate independent and identically distributed exponential interarrival times X_i. We can then represent the variables as $X_i = -(1/\lambda)\ln(U_i)$. Rewriting the expression for Y gives

$$
\begin{aligned}
Y &= \min\{n \geq 1 : \ln(U_1) + \ln(U_2) + \ldots + \ln(U_n) < -\lambda\} \\
&= \min\{n \geq 1 : \ln(U_1 U_2 \ldots U_n) < -\lambda\} \\
&= \min\{n \geq 1 : U_1 U_2 \ldots U_n > e^{-\lambda}\}.
\end{aligned}
$$

We can therefore simulate Y by generating independent uniformly distributed variates U_i and taking the product of these variates until it first falls below $e^{-\lambda}$. The number of uniformly distributed variates in the product yields Y. Then we get the desired Poisson variate as $X = N(1) = Y - 1$. Traffic traces of size 512 are generated in each simulation scenario.

Simulation of Markovian additive processes

The MAP can be used to model and simulate bursty traffic, such as video sources. The general model has a number of states, where the transition between the states is controlled by a Markov chain. In this study, we let the MAP have two states only, a silent and an active state, denoted d and a, respectively. Denoting the traffic intensity by X_t, the Markov chain controlling the transition between the two states is defined by the transition probabilities

$$
\begin{aligned}
a &= \mathbf{P}(X_t = 1 | X_{t-1} = 0), & (13.5) \\
d &= \mathbf{P}(X_t = 0 | X_{t-1} = 1). & (13.6)
\end{aligned}
$$

The Markov chain is then represented by the matrix

$$M = \begin{pmatrix} 1-a & a \\ d & d-1 \end{pmatrix},$$

so that the steady-state probabilities are given by

$$\pi_d = \frac{d}{a+d},$$
$$\pi_a = \frac{a}{a+d}.$$

The smaller the conditional probabilities a and d, the burstier the resulting traffic. At each time instant, the conditional probabilities of the sources being silent or active are given by (13.5)–(13.6), keeping track of the states of the process at each time instant.

As described in [26], the process simulation uses a training period to "burn in" and a block of arrivals, which amounts to aggregation of individual MAP sources. In the simulation, aggregation of size 100 has been used. Traffic traces of size 512 are generated in each simulation scenario.

During one time step in the queue, we can simulate the state change of the MAP by comparing the probabilities in the Markov chain with a uniformly distributed random number $u = U(0, 1)$. If the chain is in silent state and $u < d$, the process is switched to tho active state, and if it is in active state and $u < a$, it is switched to the silent state.

When in active state, the total packets of a block, say k packets, are added to the queue. When fed to the queue, a maximum amount of s packets are processed and removed from the queue in each time step. Simulations of queues with finite buffers are easily accomplished in discrete time, and various performance metrics can be calculated by introducing counters into the simulation process.

Simulation of fractional Brownian motion

We can simulate a fractional Brownian motion using the Cholesky method. It uses the Cholesky decomposition of the covariance matrix, $C = \Sigma\Sigma'$, where Σ is a lower triangular matrix of the covariance matrix given by Eq. (4.6). It can be shown that such a decomposition exists whenever the autocovariance matrix is positive definite (and symmetric, which is true by its construction).

The simulated packet arrivals are then obtained by multiplying the matrix Σ by a vector of independent normal standard random variables η of suitable size, that is,

$$\mathbf{X} = \Sigma\eta.$$

It is necessary to restrict the resulting values to be equal to or greater than zero, since traffic obviously cannot be negative.

The traffic load is chosen so that the peak rate exceeds the service rate s of the queue in order to obtain interesting node performance parameters, whereas the mean arrival rate must be lower than s. The Markov parameters are chosen as in [93] with $d = 0.072$ and $a = 0.028$. With a router capacity of $s = 60$ packets per time unit, this choice of parameters generates traffic with a peak rate exceeding s, but with a mean rate less than s.

Traffic Aggregation

The traffic aggregation logic is based on comparing the entropy for aggregated and nonaggregated traffic streams. With three traffic types and two queues, we may aggregate two traffic types, which can be done in three ways, and compare the resulting entropies.

The entropy is used in two steps. Firstly, the traffic types to aggregate are determined. The decision of which traffic types to aggregate is made so that the aggregate has an entropy as close as possible to the entropy of the nonaggregated traffic. The entropy of the single traffic types in different simulations is such that for Poisson traffic, it is high and nearly constant, whereas for FBM the average is lower but with high variability. The entropy of MAP traffic is somewhere in between. Therefore, decision of aggregation is mainly driven by the behavior of the FBM traffic.

Next, the aggregated and nonaggregated traffic is mapped onto two queues in the node. Here, the entropy is used as a scaling parameter, where the capacity is allocated in proportion to the entropy. This is a simple heuristic, which is applied both to processing capacity and buffer space. More sophisticated control logic can be devised, but this simple principle suffices to study the effect of traffic aggregation on heterogeneous traffic.

Routing Strategy

For the analysis on a network level, we assume that paths can be chosen optimally with respect to delay. This means that an end-to-end connection is set up on the route that induces the least amount of delay.

We note that a shortest path can be defined with respect to any numerical weight, not only physical distance. Such weights may be cost, delay, or some reliability measure. To find a minimum-delay path, we use the algorithm by Dijkstra.

In finding an end-to-end optimal path, we start from the first end node s and scan its neighbors j for the lowest weight. Letting d(s, j) be the total weight going from s to j, the algorithm successively uses the relation

$$d(s, j) = \min_{i \in S}(d(s, i) + d_{ij}),$$

where S is the set of nodes and d_{ij} is the weight on edge (i, j). The algorithm compares the minimum weight paths going from s to any node i, adding the weight going from i to j.

Dijkstra's algorithm uses weights on edges, whereas the delay is actually induced in the nodes, so we need to map node delays onto edges. This can be done since the delays, just like distances, are positive and additive.

The minimum-delay route is determined locally, and by using this route for a particular end-to-end connection, the node delays would change. Since we want to study the global properties of minimum-delay routing, we need an iterative procedure where minimum-delay routes are recalculated as traffic is added to the already determined minimum-delay routes.

To find the optimal end-to-end routing from a network perspective, we use a fixed-point equation

$$F(\Delta_{\min}) = \Delta_{\min}. \tag{13.7}$$

Such fixed-point equations are commonly used to analyze blocking networks [93], where $F(\cdot)$ is a function of the total blocking on a route, given by Erlang's B-formula. The blocking is a convex function of the load, just as the delay. We here assume that at least one such fixed point exists and that the equation converges to this point (or points). This assumption is partly justified by the fact that the node delays are limited and a function of traffic load.

Thus, assuming delays induced by the traffic offered to the routers, we sequentially determine the end-to-end minimum-delay paths, adding traffic to the nodes along the paths corresponding to an end-to-end traffic matrix. This gives the initial solution to the fixed-point equation (13.7). The minimum-delay routes are then successively recalculated, using the previously defined edge delays, updating the traffic on the routers along any computed path, and updating the edge delays accordingly.

Interestingly, the fixed-point equation converges in many cases to two solutions, leading to a hysteresis in the network delay. This stability phenomenon has also been observed in blocking networks [120][93]. In blocking networks, which lack centralized control, trunk reservation has been suggested as a stabilizing strategy [120]. In our case, we let the centralized logic select the best network routes, using delay measurements from the nodes.

Quality-of-Service Evaluation

A node consists of two or three queues, which are simulated independently in the sense that once resources are assigned to a specific queue, these cannot be shared by another queue in the node.

The queues are fed with the aggregate or single-type traffic, represented by vectors of arriving packets. At each time instant, the queues are governed by Lindley's equation (13.8) with the constraint of limited buffer space B, i.e.,

$$Q_{t+1} = \begin{cases} \max\{0, Q_t + A_t - S_t\} & \text{if } Q_t + A_t - S_t \le B, \\ 0 & \text{otherwise,} \end{cases} \tag{13.8}$$

which can be expressed as $Q_{t+1} = \min\{\max\{0, Q_t + A_t - S_t\}, B\}$. The arrivals A_t are given by the vectors of offered load, and $S_t = s$ is a constant number of packets processed in each time step.

In the router simulator, at each time instant we count the number of lost packets, that is, $Y = \max\{0, \hat{Q}_t - B\}$, where $\hat{Q}$ is the queue length in a queue with infinite buffer. These are the packets in excess of router capacity, that is, the packets that are being processed during the time step, and the free buffer space.

By recording the number of packets in the system at each simulation step (the state occupancy), we can determine the delay. From the recorded state occupancy, we can determine the state probabilities p_n, and the delay is given by

$$\delta_i = \sum_{n=0}^{C} 1 \cdot p_n$$

packets per time step, where C is the total router capacity.

The difference in performance is measured using the standard "error" measure

$$\Delta m = \frac{m_o - m_r}{m_r},$$

where m_o is the observed metric and m_r is the reference metric. When m_r is zero, the metric degenerates, and we set

$$\Delta m = \begin{cases} 1 & \text{if } m_r = 0 \text{ and } m_o > 0, \\ 0 & \text{if } m_r = 0 \text{ and } m_o = 0, \\ -1 & \text{if } m_r = 0 \text{ and } m_o < 0. \end{cases}$$

As a general rule, we cap the performance measure so that $0 \geq \Delta m \geq 1$ to avoid large values whenever the underlying metric is small. This also allows us to represent the difference in performance as a percentage figure.

In order to make the percentage figures more comparable on the node level, we scale by the average number of affected packets: the number of buffered, lost, and forwarded packets to the total number of packets.

13.4 The Effect of Traffic Aggregation

We analyze the effect of traffic aggregation on the node level, comparing two aggregation strategies with the case of no aggregation, and recording delay, packet loss, and throughput. Each case is simulated 1000 times, where each case consists of the three traffic types of size 512 time steps. Each case therefore consists of 512,000 simulation points.

Node Level Traffic Aggregation

Traffic aggregation on the node level includes the operations of merging two of the three traffic streams and configuring the node resources proportionally. We compare two cases of traffic aggregation: aggregation of real-time traffic, always aggregating Poisson and MAP traffic, and dynamic aggregation, where traffic types are aggregated so that the entropy of the two resulting traffic streams are as equal as possible.

The entropy-based traffic aggregation is analyzed in isolation to be able to assess its efficiency and impact on performance on the node level. The aggregation decision is based on the entropy of each of the streams of voice, video, and data traffic. Aggregation is performed so that the entropy of the traffic load in each channel is as equal as possible.

For each queue with capacity C (including buffer space), we have

$$\sum_{j=1}^{N} n_j \alpha_j \leq C,$$

where N is the number of sources in an aggregate and α_j is the effective bandwidth of traffic flow j. We do not try to estimate the statistical multiplexing gain, but the gain in the QoS triple consisting of delay, packet loss, and throughput.

In each simulation, we collect the total and maximum delay, the total and maximum packet loss, and the total and maximum throughput, and we compare that with the situation where no aggregation is performed, using three queues, and resources are allocated equally between these queues. The improvement is referred to as gain.

We compare the gain in throughput and performance by comparing the strategies:

(1) no aggregation, static mapping of traffic to equally configured queues,
(2) aggregation of real-time traffic sources (Poisson and MAP) mapped onto queues with proportionally scaled resources,
(3) dynamic aggregation, where traffic is aggregated to make the entropy as equal as possible between the two queues, and proportionally scaled resources.

By measuring the change in total and maximum delay, packet loss, and throughput when compared to no aggregation, we have descriptors of the relative performance of each strategy. Clearly, the larger the relative improvement the better. It is, however, important to justify the order of the performance statistics. We have

(1) *Maximum delay.* The statistic measures the worst end-to-end delay, which is the main performance indicator to be improved. It is, however, closely related to the total delay, and the former should not be improved at the expense of the latter.
(2) *Maximum and total packet loss.* For real-time services, packet loss can lead to severe quality degradation. For best effort traffic, however, packet loss is typically compensated by retransmissions controlled by TCP. We also note the duality between delay and packet loss: the larger the buffer, the longer the delays and the smaller the packet loss. The change in total packet loss – if negative – should be small due to its adverse effect on real-time traffic. Packet retransmission is not considered in the study, since it is traffic-dependent and leads to a bias in the throughput figures.
(3) *Total throughput.* The statistic represents the utility of shared resources, and the larger the increase, the better. The improvement is achieved through statistical multiplexing of the traffic or increased buffering.

Service Type Traffic Aggregation

It has been suggested in the literature [121] that aggregation of real-time sources would give the best QoS. The aggregation strategy is to aggregate real-time traffic, justified by their similar characteristics and QoS demands, while keeping nonreal-time traffic on a separate queue. We denote this strategy RT/NRT for brevity. On the node level, a comparison of the RT/NRT with the case of separate queues is shown in Fig. 13.3.

The results indicate a decrease in maximum packet loss and an increase in maximum throughput. Thus, the overall performance of real-time traffic improves when compared to queueing without traffic aggregation.

The increase in maximum throughput is associated with the aggregation of real-time traffic, and this can be attributed to statistical multiplexing of real-time sources. This is an expected

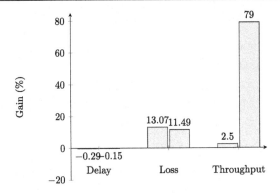

Figure 13.3: Performance gains of RT/NRT traffic aggregation.

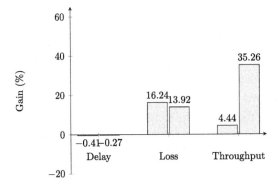

Figure 13.4: Performance gains of dynamic traffic aggregation.

result, as aggregation of Markovian sources and resources improves the efficiency. At the same time, since the change in average throughput remains very small, the other queue must experience a similar decrease in throughput. There is no or a small net gain in throughput in this case.

Dynamic Traffic Aggregation

A comparison of the dynamic aggregation strategy with no aggregation is shown in Fig. 13.4.

A fully dynamic traffic aggregation enabled by SDN/NFV is more flexible than any deterministic strategy, and it therefore has the potential to be much more efficient. In this strategy, we aggregate the two traffic streams that give an as even value of entropies as possible. There is a discernable improvement in total and maximum packet loss as well as maximum throughput. We note that the gain in total throughput is higher and the gain in maximum throughput is lower than for RT/NRT aggregation.

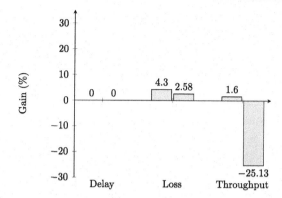

Figure 13.5: Performance gains of dynamic traffic aggregation compared with RT/NRT aggregation.

Relative difference between dynamic and RT/NRT aggregation

It is instructive to compare the dynamic with the RT/NRT strategy. The result is shown in Fig. 13.5.

The dynamic aggregation strategy shows a net improvement of packet loss and a slightly higher total throughput than the RT/NRT strategy. The increase in total throughput and decrease in maximum throughput in the comparison indicates a higher degree of fairness in QoS provisioning between the traffic streams. A higher maximum throughput is obtained when real-time traffic has precedence, as an effect of the smaller fluctuations on longer time scales.

The main idea is that resource utilization improves with aggregation, so that *similar* traffic types share a queue and thereby benefit from statistical multiplexing. For dissimilar traffic types, on the other hand, starvation of resources of one type by another may decrease performance. It is therefore suggested to separate self-similar traffic from Markovian traffic and to allocate resources in proportion to the load.

To summarize, the simulation result of traffic aggregation showed a marked improvement in terms of packet loss and/or throughput. We justify our use of these statistical metrics by the fact that the variability of self-similar traffic is decreased by the finite buffers, which affects long-range dependent traffic at all nodes along its path and produces a smoothing effect of such traffic.

13.5 The Effect of Optimal Routing

Controlled flow routing can be used to further take advantage of the resources available in the network. We investigate routing optimized with respect to the total delay, that is, the sum of delays induced at all nodes in the network and the sum of the maximum delays. For the search for an optimal flow assignment, we use the algorithm by Dijkstra.

Dijkstra's algorithm (see Section 2.4) finds the shortest path between two nodes s and t by scanning the neighbors at each step, and it builds a tree of shortest paths to all nodes from s. It uses the "distances" of the edges, where the distance can be any positive real number, which we call the edge weight. Thus, the weight can be Euclidean distance, number of hops, or, with a slight modification, delay. In Dijkstra's algorithm, we make two important observations. Firstly, the shortest path need not be unique. However, all shortest paths must have the same total weight in the network. We are interested in any such shortest path, as we look for a global optimum. Secondly, in its original form, the algorithm does not give the shortest path, but only the sum of weights along this path. The actual paths can be found by labeling each edge and adding an edge (i, j) to an edge set E each time such an edge is traversed and a node j is added to the set S. After the destination node t has been reached, we use backtracking from t through the edges in E to find the path back to s.

Now, we need to make two modifications to find the optimal flow assignment with respect to delay. Firstly, the delays, which are properties of the nodes, need to be mapped to edge weights. Secondly, since the algorithm finds the shortest path with all weights given, an iterative procedure is needed to find a global flow assignment in the network where weights depend on the assignment. The first modification is based on the additivity of delays. Since the delays are additive, the total delay of an end-to-end connection can be expressed as

$$\Delta_{st} = \sum_{i=s}^{t} \delta_i,$$

where i starts and ends at the respective end points $\{s, t\}$. In this expression, Δ_{st} is the total end-to-end delay and δ_i is the delay induced in node i. By setting the edge weights to $d_{ij} = \delta_i/2 + \delta_j/2$, we see that

$$\begin{aligned}
\Delta_{st} &= \delta_s/2 + (\delta_s/2 + \delta_i/2) + (\delta_i/2 + \delta_j/2) + \ldots + (\delta_k/2 + \delta_t/2) + \delta_t/2 = \\
&\quad \delta_s/2 + d_{si} + d_{ij} + \ldots + d_{kt} + \delta_t/2.
\end{aligned}$$

The first and the last term do not influence the shortest path, since the start and end nodes are fixed. To obtain a numerical value of the total delay, we simply add these two terms to the result from Dijkstra's algorithm. Now, the minimum-delay route can be determined locally, for a particular end-to-end connection. This choice of path, however, assumes that the delays remain fixed in the network. Since we want to study the global properties of minimum-delay routing, we need an iterative procedure where minimum-delay routes are recalculated as traffic is added to the already determined minimum-delay routes.

The end-to-end distribution of traffic is given by an $n \times n$ probability matrix, where each row signifies the node i traffic is originating from and each column the node j a percentage p_{ij} of the traffic is addressed to. This is called the *traffic matrix* in the following. For this matrix, we must have $p_{ii} = 0$ and $\sum_j p_{ij} = 1$. In each simulation, the p_{ij} are drawn from a uniform distribution with the condition that $\sum_j p_{ij} = 1$. This ensures that the assumptions on the end-to-end traffic distribution is minimal. To find the globally optimal end-to-end routing from a network perspective, we use the fixed-point equation (13.7). We have

$$F(\Delta_{\min}) = \Delta_{\min}.$$

After initial generation and dynamic aggregation of traffic, the fixed-point equation iteratively maps delays as weights onto the edges, computes the minimum-delay paths, and aggregates the traffic along these paths.

Algorithm 13.5.1 (Fixed-point optimal route algorithm).

Given a network $G = (V, E)$, $n = |V|$, n-dimensional traffic vector $\mathbf{t}$, and an $n \times n$ traffic matrix T.

STEP 0:
> Populate $\mathbf{t}^0$ by generating traffic and apply
> dynamic traffic aggregation.

STEP 1 until convergence:
> **for each node** $i \in V$ **do**
>> determine delays d_i induced
>
> **end**
> **for each link** $(i, j) \in E$ **do**
>> set weights $w(i, j) = \delta_i/2 + \delta_j/2$
>
> **end**
> **for each node** $i \in V$ **do**
>> **for each node** $j \in V$ **do**
>>
>> Find minimum-delay paths $P_{\min}^{(i,j)}$ using Dijkstra's algorithm
>> Set $\mathbf{t}(i) = \mathbf{t}^0(i) + \sum_{j \neq i} T(i, j)\mathbf{t}(j)$
>>
>> **end**
>
> **end**

Output: minimum-delay flow allocation.   ●

We assume that the fixed-point algorithm converges. Should it happen that the algorithm shows oscillating behavior, the solution with the best characteristics is chosen. Since the minimum-delay paths may be longer than the shortest paths in terms of hops, we analyze the effect of optimal routing next.

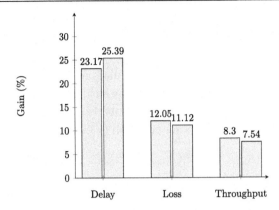

Figure 13.6: Performance gains with dynamic traffic aggregation and optimal routing.

Traffic Aggregation Under Minimum Total Delay Routing

Considering congestion control on a network level, optimal routes through the network can be chosen minimizing the number of hops (ordinary shortest paths), the total end-to-end delay, or the sum of maximum end-to-end delays. The gain is determined as the performance improvement with traffic aggregation at the nodes and optimal routing, compared to the situation with traffic aggregation at the nodes and shortest-path routing. It should be noted that in the network scenarios, nodes carry not only their own traffic, but also transit traffic, which is a random quantity determined by the random traffic matrix.

The performance gains for the case of routing minimizing the sum of end-to-end delays are shown in Fig. 13.6. The largest gain is in delay, which is expected due to the optimal routing. The distribution of gains, however, depends on the node configurations and the traffic levels. The positive gains in delay, packet loss, and throughput are significant with a p-value less than 0.3%.

By making necessary modifications to the simulation environments, we indicate the gains from dynamic traffic aggregation compared to RT/NRT aggregation and optimal routing compared to shortest-path routing, respectively. The optimally routed dynamic aggregation compared with RT/NRT aggregation is shown in Fig. 13.7. We see a marked improvement in delay, but slightly deteriorating packet loss and throughput. This shows that on a network level, dynamic traffic aggregation leads to lower total delays.

Dynamic Traffic Aggregation Under Shortest-Path Routing

The effect of dynamic traffic aggregation under optimal routing compared to simple shortest-path routing is shown in Fig. 13.8.

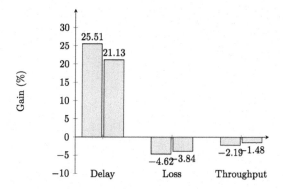

Figure 13.7: Performance gains comparing dynamic and RT/NRT aggregation under optimal routing.

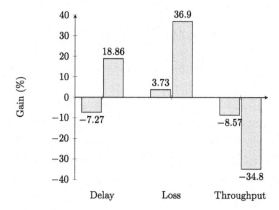

Figure 13.8: Difference in performance between optimal and shortest-path routing.

Under optimal routing, the total delay and throughput is worse than for shortest-path routing, due to the fact that the average paths are longer in the former than in the latter. However, maximum delay and packet loss are improved, indicating an improvement in the worst end-to-end QoS.

We conclude that:

(1) When using optimal routing, the overall performance gain is better for dynamic traffic aggregation than for RT/NRT routing. This follows from the better resource utilization in the nodes.

(2) Optimal routing can improve the worst end-to-end QoS at the expense of average values, following the dynamic load distribution in the network.

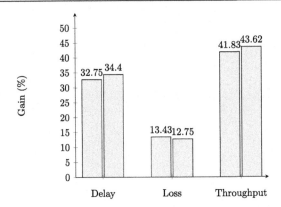

Figure 13.9: Performance gain of dynamic traffic aggregation under minimum maximum-delay routing.

Traffic Aggregation Under Minimum Maximum-Delay Routing

When the optimal routes are chosen to minimize the maximum end-to-end delay, the gain in total and maximum delay is higher than in the previous case, as shown in Fig. 13.9. By comparing Figs. 13.6 and 13.9, it can be seen that routes optimized with respect to maximum end-to-end delay give higher overall gain. Using the maximum end-to-end delay as optimization criterion has some obvious advantages. Firstly, minimizing the maximum delay leads to a higher degree of fairness in the network. Secondly, from an optimization point of view, the maximum delay is a more strict and explicit condition than the total delay, which should improve the convergence of the solution to a minimum.

To summarize, dynamic traffic aggregation provides a more flexible utilization of resources. At the same time, preserving fairness between traffic streams is imperative. On the node level, traffic aggregation leads to gains in packet loss and throughput – in particular maximum throughput. When combined with optimal routing, a load balancing effect on the network level is achieved, resulting in increased fairness without using any particular scheduling or priority principles. Routes optimized with respect to maximum delay give better results than when optimizing with respect to total delay, which is a direct consequence of improving the worst end-to-end performance. The congestion control is based on fast traffic measurements and estimation of the network states. This can be accomplished using the methods for entropy estimation and heavy hitters described in Chapter 12.

CHAPTER 14

The Internet of Things

The Internet of Things (IoT) is expected to be one of the most important applications in 5G, generating large amounts of traffic. The term can be interpreted as automatized data collection, supplied and accessed through the internet. Data collection is therefore done rather by sensors than by humans. We interpret the term as more or less independent sensor networks – specifically wireless sensor networks.

Projected applications of IoT include environmental monitoring, telemedicine, and smart metering. Since wireless sensors are suitable for deployment in inaccessible terrain and simply can be dropped into the environment from an aircraft or drone, the roll-out is fairly inexpensive. Such sensor networks can be used to monitor crops, or to detect wildfires, earthquakes, or tsunamis. Another promising use of sensor networks is in telemedicine; sensors can be used to monitor patients' health and they have a large potential of helping saving lives. Another large area of IoT – albeit rather for stationary sensors – is smart metering, where sensors and power controls allow for automatic or remote adjustments of indoor temperature to improve comfort levels and energy efficiency.

Wireless sensor networks (WSNs) have the property that the architecture can be entirely ad hoc. The wireless connectivity allows sensors to be distributed randomly to cover an area, where in the initiation phase of the networks, connectivity and routing principles are established. The challenge in WSNs is their *energy efficiency*. Even though sensors may be supplied with power from both solar cells and batteries, energy efficiency is often regarded as a key performance characteristic.

Sensor networks have been devised for environmental sensing, such as flooding and wildfires, surveillance using drones, and automotive and wearable applications, such as collision control and telemedicine. Other types of IoT are smart grids, connecting energy consumers and providing data on power consumption. The applications range from giving consumers better control of their usage to facilitating a smarter energy distribution, particularly for alternative energy sources. In the latter case, electric-car batteries may be used as temporary storage of renewable energy, which could later be discharged to provide a smoother energy supply.

We will focus on the design and properties of ad-hoc WSN, with arbitrary topology and no specific location information available. The example we focus on is WSN for the detection of forest wildfires. The reason behind this choice is that forest terrain presents the need for different energy models, due to the presence of obstacles limiting the radio coverage. The discussion is valid for most WSNs with minor modifications.

5G Networks. https://doi.org/10.1016/B978-0-12-812707-0.00019-X

14.1 Architecture

The IoT consists of networks, has naturally evolved from the traditional internet, and reuses many techniques used on the internet, such as routing protocols. The limitation, however, is that IoT networks consist of devices with weak processing power, and new specific techniques are needed to address energy efficiency and lightweight protocols.

Routing Protocols

WSNs consist of small devices – sensors – with sensing, wireless communication, and computation capabilities. However, by the necessity of being small and inexpensive to manufacture, sensors have limited processing capabilities, wireless link bandwidth, and battery power. These factors place strong demands on routing protocols.

WSNs are often deployed in a more or less uncontrolled manner, without a clear topology. We may consider the topology as either deterministic or random, where in a deterministic topology the nodes are arranged in a predefined pattern, whereas in a random topology, a sensor finds a neighbor in a random direction and at a random distance. Random types of topology are typically the outcome when the nodes are deployed from air, where there is little control over the final position of the nodes. This also means that the routing possibilities through the network somehow must be determined after its deployment.

Energy efficiency is a major concern in WSNs, since devices often by necessity have a limited supply of energy. WSN protocols must therefore be designed to minimize the energy consumptions in the devices and in the network as a whole. Since each node is part of a network, its failure not only affects its neighborhood, but may in extreme cases disconnect parts of the network. Protocols should therefore ideally be proactive and flexible in choosing routes causing minimum impact.

Device mobility adds further complexity to a WSN. In such networks, routing is more challenging due to changing topology and routing paths. The most common approach of routing protocols is finding routing paths on demand. Precomputed routing paths are usually of little use due to the changing conditions.

Mostly being unmaintained, WSNs are prone to failure for a number of reasons. WSN protocols therefore must be robust to failure situations and find efficient alternative routes or operation modes. Such measures may include preferential routing over routes having more spare energy, or restricting the use of parts of the networks.

In some cases, WSNs can be expected to expand with time by added nodes. Routing protocols should therefore be able to accommodate planned changes as well, and support rebalancing and route redefinitions.

Routing Protocols for the Internet of Things

Many protocols have been proposed for the IoT. This fast developing area is bound to become obsolete fast, although the underlying principles pertaining to IoT remain valid.

6LoWPAN

6LoWPAN – short for *IPv6 over Low-Power Wireless Personal Area Networks* – is the name of a working group in the internet area of the IETF. It is intended to extend IPv6 networks to IoT networks, with the possibility of reusing existing IPv6 technologies and infrastructure. However, these technologies were originally designed for devices with higher processing capability and memory resources and may therefore not be suitable for IoT architectures.

Zigbee

Zigbee, based on IEEE 802.15.4, is intended for personal area networks with small, low-power radio devices with low bandwidth in ad-hoc network topologies. It is specifically designed for small-scale wireless networks of devices having low cost and range.

RPL

RPL – IPv6 Routing protocols for Low Power and Lossy Network, is also based on IPv6 and established internet protocol technology. It is designed for networks of devices low in power, computation capability, and memory. The transmission in the corresponding networks is unreliable with high loss rate.

LEACH

We describe some details of the protocol LEACH – Low-Energy Adaptive Clustering Hierarchy [125][122][124]. LEACH is a hierarchical clustering protocol supporting data fusion. It is self-adaptive and self-organized. It uses *round* as unit, where each round is made up of a cluster set-up phase and a steady-state phase. To reduce unnecessary energy dissipation, the steady-state phase must be much longer than the set-up phase.

At the set-up phase, clusters are formed by randomly selecting cluster heads. The selection is performed by generating a random number uniformly between 0 and 1 and comparing this number with the threshold value $t(n)$. If the number is less than $t(n)$, then the node is set as a cluster head in this round; otherwise it functions as a common node. The threshold $t(n)$ is determined by the relation

$$t(n) = \begin{cases} \dfrac{p}{1 = p\left(r \quad \mathrm{mod}\ \frac{1}{p}\right)} & \text{if } n \in G, \\ 0 & \text{if } n \notin G, \end{cases}$$

where p is the percentage of the cluster head nodes of all nodes, r is the number of the round, and G is the collections of the nodes that have not yet been head nodes in the first $1/p$ rounds. Using this threshold, all nodes will sooner or later be assigned head nodes after $1/p$ rounds.

Each node becomes a cluster head with probability p when the round begins, and the nodes which have been head nodes in this round will not be head nodes in the next $1/p$ rounds. The number of nodes which are capable of being head nodes will gradually reduce, and so for the remaining nodes, the probability of being assigned head nodes must be increased. After $1/p - 1$ rounds, all nodes which have not been assigned head nodes are selected as head nodes with probability 1. After round $1/p$ finishes, the state of all nodes is reset.

When clusters have initialized, the nodes start to transmit the sensing data. The cluster heads receive data sent from other nodes and the received data are sent to the base station after being aggregated (data fusion).

14.2 Wireless Sensor Networks

We consider a WSN with a random topology, such as what arises when sensors are dropped randomly from an aircraft or drone. Such networks are known as ad-hoc networks. We assume that we have a base station that manages the network and collects data. Each sensor node acts as a router, so that the routing capabilities of the network are fully configurable with respect to the actual random topology.

Energy Models

The network needs to be initialized, which includes identification of sensor neighborhoods and finding routes to the base station. It is assumed that the logic resides in the base station and that sensors can be configured by downstream messages originating from the base station.

The main characteristics of the WSN we are interested in are its *lifetime* and its *coverage*. The lifetime of the network is strongly dependent on the *energy model* used.

Due to varying topology and environmental conditions, the average conditions for signal propagation change from network to network. We use radio frequency (RF) models to estimate both the signal conditions and the energy consumption of the radio unit. We study energy models for some typical environments using the LEACH simulation toolkit.

The energy models are used to simulate important properties of a WSN, specifically in tree-obstructed propagation environments, which predict the radio energy dissipation of the WSN nodes.

Aldosary and Kostanic [123] present empirical path loss models for radio propagation in sparse and dense tree environments. They study the impact of the obstructive environment on the performance of a WSN. The path loss models are given by

$$
L_p = \begin{cases} 60.844 + 33.363 \log\left(\frac{d}{d_0}\right) & \text{for sparse tree terrain,} \\ 52.14 + 40.2 \log\left(\frac{d}{d_0}\right) & \text{for dense tree terrain,} \end{cases} \tag{14.1}
$$

where L_p is the path loss at a distance d, expressed in dB, and d_0 is a reference distance in meters. The link budget, that is, the relationship between the path loss, transmitted and received power, and gains is generally written

$$
L_p = P_t - P_r + G_t + G_r, \tag{14.2}
$$

where P_t is the transmitted power, P_r is the received power, G_r is the receiving antenna gain, and G_t is the transmitting antenna gain. By substituting Eq. (14.1) into Eq. (14.2), we have

$$
P_r = \begin{cases} \frac{P_t G_t G_r}{1.21 \cdot 10^6 \cdot d^{3.33}} & \text{for sparse tree terrain,} \\ \frac{P_t G_t G_r}{0.163 \cdot 10^6 \cdot d^{4.02}} & \text{for dense tree terrain.} \end{cases}
$$

The energy model is used to estimate power dissipation by the transmitter, the power amplifier, and the receiver. The distance between two nodes determines the propagation loss, which is modeled by a factor to a power of d.

The authors report using a dissipation of 50 nJ/bit in the modeled 1-Mbps transceiver, that is, 50 mW, regardless whether the transceiver is transmitting or receiving. This gives

$$
\bar{P}_r \geq -94 \text{ dBm,}
$$
$$
E_{\text{sparse}} = 0.2047 \text{ pJ/bit/}d^{3.33},
$$
$$
E_{\text{dense}} = 0.0275 \text{ pJ/bit/}d^{4.02}.
$$

The empirical models for free space and two-ray propagation published by Chandrakasan and Heinzelman [122] are used for comparison, using the same parameters, that is,

$$
E_{\text{freespace}} = 1.10 \text{ fJ/bit/}d^2, \tag{14.3}
$$
$$
E_{\text{multipath}} = 0.027 \text{ pJ/bit/}d^4. \tag{14.4}
$$

The models (14.1), (14.3), and (14.4) are used in the LEACH simulation toolkit to derive WSN performance characteristics in different environments, under the LEACH protocol.

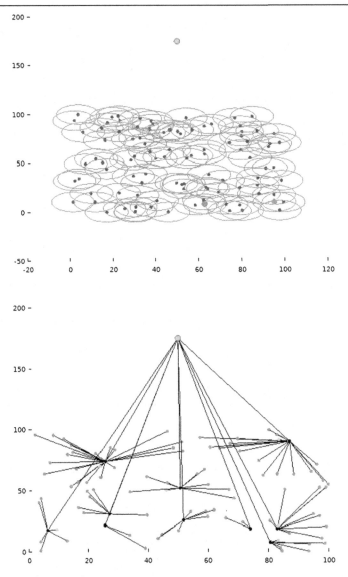

Figure 14.1: Randomly deployed WSN nodes.

Simulation Results

We specify a number of nodes N on a rectangular area A, typically distributed uniformly at random, and define a location of the base station. We define network lifetime as the failure time of the first sensor node, and we constantly measure the throughput as the cumulative number of successfully received packets at the base station. We also record the cumulative dissipated energy and estimate the coverage and connectivity ratios. See Figs. 14.1–14.4.

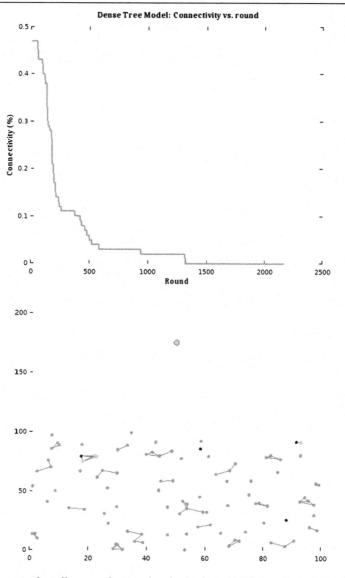

Figure 14.2: Connection diagram for randomly deployed WSN nodes under the dense tree propagation model.

The coverage for a random WSN topology is calculated using Monte Carlo simulation. We generate a large number of points uniformly in A, relate a sensing radius to the devices, and test whether the points lie within the sensing area of at least one device. The ratio of points that can be detected by at least one device to the total number of points define the relative coverage.

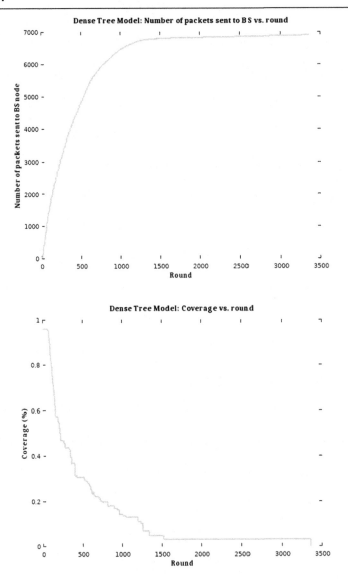

Figure 14.3: Evolution of WSN using the dense tree propagation model: coverage and throughput.

For the connectivity, we can count the number of nodes having at least one neighbor within communication range and define connectivity as the ratio of the number of nodes with at least one such neighbor to the total number of nodes.

Alternatively, we can create a graph such that there is a link between two nodes s and t whenever the node s can find a path to node t. The graph is assumed to be undirected, so the ad-

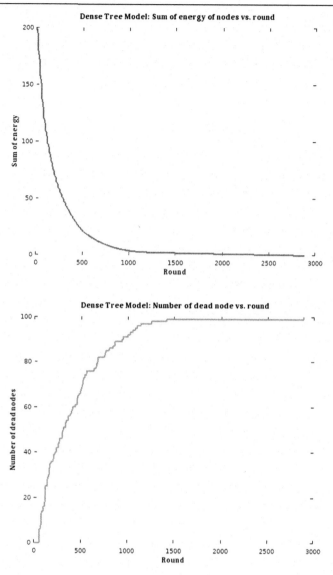

Figure 14.4: Evolution of WSN using the dense tree propagation model: failed nodes and total energy.

jacency matrix is symmetric. The connectivity is then defined as the sum of all entries in the matrix divided by twice all the possible links, $2|E| = N(N-1)$.

The path loss of a dense tree environment is larger than that of the sparse tree environment, leading to a faster depletion of energy resources.

This is caused the lack of line of sight (LOS) between almost all node pairs due to foliage, bushes, and trees, leading to extensive fading. The presence of objects such as leaves, branches, and tree trunks causes reflection, scattering, absorption, and diffraction of the signal.

The received signal may be a superposition of reflections of the signal, or small fractions of the scattered signal. Such conditions lead to large path loss, exhaustion of battery resources, and shorter lifetime of the network.

In the sparse tree environment, communicating nodes do not always have LOS, and when there is, the signal received might be a combination of direct and reflected waves due to ground reflection. In absence of LOS, the signal is again a superposition of reflected or scattered waves due to foliage, weed, or other objects.

14.3 Mobility Modeling Techniques

In IoT with moving devices, clearly mobility makes protocol design and network performance evaluation more challenging. We present some models for mobility modeling – a selection for the many models proposed in the literature. Analytical and analytical-numerical mobility models are here roughly divided into the following four classes:

- geometric models
- queueing models
- traffic flow models and
- other model types

In many cases, analytical results are supported by simulations. Some models of different types are discussed in the following subsections and analyzed with respect to their applicability from an engineering point of view.

Geometric Models

Under the class of geometric mobility models fall various approaches for modeling a two-dimensional movement of a user or a group of users. Geometric models are mathematical formulations of a user's *trajectories* in the plane.

In the paper [140], an analysis of mobility of users is performed to yield probability density functions for hand-over call rates. The *call holding times* – if there were no hand-overs – are assumed to be exponentially distributed with parameter μ_M. The mobility model assumes a uniformly distributed velocity distribution with constant velocity within a cell. At the cell boundary, the speed and direction can be changed. The direction is also uniformly distributed and independent of the speed. That is, the mobility model can be formulated as

$$f_\theta(\theta) = \begin{cases} \frac{1}{\pi} & \text{for } 0 \le \theta \le \pi, \\ 0 & \text{otherwise,} \end{cases}$$

$$f_V(v) = \begin{cases} \frac{1}{V_{max}} & \text{for } 0 \le v \le V_{max}, \\ 0 & \text{otherwise.} \end{cases}$$

The users are assumed to be spread evenly over the coverage area. Based on this model, Hong and Rappaport derive the probability density functions for the channel holding time

$$f_{T_H}(t) = \mu_M e^{-\mu_M t} + \frac{e^{-\mu_M t}}{1 + \gamma_c} \left(f_{T_n}(t) + \gamma_c f_{T_h}(t) \right) - \frac{\mu_M e^{-\mu_M t}}{1 + \gamma_c} \left(F_{T_n}(t) + \gamma_c F_{T_h}(t) \right),$$

where γ_c is the ratio of average carried hand-over attempt rate to the average carried new call origination rate, $f_{T_n}(t)$ is the probability density function of the call holding time within the cell in which the call originated (with corresponding cumulative distribution function $F_{T_n}(t)$), and $f_{T_h}(t)$ is the probability density function of the call holding time in the cell to which the call has been handed over (with corresponding cumulative distribution function $F_{T_h}(t)$). Hong and Rappaport approximate the distribution of the channel holding time by an exponential distribution. The impact of different channel reservation schemes (where channels are dedicated for hand-over traffic) is studied under this mobility model, formulated as an *M/M/c* queue.

The results in [140] show interesting results for channel reservation schemes for the particular mobility model used. However, for engineering purposes, determining the actual distributions for the mobility parameters is likely to be very difficult. Furthermore, the approximation of the channel holding time distribution with an exponential distribution seems not to be very accurate (see analyses of channel holding time measurements in Section 14.4 and [148]).

The approach in [140] is further developed in [147] and [145]. Schweigel and Zhao use the same mobility model as Hong and Rappaport and derive the same probability density distributions for square-shaped cells. In [145] the authors parameterize the mobility pattern with position, direction, and speed, and allow changes in these parameters as time progresses, so that Zonoozi and Dassanayake generalize the model in [140] to cover a mobility pattern where the users are allowed to change direction and speed within a cell. They use simulations to show that the cell sojourn time from the instant the call is originated can be described by a generalized gamma distribution. It is also shown that an increase in mobility decreases the *effective cell size* and that sojourn times become shorter. Zonoozi and Dassanayake assume that the call holding time is exponential. The cumulative distribution function for the channel holding time is expressed as

$$F_{T_{ch}}(t) = \zeta \left(F_{T_c}(t) + F_{T_n}(t) - F_{T_c}(t) F_{Tn}(t) \right) + (1 - \zeta) \left(F_{T_c}(t) + F_{T_h}(t) - F_{T_c}(t) F_{Th}(t) \right),$$

where ζ is the average relative number of handovers per call, F_{T_c} is the exponentially distributed call holding time, and F_{T_n} and F_{T_h} are the cell sojourn times for new calls and handed-over calls, respectively. When the latter are described with a generalized gamma distribution, the channel holding times are shown to be exponential. In [146] a generalized gamma distribution is also used to describe cell sojourn times for active users (users with an ongoing connection). The mobility model is similar to the one presented in [145]. The authors conclude that an exponentially distributed cell sojourn time overestimates channel holding times, compared to a gamma-distributed cell sojourn time.

The European Telecommunications Standards Institute (ETSI) defines a set of test scenarios for system simulations of the Universal Mobile Telecommunication System (UMTS) [144]. Three different mobility models are proposed: an indoor office, an outdoor pedestrian, and a vehicular environment.

The model for the outdoor pedestrian environment uses a *Manhattan grid* street structure, a rectangular grid of equally spaced streets. Users are assumed to walk along the streets and possibly change their direction at intersections with a given probability. Also speed changes are possible in given intervals. The vehicular environment is a random model without a street structure. Vehicles are assumed to move with constant speed ($v = 120$ km/h) and can change their direction every 20 m with allowed direction change of maximum $\pm 45°$, with given probabilities.

Jugl and Boche observe that the velocity of users that cross a cell boundary is different from the average velocity in a cell, a phenomenon referred to as *biased sampling* [143]. The reason for this is that users with lower speed also have a lower probability of crossing the cell boundary. From a given velocity distribution $f_V(v)$ for the users in a cell, the velocity distribution of the boundary crossing users is formulated as $f_V^*(v) = Cvf_V(v)dv$, corrected for biased sampling. This relation is useful for prediction of hand-over call rates only if the velocity can be regarded as constant for the duration of a call. With a highly variable velocity distribution, the formula cannot be applied.

Jugl and Boche propose three distributions for the random variable *traveled distance*, X, used together with a Gaussian velocity distribution in order to compute the cell sojourn time $T = X/V$ [142]. The three models represent three types of movements with different variability. For low variability (like highway traffic) the Dirac function is proposed. For high variability, a generalized gamma-type of distribution is proposed. Examples of expected cell dwelling times are computed with numerical integration. Pla and Casares use a similar model to show that the cell sojourn time is well described by a hyper-Erlang distribution [141].

The mobility model in [140] is modified in [139] to yield smooth trajectories of the users. This is achieved by the introduction of auxiliary parameters, so that a user is described by its current speed and acceleration, target speed, maximum speed, and a set of preferred speeds and maximum values for acceleration and deceleration. The time instant for a speed change is governed by a Poisson process. If a speed change occurs, the speed will continuously change from the current speed to the target speed with rate determined by the acceleration. Direction changes are modeled similarly. The proposed model is used in simulations where the preferred direction is biased towards the center of the simulation area. The simulated result shows an aggregation of users around the center point.

Empirical probability distributions for velocity and direction of users are difficult to obtain since this requires very refined measurement procedures in the network which do not exist today. Also the exact geometrical shape of the cells needs to be known, which is particularly difficult to assess for dynamic cells. The number of parameters for most geometric models is also too large for practical usage.

Queueing Models

The cell can be modeled by a queue in the following way. Users (traffic sources) arrive according to some distribution, at a certain rate, and remain in the cell (the cell sojourn time) according to some service time distribution. There is no hard limit on how many customers a cell can host (the customers are not assumed to generate any traffic at this point), so the number of servers is infinite (or very large).

Antunes [138] discusses a stochastic model taking both mobility and service call generation into account. The dissertation includes an analysis of the transient and steady-state behavior of the model. Antunes comments that a candidate model should have as few parameters as possible and that parameters should be possible to estimate from observations. In order to keep complexity down, Antunes suggests to keep the mobility part simple. Two basic models are presented – a one-dimensional one, applicable for highway traffic, and a two-dimensional one, applicable to an urban structure. The one-dimensional model is based on a highway traffic model with Poisson-distributed input in combination with a Markov-modulated fluid process. Antunes shows that the cell populations are Poisson-distributed. The two-dimensional model uses a similar approach, where a cluster of cells is assumed to be fed by Poisson-distributed inputs. The cell sojourn times are allowed to be general. Again, the resulting cell populations are shown to be Poisson-distributed. The approach proposed by Antunes is to model each individual user separately, and it imposes the assumption that all particles move independently from each other. The models allow for a generally distributed sojourn time, which, as identified by Antunes, may not be Markovian, that is, may have longer memory.

Mitchell and Sohraby use a Jackson network with multiple user classes in simulations to show the effects of channel allocation strategies [137]. In their mobility model, the call arrivals are assumed to be Poisson and the holding times exponential. They introduce routing probabilities based on hand-over rates. A similar approach is used by Ashtiani et al., who propose a multiclass Jackson network for description of user mobility [135]. The studied network is divided into regions (or cells). The interarrival times are assumed to be exponential for the cell sojourn time for active calls, which ends when a call either is terminated or handed over to another cell. The stationary distribution is found as a product form solution. The arrival rate to each cell is obtained by solving a linear system of traffic equations relating arrival rates within the system and routing probabilities, that is, the probability $r_{i,j}$ that a user will go from cell i to cell j.

Fang and Chlamtac assume a general call holding time with known distribution and distribution for the residual time of a call after successful hand-over [136]. By using Laplace–Stieltjes transforms, analytical expressions for cell sojourn times for new calls and handed-over calls are derived.

The assumption of independent users is a limitation, since the Poisson arrival process has been shown not to be a good approximation for heavy traffic situations in mobility (which often coincides with heavy utilization of the communication network). The reason is that *coupling* between elements occurs under heavy load.

As with geometric models, the main problem of using queueing methods lies in the difficulty of specifying distributions for general movement, such as velocity distributions or routing probabilities, in particular as they in general state-dependent in general, and the computationally huge demands that even modest models create.

Traffic Flow Theory

Traffic flow theory has been an active field of study since the 1930s. The objective is the analysis of flow of vehicular traffic by probabilistic means. Several classes of models have been proposed [131]. In particular, *continuum flow models* and *macroscopic models* have given inspiration for mobility modeling in wireless networks.

A continuum flow model is a model where traffic is modeled as a one-dimensional (source- and sink-free) compressible fluid. It is based on the assumptions that:

(1)   Traffic flow is conserved – this is expressed in the *continuum equation*. Let q denote the flow at a point along a line x and k the density in the same point. Then

$$\frac{\partial q}{\partial x} + \frac{\partial k}{\partial t} = 0,$$

that is, inflow is equal to outflow plus storage.

(2) There is a one-to-one relationship between speed (or flow) and density, expressed by the *equation of state*, where speed (or flow) is assumed to be a function of density, that is,

$$u = f(k). \tag{14.5}$$

With the fundamental relationship that flow equals density times speed, $q = ku$, the governing equation becomes

$$\left[f(k) + k\frac{df}{dk} \right] \frac{\partial k}{\partial x} + \frac{\partial k}{\partial t} = 0. \tag{14.6}$$

Eq. (14.6) is used in [134] to model call arrivals on a highway. The vehicles are classified into "calling" and "noncalling". Two coupled partial differential equations are formulated for the two cases. The equations thus govern a nonhomogeneous Poisson process for the call intensity rate. Rate density functions have to be specified for both types of vehicles. Numerical solutions to the equations are also presented.

As noted in [131], Eq. (14.6) is rarely used in practice due to the difficulties to find suitable initial and boundary conditions. For realistic initial and boundary conditions and complex functions (14.5), analytical solutions may not be obtainable, but Eq. (14.6) must be solved numerically. The situation is complicated further in two dimensions.

The main objection to using this approach to model general mobility is the relation (14.5). This may not be an appropriate assumption for general movement, even if it may give a reasonable approximation for vehicular traffic. Some macroscopic models are aimed at establishing relationships between certain variables on a large-scale level, based on empirical data. For example, investigations have been carried out to find a relationship between average speed and distance from a city center. Several functional relationships based on these measurements have been proposed. Examples of such functions (see [131]), where v is the average speed and r is the distance from the city center, are

$$\begin{aligned} v_1(r) &= ar^b, \\ v_2(r) &= a - be^{-cr}, \\ v_3(r) &= \frac{1 + b^2 r^2}{a + cb^2 r^2}. \end{aligned}$$

This type of models can provide valuable large-scale approximations and they are often used to estimate relationships like (14.5).

Other Model Types

Baccelli and Zuyev use *stochastic geometry* to describe user densities. Users are assumed to be distributed according to a *Poisson point process*, where the number of points in a bounded Borel set $\mathcal{B}$ has a Poisson distribution with mean $\Lambda(\mathcal{B})$ and the numbers of points in disjoint Borel sets are independent random variables. A point process is made a *marked point process* by attaching a characteristic (mark) to each point of the process, such as active in a call and inactive. The authors also use a stochastic road pattern generated by a *Poisson line process* and an assumed known velocity distribution of the users on this road pattern to derive expectation and variance of the number of active users per unit area.

The *gravity model* has been used to model traffic routing and user mobility [133]. The model formulated for traffic flow between two regions is

$$T_{i,j} = \frac{m_i m_j P_i P_j}{d_{i,j}^{\gamma_i + \gamma_j}},$$

where $T_{i,j}$ is the traffic from region i to region j, P_i and P_j are the populations of the two regions, and $d_{i,j}$ is their distance (measured between suitable points). The constants m_i, m_j, γ_i, and γ_j have to be determined from empirical data. This model can be used for user mobility modeling as well. In fact, the gravity model is incorporated in the proposed mobility model in Section 14.4.

14.4 Gibbsian Interaction Mobility Model

This section presents a novel method to estimate the fluctuations in traffic source density due to user mobility. The mobility is modeled by a Gibbs field, where the potential (and the transition kernel) is described by two simple "forces". By obtaining state probabilities associated with the Gibbs field, a better estimate of the total offered traffic can be obtained – rather than assuming a fixed number of users per area based on the network average population.

In order to describe the traffic offered to a cell, the *mobility* of users, in the sequel often referred to as *traffic sources*, has to be quantified in some way. In this section, a set of measurement data originating from an operational cellular network is analyzed for its main statistical characteristics. A model for mobility based on rather general principles is formulated. By simulation, a close resemblance of the empirical data and the simulation results are shown. It is also shown that an approximate analytical-numerical solution to the mobility model can be found for the model in the one-dimensional case. The mobility model is useful in composite models for traffic, where fluctuations due to mobility are taken into account.

The motivation for incorporating mobility in traffic engineering is given by the following observations made for the empirical data of traffic channel occupancy rates in a cell cluster:

(1) The data have higher variance than the truncated Poisson process following traditional queueing theory. Indeed, the data are very poorly modeled by the theoretically derived probability distribution.
(2) The data is short range–dependent, which motivates the description by the means of a Markov process.
(3) Mobility is responsible for fluctuation on relatively long time scales, which cannot be handled by buffers and system control mechanisms. Dimensioning should therefore be done for the peak rate of the fluctuations due to mobility.

Furthermore, it is assumed that the main reasons for the failure of traditional methods are the effects of source mobility, that is, time-varying source density and hand-over traffic. Consider a cluster of cells in a cellular network. Suppose that traffic is continuously measured in one cell. The measured traffic process can then be thought of as a composite process of traffic source density (as a result of mobility) and the traffic intensity generated by each source. These two processes are assumed to be statistically independent. The mobility pattern is difficult to assess since governing principles are not easily understood and described. The complexity of the mobility modeling suggests that a combination of methods has to be used.

Analysis of Dependence

A trace of the recorded resource utilization is shown in Fig. 14.5. The sample autocorrelation function shows positive correlation for the first few lags, which indicates the presence of short-range dependence. The initial statistical analysis on the time series data is divided into two steps. Firstly, the deviation from theoretical results is obtained by comparing the time series with the state occupancy of an $M/G/c/c$ queue, that is described by the truncated Poisson distribution,

$$p_n = \frac{\frac{(\lambda/\mu)^n}{n!}}{\sum_{i=0}^{c} \frac{(\lambda/\mu)^i}{i!}}, \quad 0 \le n \le c. \tag{14.7}$$

The Poisson distribution has been shown very successful to model independent events. For a large area, the assumption of independence *between* users is quite natural. However, a radio cell with limited coverage area will experience traffic source fluctuations and hand-over traffic, which may invalidate the Poisson assumption.

The frequency histogram of the time series is shown in Fig. 14.6. The histogram has some strong irregularities. By applying a wavelet filter, a smoothed histogram is achieved as one of the components of the corresponding multiresolution analysis. The wavelet filter is a *maximum-overlap discrete wavelet transform (MODWT)* based on Daubechies wavelet coefficients with filter width 4. The MODWT filter operates on data vectors of any length.

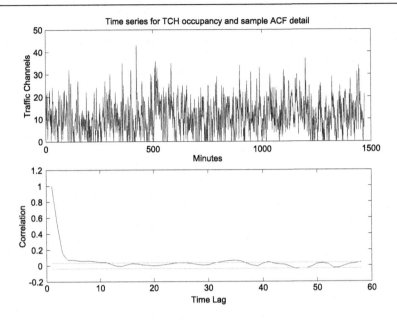

Figure 14.5: Time series and sample ACF of traffic channel occupancy data.

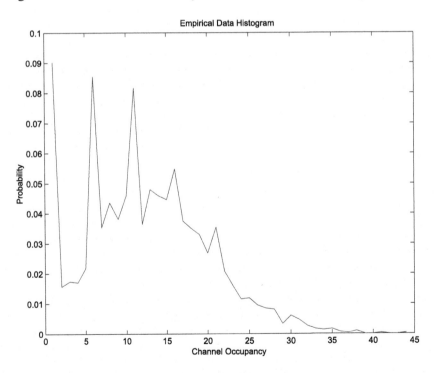

Figure 14.6: Empirical distribution of traffic channel occupancy data.

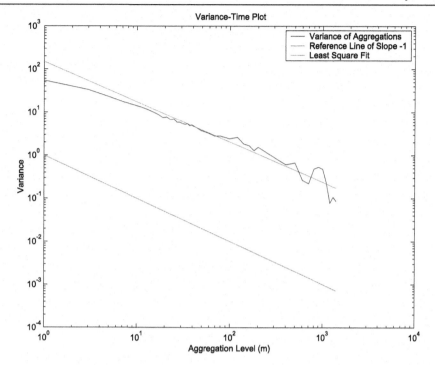

Figure 14.7: Variance-time analysis of channel occupancy data.

The choice of wavelet is often a matter of trial-and-error [132]. The main objective with the smoothing is to level out artifacts due to measurement errors or high-frequency variations. The choice of wavelet filter is therefore not very critical. The chosen *smooth* is on level $J_0 = 3$, corresponding to a time scale of 8 min.

To formulate an adequate model, it is necessary to assess whether the data are long range–dependent or short range–dependent. The variance-time analysis, using 42 different aggregation levels, is shown in Fig. 14.7. The coefficient $\hat{b}$ is estimated to be 0.93. An R/S analysis (see Section 4.3) was also performed. The result is shown in Fig. 14.8. The data are clearly located close to the line representing a value of the Hurst parameter of $H = \frac{1}{2}$, meaning that no (or very little) self-similarity is present.

Fitting a Distribution

A comparison between the channel occupancy data and the exponential distribution, the gamma distribution, and the Weibull distribution (all members of the *generalized gamma* family of distributions), is shown in Fig. 14.9.

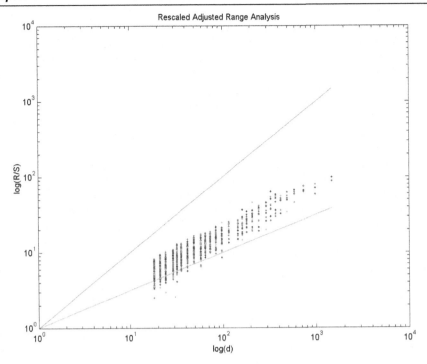

Figure 14.8: Rescaled range analysis of channel occupancy data.

For all three distributions, a maximum likelihood estimation of the parameters is performed and the result tested with the chi-square test. Here, the relative goodness of fit is of interest, so the test statistic for the three distributions with the smallest value indicates the best fitting distribution. The best fit is given by the Weibull distribution. The value of the chi-square statistic for the Weibull distribution was 0.038 compared to 0.12 and 0.36 for the gamma and exponential distributions, respectively.

Basic Assumptions

The problem of mobility is difficult as the interaction – the internal dynamics – between sources may be very complex. One therefore has to be cautious not to lose generality when choosing simplifying conjectures.

Independence

In many probabilistic models, the assumption of independence is used to achieve tractable problems. However, the results of these models are not of much use if the assumption cannot be justified.

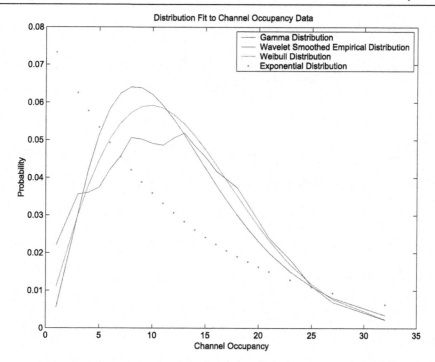

Figure 14.9: Gamma, Weibull, and exponential distributions compared to the wavelet smoothed empirical distribution.

There is also a duality between time dependence and space dependence. Consider for example a user moving from one cell to another, where the transition probabilities are assumed to be state-dependent. This dependence can be incorporated by augmenting the state space to include all possible configurations of the system. Alternatively, this dependence can be interpreted as some sort of "memory", where the system has a positive correlation with its earlier states. Dependence can therefore be seen as a measure of the amount of information that is built into the model.

Stationarity

Stationarity is a concept related to time. A stationary process has a time-invariant probability distribution. This can be interpreted as the property of a constant "intensity" or "activity" for a source or group of sources. Stationarity is closely related to ergodicity – stationarity is necessary in order to have a well-defined mean.

Ergodicity

Ergodicity can be formulated mathematically as "ensemble averages equals time averages", that is,

$$E\left(X(t)\right) = \frac{1}{T} \int_0^T X_0(t)dt.$$

The process $X(t)$ is ergodic if, with probability one, all its "measures" can be determined or well approximated from a single realization $X_0(t)$ of the process.

Indistinguishability

In quantum physics, the assumption of indistinguishability of particles leads to different principles related to *counting* of states compared to classical systems. A similar interpretation can be used for mobility analysis.

Considering two sources, a and b, and two radio cells, A and B, there are four possible scenarios:

(a) both sources a and b are in cell A
(b) both sources a and b are in cell B
(c) source a is in cell A and source b is in cell B
(d) source b is in cell A and source a is in cell B

It is clear that scenario (c) and (d) are identical from a "quantum" point of view. If the probability is equal, say p, for each of the four scenarios, the following cases can occur:

(1) both sources are in cell A with probability p
(2) both sources are in cell B with probability p
(3) one source is in cell A and the other is in cell B with probability $2p$

The state space is reduced in the latter case by collecting all identical configurations into one state. The number of degenerate configurations is given by the *binomial numbers*.

Dependence Structure

A fundamental concept in statistical physics is the *system* of a large number of interacting particles. The Gibbs–Boltzmann statistics of such a system were independently formulated by Josiah Willard Gibbs (1839–1903) and Ludwig Boltzmann (1844–1906). The following discussion on the foundation of statistical physics followed largely [130].

Consider a system, S_1, embedded in a larger system of the same type. The combination of S_1 and the larger system embedding it is denoted S_0. The embedding system without S_1 is denoted S_2. That is,

$$S_0 = S_1 \cup S_2.$$

The only assumption on the system is the *ergodic assumption*, which in physical terms states that the density is dependent only on the energy of the system.

In order to be able to model radio cells with a mechanical system, a metaproperty corresponding to the *energy* has to be assigned to the cell. By letting the *system* of radio cells be modeled as a collection of identical cells, we let the characteristic be an abstract quantity referred to as "energy" rather than the sources themselves.

Thus, each source has an energy which is equal for all sources in the same site, and the sum of the source energies is the energy of the site. The interpretation of the change of state as a result of a source moving from one site to another is therefore related to a change in energy, rather than a change in the number of sources (the "mass" of the cell). The energy is assumed to be state-dependent.

The assumption is made that the system obeys the law of conservation of the number of sources (which is also a common assumption in traffic flow theory [131]), that is,

$$\frac{d\rho}{dt} = \text{influx} - \text{outflux},$$

where ρ denotes the density. This assumption implies that sources can only move from their current site to a neighboring site and that no sources are created or destroyed. Since the instantaneous movement is restricted to neighboring sites, the energy is defined for a site with respect to its neighbors. The coupling (interaction) between sites is therefore assumed to decrease rapidly with distance. The distance d is defined so that $d = 1$ to all sources in its neighborhood, that is, in sites with which it has a common boundary, and $d = \infty$ otherwise. The energy is therefore the sum of attractive forces exerted on the sources in a site by sources in neighboring sites.

Now, assuming that the information on the number of sources in the current site and its neighbors is available at each time instant, and a fixed value, the *nominal capacity*, is known for each site, two "forces" dependent on the states can be formulated.

The first attempt to define an interaction between sources is to assume that the strength of interaction is dependent on the number of sources. An intuitive interpretation is that most activities are, in fact, interactions with other people. Understood in a wide sense, an interaction is any event where the behavior of a person is, one way or another, influenced by another person. Such events are, however, not tied to any particular location.

The interaction may be modeled as follows. Each chosen source is subject to a "gravity" force exerted by surrounding sources. This force between the site i and the site j is then

$$F_{i,j}^g \sim \alpha_g x_i x_j, \tag{14.8}$$

where α_g is a scaling parameter. The interpretation of this hypothesis is that for an arbitrary source, there is a uniform probability of interaction with any other source in its immediate neighborhood. The nature of the interaction should be interpreted in a very wide sense, that is, "being in the same cell". This force alone cannot describe the system satisfactorily.

When modeled as a Markov chain, where the state space consists of all possible states of the system, some states would become absorbing barriers. As time progresses, a system described by only (14.8) would have all sources concentrated to a few sites with neighboring sites empty, and the sources would not be able to leave these states. The gravity would force sources together in clusters, which clearly is not consistent with reality.

A second "retaining" force is therefore introduced, based on the nominal capacity, $\bar{x}$, of each site. For simplicity, the nominal capacity is assumed to be equal for all sites. The force is assumed to be proportional to the difference between the nominal capacity of the site and the instantaneous population in that site, that is,

$$F_{i,j}^r \sim (1 - \alpha_g)x_i(\bar{x} - x_j)^+,$$

where $(\bar{x} - x_i)^+ = \max(\bar{x} - x_i, 0)$. Intuitively, this force represents the fact that most people have certain locations that they visit more frequently than others. The force removes the absorbing barriers from the Markov chain so that there is always a positive transition probability to an empty site from a nonempty state.

Based on these "forces", transition probabilities between states can be formulated. Assuming that cell i is considered, the conditional transition probabilities can be seen as the normalized sum of the "derivatives" of the forces, $\partial F_{i,j}^g / \partial x_i$ and $\partial F_{i,j}^r / \partial x_i$, respectively. We have

$$p_{i,j} = \frac{\alpha_g x_j + (1 - \alpha_g)(\bar{x} - x_j)^+}{\alpha_g \sum x_k + (1 - \alpha_g) \sum (\bar{x} - x_j)^+}. \tag{14.9}$$

Simulation of Source Density

It should be possible, at least in principle, to create a Markov process if the state space is augmented enough. The concept relates to the amount of information contained in the evolution of a process. The state space is expanded to consist of *all* configurations. The state space will therefore be very large, since there are $(N + 1)^S$ configurations with $x_1 + x_2 + \ldots + x_S = N$ sources and S cells. The configurations with the same number of sources in each site are identical, so the $\frac{x_1! x_2! \ldots x_S!}{N!}$ identical configurations can be collected together and called one state, since the sources are indistinguishable. Now a state $s_{x_1, x_2, \ldots, x_S}$ can be defined as all configurations with exactly the number of users $x_1, x_2, \ldots, x_S$ in sites $1, 2, \ldots, S$. The problem is that the state space grows very rapidly with the number of sites and the number of sources, so that it becomes infeasible to list all states in order to determine the transition probabilities for all state combinations.

Gibbs Sampler Implementation

The Gibbs sampler is a very useful tool for simulations of Markov processes for which the transition matrix cannot be formulated explicitly because the state-space is too large. A Gibbs sampler for the model using conditional probabilities can be implemented as follows.

(1) Choose a source randomly by uniform sampling.
(2) Calculate the conditional transition probabilities for the chosen site, given by (14.9).
(3) Subject to the transition probabilities, generate a random number and update the site states if transition occurs.
(4) Leave the other sites unchanged and start from the beginning.

In the implementation of the simulation scheme, two parameters are introduced to describe the movement:

(1) the *mobility factor* α_m, which describes the degree of movement, or in other words, the degree of stochasticity versus determinism; and
(2) the *gravity factor* α_g, which describes the weighted relation of the gravity force to the retaining force.

Proposition 14.4.1. *The mobility model implemented as described in this section is an irreducible Markov chain. The Gibbs sampler therefore converges to a unique stationary solution as the number of iterations increases.*

To see this, let S be the state space of the chain for the one-dimensional model and let $S_{x_1,\ldots,x_{i-1},x_i,x_{i+1},\ldots,x_S} \in S$ be a state with neighboring states, i.e.,

$$S_{x_1,\ldots,x_{i-1}-1,x_i+1,x_{i+1}\ldots,x_S},$$

$$S_{x_1,\ldots,x_{i-1},x_i+1,x_{i+1}-1\ldots,x_S},$$

$$S_{x_1,\ldots,x_{i-1}+1,x_i-1,x_{i+1}\ldots,x_S},$$

$$S_{x_1,\ldots,x_{i-1},x_i-1,x_{i+1}+1\ldots,x_S}.$$

Then, since $p_{i,j} \neq 0$ for all neighboring nonempty states, these states intercommunicate. Also, $p_{i,j} = 0$ only when both sites i and j are empty. But in order to remain empty, all surrounding sites must be empty. This cannot be the case (since the sites are chosen arbitrarily), so by Definition 7.2.2 the Markov chain is irreducible. The argument is easily extended to cover the two-dimensional model. The Markov chain is also aperiodic, which is a property of the Gibbs sampler.

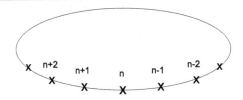

Figure 14.10: Structure of the one-dimensional Gibbs sampler.

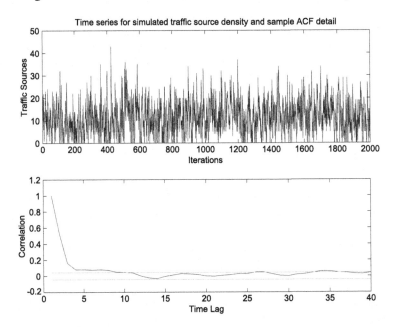

Figure 14.11: Simulated trace and sample ACF for the one-dimensional model.

Simulation Results

The one-dimensional Gibbs sampler is implemented with boundary conditions so that the S cells form a ring, depicted in Fig. 14.10. The total number of users N is constant throughout the simulation. The simulator is initiated by assigning the same number of users to all cells.

The simulation step is an arbitrarily small time step. A simulated trace of a cell in the one-dimensional model is shown in Fig. 14.11, which shows a similar short-range dependence as the empirical data. The state distribution of the simulation is shown in Fig. 14.12.

The two-dimensional Gibbs sampler is implemented so that each cell interacts with six other cells, as shown in Fig. 14.13. The simulator also needs to be defined at the boundary. In this implementation, the cells are wrapped around, so that the simulation area forms a ball. The

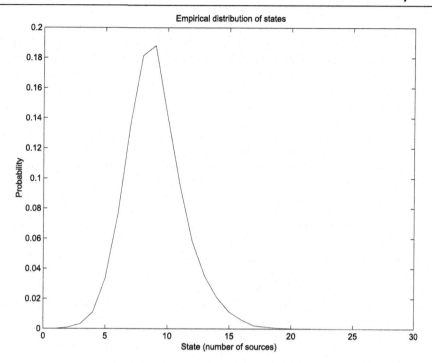

Figure 14.12: State distribution of the simulated one-dimensional system.

number of users are thereby constant throughout the simulation. This choice of boundary condition removes the need for special handling of boundary cells, which requires generation and dissipation of sources. The boundary conditions introduces dependency between the variables in that their sum N is constant. However, as the numbers of cells and users grow, this dependence becomes negligible. The simulator picks a cell, the "target cell" and computes the transition probabilities based on the interaction with the six neighboring cells.

Empirical data were obtained for 50 cells, with an average of 346 users per cell and a traffic rate of $r = 0.0331$. The simulation area was set to $S \times S$ with $S = 10$. The number of cells should be as large as possible to lessen the dependence between the cells because the number of users N is fixed. The mobility factor was set to $\alpha_m = 1$ and the gravity factor to $\alpha_g = 0.8$. In order to allow for a sufficient number of iterations for observed convergence, the following approximation was made. Each cell was assumed to host an average of 35 users generating a traffic ten times higher than originally, with $r = 0.331$. This approximation was made both to speed up the processing time and to limit the number of iterations. With a simulation area of $S \times S$ with $S = 10$ and $N = 3,500$, even 35,000 iterations only give approximately 10 "movement steps" per user. The Gibbs sampler was run for a long time period (approximately 70,000 iterations), with a "dry run" of approximately 12,000 iterations, that

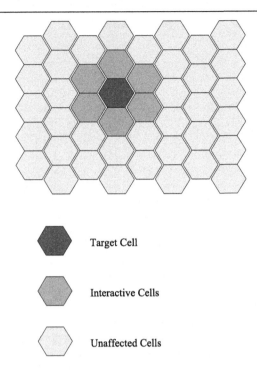

Target Cell

Interactive Cells

Unaffected Cells

Figure 14.13: Hexagonal cellular network.

is, the state information obtained during the first 12,000 iterations was discarded. The simulator counts the states in each step, giving the desired approximate probability distribution. The probability distribution is shown in Fig. 14.14, where the x-axis is scaled to represent the original number of users.

A theoretical figure for the traffic generated in the cell can be calculated based on the stationary probability distribution for the model. For each state, with m traffic sources and with $\pi(m)$ denoting the corresponding probability, the generated traffic is given by the joint probability distribution

$$p_n(m) = \pi(m) \frac{\frac{(rm)^n}{n!}}{\sum_{i=0}^{c} \frac{(rm)^i}{i!}}, \quad 0 \le n \le c,$$

where r is the generated traffic per source and c is the number of available channels. The marginal distribution of the traffic is given by $\sum_{m=0}^{N} p_n(m)$. Fig. 14.15 shows the thus obtained traffic distribution together with the empirical distribution and the classical distribution with no mobility taken into account. As can be seen in the picture, the distribution adjusted for mobility gives a much better fit to empirical data than the classical distribution. The number of allocated channels can be represented by a vertical line in Fig. 14.15, and the area cov-

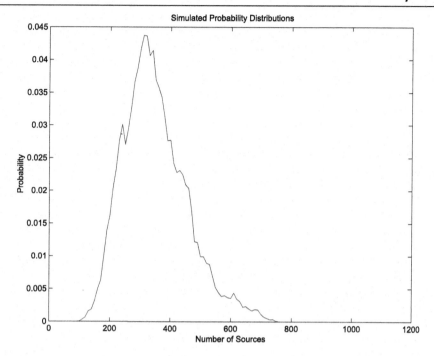

Figure 14.14: State distribution of the simulated two-dimensional system.

ered by the probability density functions to the right of that line represents the blocked (lost) traffic. Since the curve of the probability density function for the classical result is strictly lower than the empirical probability density function for an offered traffic greater than approximately 23 Erlang, the classical result will severely underestimate the blocking probability.

Mathematical Analysis of the Mobility Model

Consider the Gibbs system as outlined in Section 14.4. Let S_1 be a system embedded in larger system S_2, capable of exchanging energy with it, but not acting mechanically on any of its state variables. The system S_0 consisting of S_1 and S_2 is a large number of systems, essentially identical to S_1. Such a collection of systems is called a *canonical ensemble*. For a large ensemble, S_2 and S_0 are not essentially different. The system S_0 can be subdivided into the two subsystems S_1 and S_2, which then are *independent*.

Let the quantum (energy) states of each subsystem be labeled by a second number. Denote by $E_1^1, E_2^1, E_3^1, \ldots$ the energies of S_1 and by $E_1^2, E_2^2, E_3^2, \ldots$ the energies of S_2; denote by p_i^1 and p_k^2 the probabilities of the systems S_1 and S_2 to be in states i and k, respectively.

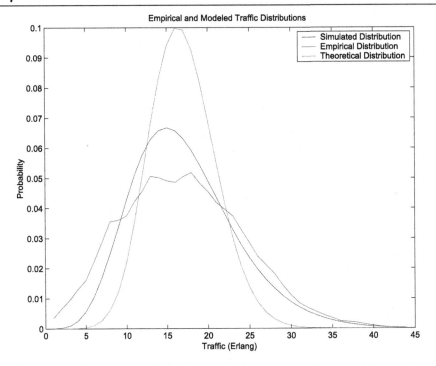

Figure 14.15: Comparison between empirical and simulated channel occupancy.

Since the two systems are independent, the total energy of S_0 is obtained by combining any two substates in pairs (i, k). The energy of such a pair state is the sum of the individual energies,

$$E_{i,k} = E_i^1 + E_k^2.$$

Because the systems are independent,

$$p_{i,k} = p_i^1 \cdot p_k^2.$$

Here, an ergodic postulate is applied, that is, it is assumed that *counting states gives the occupancy probability*. Let

$$
\begin{aligned}
p_i^1 &= p^1(E_i), \\
p_k^2 &= p^2(E_k), \\
p_{i,k} &= p^0(E_i + E_k) = p^1(E_i) \cdot p^2(E_k).
\end{aligned}
$$

Assume that S_2 and S_0 are ensembles with an extremely dense energy spectrum, so that the variable E_k can be approximated by a continuous variable and $p^2(E_k)$ can be approximated by a differentiable function. Logarithmic differentiation with respect to E_k yields

$$\frac{\partial p^0(E_i + E_k)}{\partial E_k} = \frac{d \ln p^2(E_k)}{d E_k} = -\beta_2.$$

However, β_2 does not depend on S_1. Furthermore, due to symmetry,

$$\frac{\partial \ln p^0(E_i + E_k)}{\partial E_i} = \frac{\partial \ln p^0(E_i + E_k)}{\partial E_k}.$$

Therefore,

$$\frac{d \ln p^1(E_i)}{d E_i} = -\beta_2.$$

The interpretation of this formula is that it yields the probability of occupation of a quantum state as a function of its energy. Integration gives

$$p^1(E_i) = ce^{-\beta_2 E_i}.$$

The selection of S_1 out of the ensemble S_0 was entirely arbitrary, so the argument can be repeated for every other system of the ensemble. The βs so obtained must all be identical. Thus, β comes out to be a characteristic of the ensemble S_0 only.

With the requirement that

$$\sum_i p^1(E_i) = 1,$$

the probability becomes

$$p(E_i) = \frac{\exp(-\beta E_i)}{\sum_v \exp(-\beta E_v)},$$

which is referred to as the Gibbs distribution law. If a system obeying the Gibbs distribution law is in equilibrium with another system with which it can exchange energy, then the two systems will share the parameter β. If β is a positive number, the probability for a given quantum state being occupied decreases as the energy increases. The following is true:

(1) Two systems which cannot exchange energy have in general different βs.
(2) If two systems are capable of exchanging energy their values of β become equal when equilibrium is reached.
(3) The mean energy of any system increases as β decreases.

The average energy is given by

$$U = \frac{\sum_v E_v \exp(-\beta E_v)}{\sum_v \exp(-\beta E_v)}.$$

The sum of states, also called the *partition function F*, is defined by

$$F = \sum_v \exp(-\beta E_v).$$

The following discussion on methods to find an analytical solution of the state probabilities of the one-dimensional model follows closely the mathematical solution of the one-dimensional *Ising model*; see for example [130].

Analysis-Numerical Solution to the One-Dimensional Mobility Model

Consider a number of S sites (the radio cells in this case), numbered $1, 2, \ldots, S$ and connected in a ring, so that the Sth site has the neighbors labeled $S - 1$ and 1. Let the states of the sites be described by the variables $x_1, x_2, \ldots, x_S$; see Fig. 14.10. These variables are binary variables in the case of the Ising model and discrete $(x_i \in \mathbb{N})$ in the present case.

The assumption of nearest-neighbor interaction is imposed on the system, so that each site interacts only with its two direct neighbors. Denote by $V(x, y)$ the interaction between two neighboring sites. Then the probability for a given state of the system is given by the Boltzmann potential,

$$\exp(-\beta(V(x_1, x_2) + V(x_2, x_3) + \ldots + V(x_S, x_1))), \tag{14.10}$$

from which the partition function is formed by summation (or integration), that is,

$$F = \sum_{x_1} \sum_{x_2} \ldots \sum_{x_S} \exp(-\beta(V(x_1, x_2) + V(x_2, x_3) + \ldots + V(x_S, x_1))).$$

In particular, the Markov chain with state space S has the transition probabilities

$$\mathbf{P}(s_{x_1,x_2,\ldots,x_i-1,x_{i+1}+1,\ldots,x_S} \mid s_{x_1,x_2,\ldots,x_i,x_{i+1},\ldots,x_S} =$$

$$= \frac{\alpha_g x_{i+1} + (1 - \alpha_g)(\bar{x} - x_{i+1})^+}{\alpha_g(x_{i-1} + x_{i+1}) + (1 - \alpha_g)((\bar{x} - x_{i-1})^+ + (\bar{x} - x_{i+1})^+)},$$

whereas the state probabilities are

$$\mathbf{P}(s_{x_1,x_2,\ldots,x_i-1,x_{i+1}+1,\ldots,x_S})$$
$$= e^{-\beta(V(x_1,x_2)+V(x_2,x_3)+\ldots+V(x_{i-1},x_i-1)+V(x_i-1,x_{i+1}+1)+V(x_{i+1}+1,x_{i+2})+\ldots+V(x_S,x_1))}.$$

Consider a transition matrix P with conditional probabilities. The stationary probability distribution $\pi(i)$ can be found from

$$\sum_y \exp(-\beta V(y, z))\pi(y) = \lambda_1 \pi(z), \tag{14.11}$$

where λ_1 is an eigenvalue and $\pi(\cdot)$ is the corresponding eigenvector of the matrix $\exp(-\beta V(y, z))_{y,z}$. This result is given by the *Perron–Frobenius theorem*.

Theorem 14.4.2 (Perron–Frobenius). *If P is the transition matrix of a finite irreducible chain with period d and cardinality N, then:*

(1) $\lambda_1 = 1$ *is an eigenvalue of P;*
(2) *the d complex roots of unity*

$$\lambda_1 = \omega^0, \lambda_2 = \omega^1, \ldots, \lambda_d = \omega^{d-1}, \quad \text{where } \omega = \exp(2\pi i/d),$$

are eigenvalues of P;
(3) *the remaining eigenvalues $\lambda_{d+1}, \ldots, \lambda_N$ satisfy $|\lambda_j| < 1$.*

Since the period of the Markov chain is $d = 1$, λ_1 has multiplicity 1 and for all other eigenvalues λ_i we have $|\lambda_i| < \lambda_1$. This eigenvalue is called the *Perron–Frobenius eigenvalue*. For the matrix $\exp(-\beta V(y, z))$, which is not normalized to be a probability matrix, the magnitude of λ_1 is greater than one, that is, λ_1 is the normalizing constant (partition function). The eigenvectors corresponding to different eigenvalues are orthogonal, so that

$$\sum_y \pi_\mu(y)\pi_\nu(y) = \delta_{\mu\nu},$$

where μ and ν are two different eigenvalues. Then by the spectral decomposition [30], the matrix $\exp(-\beta V(y, z))_{y,z}$ can be represented as

$$\exp(-\beta V(y, z)) = \sum_i \lambda_i \pi_i(y)\pi_i(z), \tag{14.12}$$

where $\pi_i(y)$ is the right and $\pi_i(z)$ is the left eigenvalue, respectively, associated with the eigenvalue λ_i. The partition function is given by

$$F = \sum_{x_1}\sum_{x_2}\cdots\sum_{x_S} \exp[-\beta(V(x1, x_2) + V(x_2, x_3) + \ldots + V(x_S, x_1))]. \tag{14.13}$$

Substitution of (14.12) into (14.13) gives

$$F = \sum_{x_1}\sum_{x_2}\cdots\sum_{x_S} \exp[-\beta V(x1, x_2)]\exp[-\beta V(x_2, x_3)] + \ldots + \exp[-\beta V(x_S, x_1)] =$$

$$\sum_{x_1}\sum_{x_2}\cdots\sum_{x_S}(\sum_i \lambda_i\pi_i(x_1)\pi_i(x_2))(\sum_i \lambda_i\pi_i(x_2)\pi_i(x_3))\ldots(\sum_i \lambda_i\pi_i(x_S)\pi_i(x_1)) =$$

$$\sum_{x_1}\sum_{x_2}\cdots\sum_{x_S}[(\lambda_1)^S\pi_1(x_1)\pi_1(x_2)\ldots\pi_1(x_S)\pi_1(x_1) +$$

$$+ \quad (\lambda_1)^{S-1}\lambda_2\pi_2(x_1)\pi_1(x_2)\ldots\pi_1(x_S)\pi_1(x_1) + \ldots] \approx \sum_i \lambda_i^S.$$

If the summation over x_i is stopped before the last variable, the term $\pi_{\lambda_1}^2(x)$ which is left is the probability distribution for the variable x_i. This relation becomes particularly useful in the case when the number S of systems is very large. Then all but the largest eigenvalue λ_1 can be neglected, and

$$F \approx \lambda_1^S.$$

Stochastic Fields

A *stochastic field* is a generalization of a stochastic process, where the index set has dimension greater than one. The definitions and results below can be found in [149].

Definition 14.4.1. Let $S \in \mathbb{Z}^d$ be a numerable set of *sites*, let Ω be a Polish space of states with σ-algebra $\mathcal{F}$. A configuration over S is an element ω of the measurable space of configurations $(\Omega, \mathcal{F})^S$. A *stochastic field* over S is a probability measure $\mathbf{P}(\omega)$ over $(S, \mathcal{E})^S$.

A stochastic field can be specified as a family of conditional probabilities. Let $\mathcal{S}$ be a collection of nonempty subsets of S.

Definition 14.4.2. A family of kernels $\pi = \{\pi_V, V \in \mathcal{S}\}$ that satisfies

$$\mu_{V'}\mu_V = \mu_{V'},$$

for $V \subset V'$, is called a *conditional specification*.

Definition 14.4.3. An *interaction potential* is defined by a family $\phi = (\phi_V, V \in \mathcal{S})$ of maps

$$\phi_V : \Omega \to \mathbb{R},$$

so that ϕ_V is $\mathcal{F}(V)$-measurable and the sum

$$U_V^\phi(\omega) = \sum_{V \in \mathcal{S}} \phi_V(\omega)$$

exists. This quantity (with negative sign) is referred to as the *energy* of ω in V.

The potential is called μ-*admissible* if for $V \in S$ and $\omega \in \Omega$,

$$Z_V^\phi(\omega) = \int_\Omega \exp(U_V^\phi(\omega, v))\mu^V(d\omega) < \infty, \tag{14.14}$$

where ω and v are the configurations over V and $S \backslash V$. If ϕ is admissible, the family

$$\pi_V^\phi(\omega, v) = Z_V^\phi(\omega) \exp(U_V^\phi(\omega, v)), \quad V \in S, \tag{14.15}$$

is coherent, where ω and v are configurations over V and $S \backslash V$, respectively; $\{\pi_V^\phi, V \in S\}$ is called the Gibbs specification associated with the potential ϕ. The expression (14.14) is called the partition function.

Definition 14.4.4. Let $V \in S$, let X be an event in $\mathcal{F}_V$ and ω be a configuration of Ω. Then a probability measure μ_V over $(\Omega, \mathcal{F})$ is called a *Markovian field* if

$$\mu_V(X|\omega(S \backslash V)) = \mu_V(X|\omega(\partial V)),$$

where ∂V is the neighborhood of V.

Theorem 14.4.3. *Let μ be transition probabilities for a Markovian field over $(\Omega, \mathcal{F})$, where*

$$\mu_V(x|y) > 0, \quad x \in V, y \in S \backslash V.$$

Then there is an interaction potential with bounded support $\phi = \{\phi_X, X \in C\}$ where all elements in C are single points or pairwise neighbors and $\mu_V = \pi_V^\phi$. The converse is also true.

Estimation for stochastic fields is difficult. In particular for maximum likelihood estimators, the numerical complexity grows fast with the number of sites and elements. Rather, the *conditional pseudolikelihood* is used, that is,

$$U_n = -|D_n|^{-1} \sum_{i \in D_n} \log(\pi_i(x_i|\partial x_i)), \tag{14.16}$$

where D_n is a sequence of squares, whose sides tend to infinity with n.

Estimation for the One-Dimensional Mobility Model

By Theorem 14.4.3 there exists a Gibbs potential $\phi = V(y, z)$ so that (14.15) is satisfied, which is equal to (14.10). The theorem only asserts the existence of a potential, and indeed there are many possible choices [149]. First, the conditional pseudolikelihood (14.16) gives some idea for construction of the potential; see Fig. 14.16.

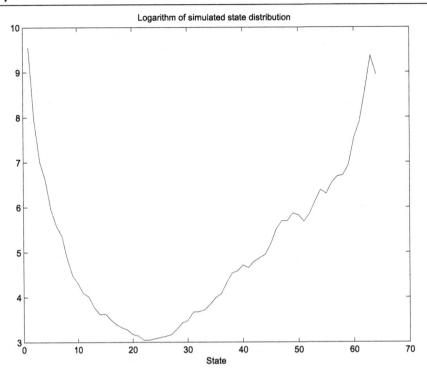

Figure 14.16: Conditional pseudolikelihood.

It turns out that it is not possible to find a global polynomial fit only in the state x. The restriction on the potential is

$$\frac{\partial V(x, y)}{\partial x} = 0$$

for $x = \bar{x}$, the average density in the site, and similarly for the derivative with respect to y.

Let the total number of sources be N. Define the *interaction potential* between any pair of sites i and $i + 1$, say, as

$$V(x_i, x_{i+1}) = \frac{(\bar{x} - x_i)^2}{2(x_i + x_{i+1})} \cdot (x_{i+1}\alpha_g + n_{i+1}(1 - \alpha_g)) + \tag{14.17}$$

$$+ \frac{(\bar{x} - x_{i+1})^2}{2(x_i + x_{i+1})} \cdot (x_i\alpha_g + n_i(1 - \alpha_g)). \tag{14.18}$$

Here, the scaling parameter $0 \leq \alpha_g \leq 1$ is the gravity factor. The divisor $\frac{1}{2}$ has been introduced since the potential is taken as the average of two force components. It turns out that the interaction potential (14.17)–(14.18) in Eq. (14.11) gives a relatively close fit, as shown in

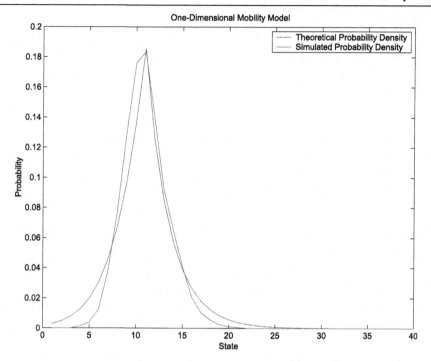

Figure 14.17: Theoretical and simulated steady-state distributions for the one-dimensional model.

Fig. 14.17. The parameter β is chosen freely to scale the distribution. The parameter α_g determines the spread of the distribution. This is consistent with intuition: the higher the value α_g, the more likely users are clustered together in certain sites.

For large x_i with $x_1 \approx x_{i+1}$, the potential is approximately quadratic in x_i, and for small x_i, x_{i+1} the function is a rational function in x_i, x_{i+1}.

It should be noted that other potentials may possibly give a better fit. The parameter α_g determines the variability in the state distribution. The larger the value of α_g, the higher is the probability of sources clustering in a few sites. The parameter β can be regarded as a shape parameter, and it is not used in the simulations. Therefore, β depends only on the chosen potential. In this investigation, the parameters were chosen based on "trial-and-error". After finding a proper *prior distribution*, that is, an interaction potential, the parameters need to be estimated. A possible method for estimation is again by simulations. In [129], a Markov chain Monte Carlo method for parameter estimation of a Gibbs field is proposed.

Unfortunately, finding the eigenvalues of the matrix is a very ill-posed problem. The conditional number of the matrix increases rapidly as the number of variables increases.

Concluding Remarks

In the plane, the interaction potential becomes

$$V(y, u) + V(y, v) + V(y, w),$$

where u, v, and w are interactions along the three axes in a hexagonal cell structure. The kernel now has four variables and cannot be readily represented by a matrix as in the one-dimensional case. There is no known analytical solution to the Ising model in three dimensions. The problem seems even to be NP-complete; see [128]. There is therefore little hope to find an analytical solution for the proposed mobility model in two dimensions. As with many other complex models, simulations seem to be the only way to determine the qualitative characteristics of the model. Indeed, simulations have been used to obtain a solution to the three-dimensional Ising model [127].

Bibliography

[1] N. Feamster, J. Rexford, E. Zegura, The road to SDN: an intellectual history of programmable networks, ACM Networks 11 (12) (2013).

[2] F. Keti, S. Askar, Emulation of software defined networks using mininet in different simulation environments, in: 6th International Conference on Intelligent Systems, Modelling and Simulation, IEEE, 2015.

[3] Open Networking Foundation, OpenFlow Switch Specification, Version 1.3.3, 2013, Accessed on 10 May 2016.

[4] K. Park, W. Willinger (Eds.), Self-Similar Network Traffic and Performance Evaluation, John Wiley & Sons, Inc., USA, 2000.

[5] S. Walukiewicz, Integer Programming, PWN—Polish Scientific Publishers, Warszawa, 1991.

[6] D.R. Karger, Global min-cuts in RNC, and other ramifications of a simple min-cut algorithm, in: Proc. 4th Annual ACM-SIAM Symposium on Discrete Algorithms, 1992.

[7] D.R. Karger, C. Stein, A new approach to the minimum cut problem, Journal of the ACM 43 (4) (1996).

[8] R. Motwani, P. Raghavan, Randomized Algorithms, Cambridge University Press, UK, 1995.

[9] Charles J. Colbourn, Some Open Problems on Reliability Polynomials, DIMACS Technical Report, No. 93-28, Canada, 1993.

[10] Jesper Nederlof, Inclusion Exclusion for Hard Problems, M.Sc. thesis, Utrecht University, 2008.

[11] J. Galtier, New Algorithms to Compute the Strength of a Graph, INRIA Rapport de recherche, No. 6592, ISSN 0249-6399, July 2008.

[12] I. Beichl, B. Cloteaux, F. Sullivan, An approximation algorithm for the coefficients of the reliability polynomial, Congressus Numerantium 197 (2009) 143–151.

[13] D.E. Knuth, Estimating the efficiency of backtrack programs, Mathematics of Computation 29 (1975) 121–136.

[14] O.K. Rodionova, A.S. Rodionov, H. Choo, Network Probabilistic Connectivity: Optimal Structures, Springer-Verlag, 2004.

[15] Barabási Lab, Accessed on 30 Dec. 2017.

[16] R. Albert, A.-L. Barabási, Statistical mechanics of complex networks, Reviews of Modern Physics 74 (2002).

[17] P. Erdös, A. Rényi, On random graphs I, Publicationes Mathematicae 6 (1959) 290–297.

[18] P. Erdös, A. Rényi, On the evolution of random graphs, Magyar Tudományos Akadémia Matematikai Kutató Intézetének Közleményei (Publications of the Mathematical Institute of the Hungarian Academy of Sciences) 5 (1959) 17–61.

[19] B. Bollobás, O. Riordan, The diameter of a scale-free random graph, Combinatorica 24 (1) (2004) 5–34.

[20] A.-L. Barabási, R. Albert, Emergence of scaling in random networks, Science 286 (5439) (1999) 509–512.

[21] J. Beran, Statistics for Long-Memory Processes, Chapman & Hall/CRC, USA, 1994.

[22] A. Pulipaka, P. Seeling, M. Reisslein, Traffic models for H.264 video using hierarchical prediction structures, in: IEEE Global Communications Conference, GLOBECOM 2012, 2012.

[23] B. Ryu, Modeling and simulation of broadband satellite networks – part II: traffic modeling, IEEE Communications Magazine 37 (7) (1999).

[24] H.E. Egilmez, S.T. Dane, K.T. Bagci, A.M. Tekalp, OpenQoS: an OpenFlow controller design for multimedia delivery with end-to-end quality of service over software-defined networks, in: Signal & Information Processing Association Annual Summit and Conference (APSIPA ASC) 2012, 2012.

[25] E. Buffet, N.G. Duffield, Exponential Upper Bounds via Martingales for Multiplexers with Markovian Arrivals, Report DIAS-APG-92-16, 1992.

[26] N.G. Duffield, Exponential Bounds for Queues with Markovian Arrivals, Report DIAS-APG-93-01, 1993.

[27] J. Ni, T. Yang, D.H.K. Tsang, Source Modeling, Queueing Analysis, and Bandwidth Allocation for VBR MPEG-2 Video Traffic in ATM Networks, 1997.

[28] W.E. Leland, M.S. Taqqu, W. Willinger, D.V. Wilson, On the self-similar nature of Ethernet traffic (extended version), Transactions on Networking 2 (1) (February 1994).

[29] T. Tuan, K. Park, Congestion Control for Self-Similar Network Traffic, Technical Report CSD-TR 98-014, 1998.

[30] G.R. Grimmett, D.R. Stirzaker, Probability and Random Processes, Oxford University Press, Hong Kong, 1992.

[31] P.J. Brockwell, R.A. Davis, Time Series: Theory and Methods, Springer, USA, 1987.

[32] W.E. Leland, M.S. Taqqu, W. Willinger, D.V. Wilson, On the self-similar nature of Ethernet traffic – extended version, IEEE/ACM Transactions on Networking 2 (1) (1993) 1–15.

[33] M.E. Crovella, A. Bestavros, Self-similarity in World Wide Web traffic: evidence and possible causes, IEEE/ACM Transactions on Networking 5 (6) (1997) 835–846.

[34] A. Popescu, Traffic self-similarity, Tutorial, IEEE International Conference on Telecommunications, ICT2001, 2001.

[35] J.D. Petruccelli, B. Nandram, M. Chen, Applied Statistics for Engineers and Scientists, Prentice-Hall, USA, 1999.

[36] V. Paxson, S. Floyd, Wide-area traffic: the failure of Poisson modeling, IEEE/ACM Transactions on Networking 3 (3) (1995) 226–244.

[37] M.S. Taqqu, V. Teverovsky, W. Willinger, Estimators for long-range dependence: an empirical study, Fractals 3 (1995) 785–798.

[38] N. Rushin-Rimini, I. Ben-Gal, O. Maimon, Fractal geometry statistical process control for non-linear pattern-based processes, IIE Transactions 45 (2012) 373–391.

[39] G.A. Miller, W.G. Madow, On the Maximum Likelihood Estimate of the Shannon-Weaver Measure of Information, Technical Report, Air Force Cambridge Research Center, 1954.

[40] P. Raghavan, C.D. Tompson, Randomized rounding: a technique for provably good algorithms and algorithmic proofs, Combinatorica 7 (1987) 365.

[41] J.H. Lin, J.S. Vitter, ϵ-approximations with minimum packing constraint violation, in: Proc. 24th ACM Symp. on Theory of Computing, 1992, pp. 771–782.

[42] M. Dorigo, Optimization, Learning and Natural Algorithms, PhD thesis, Politecnico di Milano, Italy, 1992.

[43] M. Dorigo, G. Di Caro, L.M. Gambardella, Ant algorithms for discrete optimization, Artificial Life 5 (2) (1999) 137–172.

[44] C. Blum, Ant colony optimization: introduction and recent trends, Physics of Life Reviews 2 (2005) 353–373.

[45] M. Dorigo, V. Maniezzo, A. Colorni, The ant system: optimization by a colony of cooperating agents, IEEE Transactions on Systems, Man, and Cybernetics–Part B 26 (1) (1996) 1–13.

[46] C.H. Papadimitriou, K. Steiglitz, Combinatorial Optimization: Algorithms and Complexity, Prentice-Hall, USA, 1982.

[47] J. Kennedy, R. Eberhart, Particle swarm optimization, in: Proceedings of IEEE International Conference on Neural Networks. IV, 1995, pp. 1942–1948.

[48] R. Poli, J. Kennedy, T. Blackwell, Particle swarm optimization – an overview, Swarm Intelligence 1 (2007) 33.

[49] Y. Shi, R.C. Eberhart, A modified particle swarm optimizer, in: Proceedings of the IEEE International Conference on Evolutionary Computation, 2000, pp. 69–73.

[50] X.-S. Yang, Nature-Inspired Metaheuristic Algorithms, Luniver Press, 2008.

[51] R.B. Francisco, M.F.P. Costa, A.M.A.C. Rocha, Experiments with firefly algorithm, in: B. Murgante, et al. (Eds.), Computational Science and Its Applications – ICCSA 2014, ICCSA 2014, in: Lecture Notes in Computer Science, vol. 8580, Springer, Cham, 2014.

[52] N. Goyal, L. Rademacher, S. Vempala, Expanders via random spanning trees, in: Proc. 20th Annual ACM-SIAM Symposium on Discrete Algorithms, 2008.

[53] S.E. Schaeffer, Graph clustering, Computer Science Review 1 (1) (2007) 27–64.

[54] R. Kannan, S. Vempala, A. Vetta, On clusterings: good, bad and spectral, in: Proc. Annu. IEEE Symp. Foundations of Comput. Sci. (FOCS), 2000, pp. 367–377.

[55] U. von Luxburg, A tutorial on spectral clustering, Statistics and Computing 17 (4) (December 2007) 395–416.

[56] A. Pothen, H.D. Simon, K.P.P. Liu, Partitioning sparse matrices with eigenvectors of graphs, SIAM Journal on Matrix Analysis and Applications 11 (3) (1989) 430–452.

[57] G. Fung, A Comprehensive Overview of Basic Clustering Algorithms, Technical Report, University of Wisconsin–Madison, 2001, http://pages.cs.wisc.edu/~gfung/clustering.pdf.

[58] U. Brandes, M. Gaertler, D. Wagner, Experiments on graph clustering algorithms, in: Proc. 11th Europ. Symp. Algorithms (ESA '03), in: LNCS, vol. 2832, 2003, pp. 568–579.

[59] G. Dahlqvist, Å. Björk, Numerical Methods, Prentice-Hall, Englewood Cliffs, 1974.

[60] W.R. Gilks, S. Richardson, D.J. Spiegelhalter (Eds.), Markov Chain Monte Carlo in Practice, Chapman & Hall/CRC, USA, 1996.

[61] O. Häggström, Finite Markov Chains and Algorithmic Applications, Chalmers Tekniska Högskola, Sweden, 2001.

[62] A.P. Dempster, N.M. Laird, D.B. Rubin, Maximum likelihood from incomplete data via the EM algorithm, Journal of the Royal Statistical Society. Series B (Methodological) 39 (1) (1977) 1–38.

[63] M.A. Tanner, Tools for Statistical Inference, Springer, USA, 1998.

[64] C.B. Do, S. Batzoglou, What is the expectation maximization algorithm?, Nature Biotechnology 26 (8) (2008) 897–899.

[65] A. Juan, E. Vidal, Bernoulli mixture models for binary images, in: Proc. of the ICPR 2004, 2004.

[66] The MNIST database of handwritten digits, Retrieved from http://yann.lecun.com/exdb/mnist/.

[67] L. van der Maaten, G. Hinton, Visualizing data using t-SNE, Journal of Machine Learning Research 9 (2008) 2579–2605.

[68] J. Krijthe, Package 'Rtsne', Retrieved from https://cran.r-project.org/web/packages/Rtsne/index.html, 2017.

[69] D.B. Shmoys, E. Tardos, K. Aardal, Approximation algorithms for facility location problems, in: Proceedings of the 29th Annual ACM Symposium on Theory of Computing, 1997, pp. 265–274.

[70] K. Jain, V.V. Vazirani, Approximation algorithms for metric facility location and k-median problems using the primal-dual schema and Lagrangian relaxation, Journal of the ACM 48 (2) (March 2001) 274–296.

[71] J. Chuzhoy, S. Guha, S. Khanna, J. Naor, Machine minimization for scheduling jobs with interval constraints, in: Proc. of the 45th Annual IEEE Symposium on Foundations of Computer Science (FOCS), 2004, pp. 81–90.

[72] J.T. Tsai, J.C. Fang, J.H. Chou, Optimized task scheduling and resource allocation on cloud computing environment using improved differential evolution algorithm, Journal Computers and Operations Research 40 (12) (2013) 3045–3055.

[73] Raju, CloudSim example with Round Robin Data center broker & Round Robin Vm allocation policy with circular hosts list, Retrieved from https://github.com/AnanthaRajuC/CloudSim-Round-Robin, 2016.

[74] H.M. Lee, Y.-S. Jeong, H.J. Jang, Performance analysis based resource allocation for green cloud computing, The Journal of Supercomputing 69 (2014) 1013–1026, Springer.

[75] D.B. Shmoys, E. Tardos, An approximation algorithm for the generalized assignment problem, Mathematical Programming 62 (1993) 461–474.

[76] D.S. Johnson, A. Demers, J.D. Ullman, M.R. Garey, R.L. Graham, Worst-case performance bounds for simple one-dimensional packing algorithms, SIAM Journal on Computing 3 (4) (1974) 299–325.

[77] R.E. Korf, A new algorithm for optimal bin packing, in: AAAI-02 Proc., 2002.

[78] K. Mills, J. Filliben, C. Dabrowski, Comparing VM-placement algorithms for on-demand clouds, in: Third IEEE International Conference on Cloud Computing Technology and Science, 2011.

[79] S. Guha, A. Meyerson, K. Munagala, A constant factor approximation algorithm for the fault-tolerant facility location problem, Journal of Algorithms 48 (2) (2003) 429–440.

[80] S. Voss, Capacitated minimum spanning trees, in: C.A. Floudas, P.M. Pardalos (Eds.), Encyclopedia of Optimization, Springer, Boston, MA, 2001.

[81] R. Bera, R. Lanjewar, D. Mandal, R. Kar, S.P. Ghoshal, Comparative study of circular and hexagonal antenna array synthesis using improved particle swarm optimization, Procedia Computer Science 45 (2015) 651–660.

[82] N. Pathak, P. Nanda, G.K. Mahanti, Synthesis of thinned multiple concentric circular ring array antennas using particle swarm optimization, Journal of Infrared, Millimeter, and Terahertz Waves 30 (7) (July 2009) 709–716.

[83] G. Ram, D. Mandal, R. Kar, S.P. Ghoshal, Design of non-uniform circular antenna arrays using firefly algorithm for side lobe level reduction, International Journal of Electrical, Computer, Energetic, Electronic and Communication Engineering 8 (1) (2014).

[84] B. Basu, G.K. Mahanti, Thinning of concentric two-ring circular array antenna using fire fly algorithm, Scientia Iranica D 19 (6) (2012) 1802–1809.

[85] P. Tummala, K. Sravan, Synthesis of hexagonal antenna array using firefly algorithm, in: ECBA-16, in: Academic Fora, vol. 3, 2016, p. 18.

[86] P. Saxena, A. Kothari, Optimal pattern synthesis of linear antenna array using grey wolf optimization algorithm, International Journal of Antennas and Propagation 2016 (2016), Hindawi.

[87] J.P.G. Sterbenz, D. Hutchinson, E.K. Cetinkaya, A. Jabbar, J.P. Rohrer, M. Schöller, P. Smith, Resilience and survivability in communication networks: strategies, principles, and survey of disciplines, Computer Networks 54 (2010) 1245–1265.

[88] K. Steiglitz, P. Weiner, D.J. Kleitman, The design of minimum cost survivable networks, IEEE Transactions on Circuit Theory 16 (4) (1969) 455–460.

[89] H.N. Gabow, M.X. Goemans, D.P. Williamson, An efficient approximation algorithm for the survivable network design problem, Mathematical Programming 82 (1998) 13–40.

[90] M.X. Goemans, D.P. Williamson, A general approximation technique for constrained forest problems, SIAM Journal on Computing 24 (2) (1995) 296–317.

[91] J.Y. Yen, Finding the k shortest loopless paths in a network, Management Science 17 (11) (1971).

[92] J.L. Marzo, E. Calle, C.M. Scoglio, T. Anjali, QoS online routing and MPLS multilevel protection: a survey, IEEE Communications Magazine 41 (10) (Oct. 2003) 126–132.

[93] C. Larsson, Design of Modern Communication Networks – Methods and Applications, Academic Press, 2014.

[94] R.A. Guérin, A. Orda, D. Williams, QoS routing mechanisms and OSPF extensions, in: IEEE GLOBECOM '97, vol. 3, 1997.

[95] T. Leighton, F. Makedon, S. Plotkin, C. Stein, É. Tardos, S. Tragoudas, Fast approximation algorithms for multicommodity flow problems, Journal of Computer and System Sciences 50 (2) (April 1995) 228–243.

[96] T. Leighton, S. Rao, Multicommodity max-flow min-cut theorems and their use in designing approximation algorithms, Journal of the ACM 46 (6) (Nov. 1999) 787–832.

[97] T. Leong, P. Shor, C. Stein, Implementation of a Combinatorial Multicommodity Flow Algorithm, DIMACS Series in Discrete Mathematics and Theoretical Computer Science, 1992.

[98] A. Conte, Review of the Bron-Kerbosch Algorithm and Variations, Project Report, University of Glasgow, 2013.

[99] F.J.A. Artacho, R. Campoy, Solving graph coloring problems with the Douglas-Rachford algorithm, Set-Valued and Variational Analysis 26 (2) (June 2018) 277–304.

[100] G. Peyre, Douglas Rachford proximal splitting, http://www.numerical-tours.com/matlab/optim_4b_dr/, 2010.

[101] D.B. Johnson, Finding all the elementary circuits of a directed graph, SIAM Journal on Computing 4 (1) (March 1975).

[102] A. Nijhawan, P Cycles: Design and Applications, Project Report, North Carolina State University, 2009, http://dutta.csc.ncsu.edu/csc772_fall09/wrap/Res_project/pcycle_CSC772_Final_Paper.pdf.

[103] Z. Zhang, W.-D. Zhong, A heuristic method for design of survivable WDM networks with p-cycles, IEEE Communications Letters 8 (7) (July 2004).

[104] Wayne D. Grover, Mesh-Based Survivable Networks, chapter 10, Prentice Hall, 2003.

[105] R. Asthana, Study of p-Cycle Based Protection in Optical Networks and Removal of Its Shortcomings, Dissertation, Indian Institute of Technology, Kanpur, India, 2007, http://home.iitk.ac.in/~ynsingh/phd/Y110492_IITK.pdf.

[106] D. Laney, 3D Data Management: Controlling Data Volume, Velocity, and Variety, Tech. Rep. 2001, META Group, 2001.

[107] J.L. Carter, M.N. Wegman, Universal classes of hash functions, Journal of Computer and System Sciences 18 (2) (April 1979) 143–154.

[108] R. Morris, Counting large numbers of events in small registers, Communications of the ACM 21 (10) (1977) 840–842.

[109] P. Flajolet, G.N. Martin, Probabilistic counting algorithms for data base applications, Journal of Computer and System Sciences 31 (2) (1985) 182–209.

[110] N. Alon, Y. Matias, M. Szegedy, The space complexity of approximating the frequency moments, in: Proceedings of the 28th ACM Symposium on Theory of Computing (STOC 1996), 1996, pp. 20–29.

[111] W.B. Johnson, J. Lindenstrauss, Extensions of Lipschitz mappings into a Hilbert space, in: R. Beals, A. Beck, A. Bellow, et al. (Eds.), Conference in Modern Analysis and Probability, New Haven, Conn., 1982, in: Contemporary Mathematics, vol. 26, American Mathematical Society, Providence, RI, 1984, pp. 189–206.

[112] G. Cormode, S. Muthukrishnan, An improved data stream summary: the count-min sketch and its applications, Journal of Algorithms 55 (1) (2005) 58–75.

[113] J.N. Hovmand, M.H. Nygaard, Estimating Frequencies and Finding Heavy Hitters, Master's Thesis, Aarhus University, 2016.

[114] M. Charikar, K. Chen, M. Farach-Colton, Finding frequent items in data streams, in: P. Widmayer, S. Eidenbenz, F. Triguero, R. Morales, R. Conejo, M. Hennessy (Eds.), Automata, Languages and Programming, ICALP 2002, in: Lecture Notes in Computer Science, vol. 2380, Springer, Berlin, Heidelberg, 2002.

[115] V.V. Uchaikin, V.M. Zolotarev, Chance and Stability: Stable Distributions and Their Applications, De Gruyter, 1999.

[116] A. Lall, V. Sekar, M. Ogihara, J. Xu, H. Zhang, Data streaming algorithms for estimating entropy of network traffic, in: Proceedings of the Joint International Conference on Measurement and Modeling of Computer Systems (SIGMETRICS'06), 2006, pp. 145–156.

[117] A. Kumar, M. Sung, J. Xu, J. Wang, Data streaming algorithms for efficient and accurate estimation of flow size distribution, ACM SIGMETRICS Performance Evaluation Review 32 (1) (June 2004) 177–188.

[118] Q. Zhao, A. Kumar, J. Wang, J. Xu, Data streaming algorithms for accurate and efficient measurement of traffic and flow matrices, in: SIGMETRICS '05 Proceedings of the 2005 ACM SIGMETRICS International Conference on Measurement and Modeling of Computer Systems, pp. 350–361.

[119] M. Gerla, L. Kleinrock, Flow control: a comparative survey, IEEE Transactions on Communications 28 (4) (1980).

[120] R.J. Gibbens, P.J. Hunt, F.P. Kelly, Bistability in communication networks, in: Physical Systems, Oxford University Press, Disorder, 1990, pp. 113–128.

[121] K. Dolzer, W. Payer, M. Eberspächer, A simulation study on traffic aggregation in multi-service networks, in: COST 257TD(00)05, 2000.

[122] W.R. Heinzelman, A.P. Chandrakasan, H. Balakrishnan, An application-specific protocol architecture for wireless microsensor networks, IEEE Transactions on Wireless Communications 1 (4) (2002).

[123] A. Aldosary, I. Kostanic, The impact of tree-obstructed propagation environments on the performance of wireless sensor networks, in: IEEE 7th Annual Computing and Communication Workshop and Conference (CCWC 2017), 2017.

[124] W.R. Heinzelman, A.P. Chandrakasan, H. Balakrishnan, Energy-efficient communication protocol for wireless microsensor networks, in: Proceedings of the Hawaii International Conference on System Sciences, 2000.

[125] C. Fu, Z. Jiang, W. Wei, A. Wei, An energy balanced algorithm of LEACH protocol in WSN, IJCSI International Journal of Computer Sciences Issues 10 (2013) 1.

[126] M.B. Yassein, A. Al-zou'bi, Y. Khamayseh, W. Mardini, Improvement on LEACH protocol of wireless sensor network (VLEACH), International Journal of Digital Content Technology and Its Applications (JDCTA) 3 (2) (2009) 132–136.

[127] W. Janke, R. Villanova, Ising model on three-dimensional random lattices: a Monte Carlo study, The American Physical Society: Physical Review 66 (2002).

[128] B.A. Cipra, The Ising model is NP-complete, SIAM News 33 (6) (2000).

[129] X. Descombes, R.D. Morris, Estimation of Markov random field prior parameters using Markov chain Monte Carlo maximum likelihood, IEEE Transactions on Image Processing 8 (7) (1999) 954–963.

[130] G.H. Wannier, Statistical Physics, Wiley & Sons, USA, 1966.

[131] Transportation Research Board (TRB), Traffic Flow Theory, Special Report 165, U.S. Department of Transportation, Federal Highway Administration, 1975, Retrieved from http://www.tfhrc.gov/its/tft/tft.htm.

[132] D.B. Percival, A.T. Walden, Wavelet Methods for Time Series Analysis, Cambridge University Press, USA, 2000.

[133] D. Lam, D.C. Cox, J. Widom, Teletraffic modeling for personal communications services, IEEE Communications Magazine: Special Issues on Teletraffic Modeling Engineering and Management in Wireless and Broadband Networks 35 (1997) 79–87.

[134] K.K. Leung, W.A. Massey, W. Whitt, Traffic models for wireless communication networks, IEEE Journal on Selected Areas in Communications 12 (8) (Oct. 1994) 1353–1364.

[135] F. Ashtiani, J.A. Salehi, M.R. Aref, Analytical computation of spatial traffic distribution in a typical region of a cellular network by proposing a general mobility model, in: IEEE ICT Proceedings, Tahiti, 2003, pp. 295–301.

[136] Y. Fang, I. Chlamtac, Analytical generalized results for handoff probability in wireless networks, IEEE Transactions on Communications 50 (3) (2002) 396–399.

[137] K. Mitchell, K. Sohraby, An analysis of the effects of mobility on bandwidth allocation strategies in multi-class cellular wireless networks, in: IEEE INFOCOM 2001, vol. 2, 2001, pp. 1075–1084.

[138] N.G.R. Antunes, Modeling and Analysis of Wireless Networks, Doctoral Dissertation, Universidade Técnica de Lisboa, Instituto Superior Técnico, 2001.

[139] C. Bettstetter, Smooth is better than sharp: a random mobility model for simulation of wireless networks, in: Proc. ACM Workshop on Modeling, Analysis and Simulation of Wireless and Mobile Systems, 2001, pp. 19–27.

[140] D. Hong, S.S. Rappaport, Traffic model and performance analysis for cellular mobile radio telephone systems with prioritized and non-prioritized handoff procedures, IEEE Transactions on Vehicular Technology 35 (3) (1986) 77–92.

[141] V. Pla, V. Casares, Analytical-numerical study of the handoff area Sojourn time, in: 4th IEEE Globecom, 2002.

[142] E. Jugl, H. Boche, Dwell time models for wireless communication systems, in: Proc. IEEE Vehicular Technology Conference VTC'99, vol. 5, 1999, pp. 2984–2988.

[143] E. Jugl, H. Boche, Analysis of analytical mobility models with respect to the applicability for handover modeling and to the estimation of signaling cost, in: Proceedings of the 6th Annual International Conference on Mobile Computing and Networking, 2000, pp. 68–75.

[144] ETSI, Universal Mobile Telecommunications System (UMTS); Selection Procedures for the Choice of Radio Transmission Technologies of the UMTS (UMTS 30.03, Version 3.2.0), Technical report, European Telecommunication Standards Institute, 1998.

[145] M.M. Zonoozi, P. Dassanayake, User mobility modeling and characterization of mobility patterns, IEEE Journal on Selected Areas in Communications 15 (7) (1997) 1239–1252.

[146] S. Thilakawardana, R. Tafazolli, Impact of service and mobility modelling on network dimensioning, in: European Wireless Conference, Italy, 2002.

[147] M. Schweigel, R. Zhao, Cell residence time in square shaped cells, 2nd Polish-German Teletraffic Symposium PGTS (2002) 213–220.

[148] F. Barceló, J. Jordán, Channel holding time distribution in public cellular telephony, in: Proc. 9th Int. Conf. on Wireless Communications, vol. 1, 1999, pp. 125–134.

[149] X. Guyon, Random Fields on a Network: Modeling, Statistics, and Applications, Springer-Verlag, USA, 1995.

[150] E. Dimitrova, P. Vera-Licona, J. McGee, R. Laubenbacher, Discretization of time series data, Journal of Computational Biology 17 (6) (2010) 853–868.

Index

Printed in the United States
By Bookmasters